Heinrich Klefenz

Industrial Pharmaceutical Biotechnology

KU-281-341

Heinrich Klefenz

Industrial Pharmaceutical Biotechnology

U.WEL.
LEARNING RESOURCES
ACC. N 2392020
CONTROL
3527299SS
DATE
27 JUN 2006
SITE
WV
615.
19
KLE
SAS
WITHDRAWN

Dr. Heinrich Klefenz
Hauptstr. 35
D-76879 Bornheim
Germany

This book was carefully produced. Nevertheless, author and publisher do not warrant the information contained therein to be free of errors. Readers are advised to keep in mind that statements, data, illustrations, procedural details or other items may inadvertently be inaccurate.

Cover illustration: Design by 'das trio kommunikation und marketing gmbh; Mannheim, München'

Copyright of and reprint permissions granted by
American Society for Microbiology (Tables 7.1, 7.2; ref. 502)
American Association for the Advancement of Science (Tables: 4.7, ref. 219; 5.1, ref. 224; Figures: 4.1, ref. 154; 4.2, ref. 510; 6.1, ref. 301)
Nature Publishing Group (Fig. 1.3, ref. 432; Tables: 1.6, ref. 432; 1.7, ref. 433; 1.8, ref. 436; 1.9, ref. 437; 1.10, ref. 439).

1st Edition 2002
 1st Reprint 2005

Library of Congress Card No.: Applied for.

British Library Cataloguing-in-Publication Data:
A catalogue record for this book
is available from the British Library

Die Deutsche Bibliothek Cataloguing-in-Publication Data:
A catalogue record for this publication is available from Die Deutsche Bibliothek
ISBN 3-527-29995-5

© WILEY-VCH Verlag GmbH, Weinheim (Federal Republic of Germany), 2002

Printed on acid-free paper.

All rights reserved (including those of translation into other languages). No part of this book may be reproduced in any form – by photoprinting, microfilm, or any other means – nor transmitted or translated into a machine language without written permission from the publishers. Registered names, trademarks, etc. used in this book, even when not specifically marked as such, are not to be considered unprotected by law.

Composition: Manuela Treindl, Regensburg
Printing: Strauss Offsetdruck GmbH, Mörlenbach
Bookbinding: J. Schäffer GmbH & Co. KG, Grünstadt

Printed in the Federal Republic of Germany.

3/97

UNIVERSITY OF
WOLVERHAMPTON
ENTERPRISE LTD.

LR/LEND/002

Harrison Learning Centre
Wolverhampton Campus
University of Wolverhampton
St Peter's Square
Wolverhampton WV1 1RH
Telephone: 0845 408 1631

ONE WEEK LOAN

- 6 MAY 2011

1 3 JUN 2013

Telephone Renewals: 01902 321333
Please RETURN this item on or before the last date shown above.
Fines will be charged if items are returned late.
See tariff of fines displayed at the Counter.   (L2)

Preface

Biotechnology and its applications in medicine, pharma, and related industries represent one of the most influential developments and pose one of the greatest challenges of the 21st century, both with respect to its political, societal, and ethical implications and in the search for the fulfillment of its promises for health.

Biotechnology is stepping beyond previously insurmountable boundaries in understanding and manipulating life, in the efforts to understand biology, to eradicate disease, to maintain health and vigor, and to endow humans and life forms with desired properties.

This book aims to describe a fast-moving subject (or rather a whole interconnected system of subjects) and, like in optics, some parts of the picture may be blurred and will require further refining. It pulls together topics, which are essential for the realization of the promises of biomedicine – the repertoire of genomics, proteomics, cytomics, bioinformatics, and the interaction of networks – and combines with pertinent methods in nanotechnologies, such as engineering tools to design and construct devices, artificial intelligence and vision processing for nano-devices, implantates, and for the envisioned swarms of remedial nano-robots.

Crucial topics for future therapies are regenerative medicine and the cultivation of tissues and organs as well as the underlying genetics and regulatory, developmental, biochemical networks.

Complex traits, critical in multifactor and degenerative diseases, are being dealt with, with a focus on senescence which forms the background against which numerous degenerative and acute diseases develop, the elucidation of which will facilitate the strengthening of immune responses, the maintenance of homeostasis and biochemical networks, the preservation of the integrity of genetic and cellular structures.

Drug discovery encompasses the identification of molecular structures, the creation of active molecules, and the development of novel comprehensive therapies like immunotherapy and cellular or organismal therapy with genetically engineered cells.

Biotechnology, chemistry, physics provide the tools for target identification, for the creation of new molecular structures, and for the recovery of biologically active molecules provided by the biosphere and efficiency-honed during continuous evolutionary processes.

The huge amounts of data and information alone will not be sufficient to lead to new molecular entities and novel therapies, since synthesizing millions of compounds will neither fill the universe of potential molecular structures nor allow the identification of those three-dimensional structures specifically interacting with targets. The knowledge of the biological processes and structures as the templates and targets for the identification of active molecules is indispensable.

Biological plus chemical functional information and knowledge of interactions and networks will be the foundation to which the essential components of creativity and innovation (and chance) are to be added as keys for the successful application of the pertinent technologies.

The reference list of more than 700 literature citations is meant to underpin the contents and the conclusions of the book's theme, and to serve as a starting point for delving deeper into individual subjects.

Special thanks go to Dr. Hovsep Sarkissian for his support in layout, in the production of figures and tables, in proofreading and the generation of a readable manuscript. Thanks are also due to the staff of Wiley-VCH for their organization, continuous encouragement, and stimulation; and to the 'Muttersprachler' who critically read the English manuscript provided and contributed to the professionalism of the writing.

Utmost to be thankful for is the patience, understanding, and support of my family and our two children who have tolerated extended periods of negligence.

Undoubtedly the rapid development in biotechnology, biomedicine, and supporting technologies, will affect many topics of the book's field and will necessitate modifying, changing, or complementing the subjects.

I have no doubt that in our efforts to fulfill the potential of pharmaceutical biotechnology, we are on a steep uphill slope and the top of the mountain (control of health, disease, and desired properties) is in the clouds at incalculable but reachable distance.

I welcome critical comments or suggestions about the book, proposals for areas to be dealt with in the future, and I am ready to provide further details, information, or references about the various topics upon request.

H. Klefenz   Bornheim, December 2001
e-mail address: sarkle@t-online.de

Contents

1 Introduction to Functional Biotechnology

1.1 Scientific and Technological Foundations

Pharmaceutical biotechnology focuses on biotechnology with pharmaceutical relevance as the central science and technology of the 'Life Sciences' with its fundamentals, developments, influences and effects.

This monograph demonstrates the paradigmatic changes effected by biotechnology in combination with pharmaceutical science, cell biology, chemistry, electronics, materials science and technology, plus organizational changes on pharmaceutical research, development and industry as well as pharmaceutical-related animal and plant biotechnology ('Life Sciences').

Pharmaceutical biotechnology exemplifies the transformation towards a knowledge-based society with innovation as the essential basis of activity in an age of globalization, increased competition, and accelerated speed of development, changes and decisions.

The total spectrum of concepts, processes and technologies of biotechnology, chemistry and electronics is being applied in modern industrial pharmaceutical research, development and production.

In pharmaceutical and medical research, diagnostics, production and therapy, the results of genome sequencing and studies of biological–genetic function (functional genomics) are combined with chemical, microelectronic and micro system technologies to produce medical devices, known as diagnostic 'Biochips'.

In chemical, pharmaceutical and biotechnological production processes the multitude of biologically active molecules is expanded by additional novel structures created with newly arranged 'gene clusters' and (bio-) catalytic chemical processes.

Materials synthesized with chemical and biotechnological processes support novel implantates, tissue engineering and even competitors to silicon-based computing, as well as analytics, diagnostics, medical devices, electronics, data processing and energy conversion.

New organizational structures in the cooperation of institutes, companies and networks enable faster knowledge and product development, and immediate application of scientific research and process developments.

Target groups of readers are biotechnologists, pharmaceutical scientists, biochemists, biologists, physicians, pharmacologists, chemists, reproductive biologists, genetic engineers, agro-scientists, and animal and plant breeders.

Organizationally, this monograph is addressed to scientists, technicians and managers of biotechnology, pharmaceutical and chemical companies, research institutes, and biotech ventures, and decision makers in industry, science, venture capital/finance and politics.

This monograph aims to present an integrated view of the manifold and diverse developments and their impact on the discovery of new drugs and therapies. Specifically, the topics deal with:

- The integration of genomics, proteomics, cytomics, structural and functional biology.
- Studies of networks and multi-gene traits at the molecular, genetic, biochemical, cellular and organism levels.
- Micro- and nanotechnologies for R & D and therapy.
- Stem cell research, therapeutic cloning and regenerative medicine.
- Drug discovery and therapy development from genomics, proteomics to small molecules, biopharmaceuticals to systems.
- Organizational solutions and core competencies for the pharmaceutical industry.
- Bioinformatics, functional genomic, structural analysis and computational biology.
- Scientific and technological foundations.

1.2   Genomics

Functional genomics is the scientific field dealing with extracting or synthesizing biologically relevant and therapeutically useful information from sequences, genomics, proteomics, expression profiles and linkage studies. The analysis of genomic, expression and proteomic data produces networks of functional interactions and linkages between proteins, cells, tissues and organs.

Proteins are the main catalysts, structural elements, signaling messengers and molecular machines of biological tissues. Phylogenetic profile generation and two-hybrid screen methods are the major techniques used to study protein–protein interactions.[1]

Gene-based diagnostics is rapidly expanding in the medical/industrial sector. It involves the study of DNA and RNA as compared to 'classical' medical diagnostics, which deals with enzymes, hormones, proteins and metabolic intermediates. The total business volume in medical diagnostics is about US$ 18 billion (1998), out of which gene-based diagnostics comprises US$ 500–700 million, with annual growth rates of 25%.

The pharmaco-genomics market (products and services) is estimated to grow from US$ 47 million in 1998 to US$ 795 million in 2005, with the major areas being cardiovascular diseases (US$ 139 million), infectious diseases (US$ 123 million), central nervous system (CNS)-related disorders (US$ 72 million) and cancer (US$ 41 million). In 1999, 28 pharmaco-genomic collaborations had been formed, 20 concerning the application of pharmaco-genomics to drug development; seven were involved in drug discovery and four in marketed drugs.

There are conceptual and real developments aimed at bringing the fields of genomics, functional genomics, pharmaco-genomics, single-nucleotide polymorphism (SNP) studies, imprinting, metabolic networks, genetic hierarchies in embryonic development and epigenetic mechanisms of cancer together under the conceptual umbrella of 'epigenomics', studying complex phenotypes from the genomic level down. The focus of scientific efforts is genome-scale mapping of the methylation status of CpG dinucleotides, the identification and analysis of epigenomic loci in the major histocompatibility complex (MHC), and the comparative analysis of epigenomic information from different organisms.[2]

The flow of novel genes from efforts in genomics provides the opportunity to greatly expand the number of therapeutic targets – the limited resource in drug discovery. Strategies to accelerate the evaluation of candidate molecules as disease-relevant targets involve the establishment of pertinent models (e.g., mice, cells, organs, zebra fish, nematodes and yeast).

The challenge of transforming DNA sequences into disease-relevant targets will continue to be a major requirement in drug discovery.[3] Genomics stretches from gene sequencing, gene analysis and trait analysis via structural genomics to functional genomics.

Structural genomics aims to experimentally determine the structures of all possible protein folds. Such efforts entail a conceptual shift from traditional structural biology in which structural information is obtained on known proteins to one in which the structure of a protein is determined first and the function assigned later. Whereas the goal of converting protein structure into function can be accomplished by traditional sequence motif-based approaches, recent studies have shown that assignment of a protein's biochemical function can also be achieved by scanning its structure for a match to the geometry and chemical identity of a known active site. This approach can use low-resolution structures provided by contemporary structure prediction methods. When applied to genomes, structural information (either experimental or predicted) is likely to play an important role in high-throughput function assignment.

Sequence genomics is the starting point for structural and functional genomics which provide the experimental structural data for the molecular design of antagonists, agonists and biologically (respectively, pharmacologically) active substances.[4]

Table 1.1 shows a compilation of projects, sources and databases for structural data to facilitate access to these fundamental sources for pharmaceutical development.

Genomics, the study of the whole genome, requires ever-increasing efficiency in the methods used for gene analysis.

An automated, high-throughput, systematic cDNA display method called total gene expression analysis (TOGA) was developed. TOGA utilizes 8-nucleotide sequences, comprised of a 4-nucleotide restriction endonuclease cleavage site and adjacent 4-nucleotide parsing sequences, and their distances from the 3′ ends of mRNA molecules to give each mRNA species in an organism a single identity. The parsing sequences are used as parts of primer-binding sites in 256 polymerase chain reaction (PCR)-based assays performed robotically on tissue extracts to determine simultaneously the presence and relative concentration of nearly every mRNA in the extracts, regardless of whether the mRNA has been discovered previously. Visualization of the electrophoretically separated fluorescent assay products from different extracts displayed via a Netscape browser-based graphical user interface allows the status of each mRNA to be compared among samples and its identity to be matched with sequences of known mRNAs compiled in databases.[5]

Methods for gene expression analysis include transcript sampling by sequencing or by hybridization signature, transcript amplification and imaging, and hybridization to gene arrays. Serial analysis of gene expression (SAGE), one of the most effective methods, is

Table 1.1. Structural genomics resources [Refs in 4].

At present, several pilot structural genomics projects are underway (see Table 1.1). As a proof of principle, Kim and coworkers[58] have solved the crystal structure of *Methanococcus jannachii* Mj0577 protein, for which the function was previously unknown. The structure contains a bound ATP, suggesting Mj0577 is an ATPase or an ATP-mediated molecular switch; this was subsequently confirmed by biochemical experiments[58]. Importantly, efforts are also underway to minimize a duplication of efforts among the various structural genomics groups. For example, a very useful database, PRESAGE, has been assembled by Brenner and coworkers[59] that provides a collection of annotations reflecting current experimental status, structural assignments, models, and suggestions. Another similar resource is provided by the Protein Structural Initiative (http://www.structuralgenomics.org/).

URLs for structural genomics pilot projects, computational tools, and key databases

Resource	Description	URL
Projects		
Center for Advanced Research in Biotechnology (Rockville, MD) and the Institute for Genomic (Rockville)	Solve structures of unknown function in *Haemophilus influenzae*	http://structuralgenomics.org/
Brookhaven National Laboratory (Upton, NY), Rockefeller University (New York, NY), and Albert Einstein School of Medicine (New York, NY)	Pilot genomics project on yeast	http://proteome.bnl.gov/ targets.htm
New Jersey Commission on Science and technology, and Rutgers University (Piscataway, NJ)	Metazoan organisms, human pathogen proteins	http://www-nmr.cabm. Rutgers.edu/
Los Alamos National Laboratory and The University of California, Los Angeles	Thermophilic archeon *Pyrobaculum aerophilum*	http://www-structure. llnl.gov/PA/PA_intro.html
Argonne National Laboratory (Argonne, IL)	Technology for high throughput structure determination	http://www.bio.ani.gov/ research/ structural_genomics.htm
PRESAGE	Structural genomics clearing house; coordination of efforts	http://presage.Stanford.edu
Protein structure initiative	structural genomics clearing house	http://structuralgenomics.org/

limited by the small amount of sequence information obtained for each gene. Transcript sequencing following subtractive hybridization is limited to binary comparisons. Transcript imaging approaches such as differential display, partitioning by type IIS restriction enzymes, representational difference analysis (RDA) and amplified fragment length polymorphism (AFLP) are rapid and theoretically comprehensive since they use fragment patterns on gels to infer gene expression. The development of microarrays has significantly enhanced the capacity of hybridization techniques to identify differences in gene expres-

Table 1.1. Structural genomics resources (Cont'd).

URLs for structural genomics pilot projects, computational tools, and key databases

Resource	Description	URL
Tools		
Eisenberg group	Threading tools	http://www.doe-mbi.ucla.edu/ PeopleEisenberg/Projects
Expasy	Swiss-Prot site contains many sequence and structure searching tools	http://www.expasy.ch/
Gerstein group	Structure prediction of eight genomes, comparative genomics	http://bioinfo.mbb.yale.edu/ genome/
National Center for Biotechnology Information (Bethesda, MD)	BLAST sequence similarity search tools	http://www.ncbi..nml.nih.gov /BLAST/
Sali group	Tools for protein structure modeling, incl. MODELLER	http://guitar.Rockefeller.edu/ subpages/programs/ programs.html
Skolnick-Kolinski group	Threading tools, *ab initio* folding tools, FFF library	http://bioinformatics. danforthcenter.org
Thornton group	Library of three-dimensional active site motifs	http://www.biochem.ucl.ac. uk/bsm/PROCAT/PROCAT. html
Databases		
Protein Data Bank	Database of solved protein structures	http://nist.rcsb.org/pdb
Expasy	Swiss-Prot protein sequence and structure database	http://www.expasy.ch/
CATH	Protein structure classification database	http://www.biochem.ucl.ac. uk/bsm/cath/
SCOP	Murzin's database of protein structure classification	http://scop.mrc-lmb.cam.ac. uk.scop/

sion. In practice, however, hybridization methods are limited by an inability to detect genes with no expressed sequence tag (EST) representation.

A methodological variation to expression analysis was developed which provides rapid, comprehensive sampling of cDNA populations together with sensitive detection of differences in mRNA abundance for both known and novel genes. By using this method, the gene expression in a rat model of pressure overload-induced cardiac hypertrophy was analyzed.

This mRNA profiling technique for determining differential gene expression utilizes, but does not require, prior knowledge of gene sequences. The method permits high-throughput reproducible detection of most expressed sequences with a sensitivity of greater than 1 part in 100,000. Gene identification by database query of a restriction endonuclease fingerprint, confirmed by competitive PCR using gene-specific oligonucleotides, facilitates gene discovery by minimizing isolation procedures. This process, called Gene Calling, was validated by analysis of the gene expression profiles of normal and hypertrophic rat hearts following *in vivo* pressure overload.[6]

Efficiency improvements in the development process for the next generation of therapeutic products require a strategy to overcome the 96% attrition rate between drug discovery projects at the laboratory level and new drugs in the marketplace. The required new strategies need to be directed towards the identification of therapeutic targets and their validation while addressing the milestones of the development process.

In order to fulfil these requirements, an improved understanding of the pathophysiology of human disease at the molecular level is necessary to elucidate alterations in biochemical pathways associated with disease phenotypes. These pathway changes reflect the genetic and biochemical alterations in expression resulting in the disease phenotype. Elucidating these changes can reveal disease-associated processes, and focus diagnostic and therapeutic development efforts on relevant disease markers and targets. Both gene and protein expression profiling methodologies are necessary to monitor and record changes in the expression of genes and gene products.

SAGE is a sequence-based genomics tool that features comprehensive gene discovery and quantitative gene expression capabilities. An experimentally and conceptually open system, SAGE can reveal which genes are expressed and their level of expression, rather than just quantifying the expression level of a predetermined and presently incomplete set of genes such as in experiments carried out by closed-system gene expression profiling platforms like microarrays. These superior aspects enable SAGE to be used as a primary discovery engine to characterize human disease at the molecular level while pinpointing potential targets and markers for therapeutic and diagnostic development.[7]

The study of gene expression profiles for identifying multi-effect phenomena supports the identification of causal genes or gene networks.

The molecular mechanisms of pulmonary fibrosis, which are as yet poorly understood, provide a suitable target system to analyze the genetic basis of the disease. Oligonucleotides were used to analyze gene expression programs that underlie pulmonary fibrosis in response to bleomycin, a drug that causes lung inflammation and fibrosis, in two strains of susceptible mice (129 and C57BL/6). The gene expression patterns were compared in these mice with 129 mice carrying a null mutation in the epithelial-restricted integrin β_6 subunit ($\beta_6^{-/-}$), which develop inflammation but are protected from pulmonary fibrosis. Cluster analysis identified two distinct groups of genes involved in the inflammatory and fibrotic responses. Analysis of gene expression at multiple time points after bleomycin administration showed sequential induction of subsets of genes that characterize each response. The availability of this comprehensive data set allows the accelerated development of active compounds and of strategies for intervention at various stages in the development of fibrotic diseases of the lungs and other organs.[8]

In view of the increasing requirements for analyzing gene function on a genomic scale, there is a clear need to develop methods that allow this analysis do be done in an economically efficient way.

A transposon-tagging strategy for the genome-wide analysis of disruption phenotypes, gene expression and protein localization was developed and applied to the large-scale analysis of gene function in the budding yeast *Saccharomyces cerevisiae*. A large collection of defined yeast mutants within a single genetic background was generated (over 11,000 strains), each carrying a transposon inserted within a region of the genome expressed during vegetative growth and/or sporulation. These insertions affect nearly 2000 annotated genes, thus representing about one-third of the 6200 predicted genes in the yeast genome. This collection was used to determine disruption phenotypes for almost 8000 strains using 20 different growth conditions. The data sets thus obtained were clustered and allowed the clear identification of groups of functionally related genes. More than 300 previously non-annotated open reading frames (ORFs) were discovered and analyzed by indirect immunofluorescence of more than 1300 transposon-tagged proteins. The study comprises more than 260,000 data points and represents a useful functional analysis of the yeast genome.[9]

A powerful technique for the identification of differentially expressed genes without cloning and amplification in a biological host has been developed. The method involves the cloning of nucleic acid molecules onto the surface of 5-µm beads rather than biological hosts, whereby a unique tag sequence is attached to each molecule. The tagged library is subsequently amplified. The unique tagging of the molecules is achieved by sampling a small fraction (1%) of a very large repertoire of tag sequences. The resulting library is hybridized to microbeads that each carries about 10^6 strands complementary to one of the tags. About 10^5 copies of each molecule are collected on each microbead. Since the clones are segregated on microbeads, they can be handled simultaneously and subsequently assayed separately. The broad utility of this approach was demonstrated by labeling and extracting microbead-bearing clones differentially expressed between two libraries by using a fluorescence-activated cell sorter (FACS). As no prior information about the cloned molecules is required, the method is especially useful where sequence data are incomplete or non-existent. The technique also permits the isolation of clones that are expressed only in certain tissues or that are differentially expressed between normal and diseased states. Clones of specific interest may then be spotted on other more cost-effective, low-density planar microarrays, which are focused on target tissues or diseases.[10]

The crucial experimental tools for measuring complex differential expression profiles are microarrays (DNA arrays). Experimental genomics in combination with the growing body of sequence information promises to thoroughly advance the studies of cells and cellular processes. Information on genomic sequence can be used experimentally with high-density arrays that allow complex mixtures of RNA and DNA to be tested in a parallel and quantitative way. DNA arrays can be used for many different purposes, such as to measure levels of gene expression (mRNA abundance) for tens of thousands of genes simultaneously. Measurements of gene expression and other applications of microarrays constitute a major thrust of genomics, and facilitate the use of sequence information for experimental design and data interpretation to understand function.[11]

The high-throughput technologies enable researchers to study gene expression for thousands of genes simultaneously, thus involving a huge repertoire of data. The resulting output of microarray studies is subject to experimental bias and substantial variability, thus requiring statistical analysis and the replication of studies.

Statistical methods for analyzing replicated cDNA microarray expression data and results of controlled experiments have provided valuable arguments for statistically controlled and validated experimentation. A study was conducted to investigate inherent variability in gene expression data, and the extent to which replication in an experiment produces more consistent and reliable findings. A statistical model was applied that describes the probability that mRNA is contained in the target sample tissue, subsequently converted to probe and ultimately detected on the slide. An analysis of the combined data from all replicates was also carried out. Of the 288 genes studied in this controlled experiment, 32 would be expected to produce strong hybridization signals because of the known presence of repetitive sequences within those genes. Results based on individual replicates show that there are 55, 36 and 58 highly expressed genes in replicates 1, 2 and 3, respectively. An analysis using the combined data from all three replicates reveals that only two of the 288 genes are incorrectly classified as expressed. The experiment demonstrates that any single microarray output is subject to substantial variability. By pooling data from replicates, a more reliable analysis of gene expression data can be achieved. Thus, designing experiments with replications will greatly reduce misclassification rates. At least three replicates should be used in designing experiments when using cDNA microarrays, particularly when gene expression data from single specimens are being analyzed.[12]

Functional genomic studies of a particular species depend on the identification of all of the expressed genes from the genome under investigation. The difficulty of genome-wide gene identification is proportional to the number of genes expressed in a particular genome. The number of expressed genes in the human genome is estimated at between 60,000 and 150,000 (references 1–4 in Wang *et al.*[13]). The EST (Expressed Sequence Tag) project and CGAP (Cancer Genome Anatomy Project) constitute major efforts to identify all of the expressed human genes. These efforts have resulted in the identification of 38,039 human genes from 886,936 human EST sequences through the EST project and 44,391 human genes from 804,804 EST sequences through the CGAP (reference 7 in Wang *et al.*[13]; also www. ncbi.nlm.hih.gov). The rate of novel gene identification through the EST project declined from 10.6% of EST sequences in 1996 (36,000 novel sequences from 340,000 EST sequences) to only 2.7% of EST sequences collected in 1998 (638 novel sequences identified from 23,038 EST sequences, and UniGene and dbEST databases), despite the fact that many expressed genes still were unidentified. Since most of the procedures in the current CGAP are similar to the EST project, the rate of novel gene identification in the CGAP may decline at some point from its current rate (5.4%), leaving many expressed human genes unidentified.

A possible explanation for this decline in gene identification is that genes expressed at a low level have a lower probability of being identified than those expressed at a higher level. There could also be systematic flaws in the current approaches, leading to difficulties in identifying novel genes. An analysis of the current technologies for genome-wide gene identification indicates that the existence of poly(dA/dT) sequences in cDNA clones is significantly responsible for the problem.

All cDNA libraries currently used for genome-wide gene identification are generated through oligo(dT) priming for reverse transcription. Since human mRNAs contain an average of 200 adenosine (A) residues at their 3′ end, oligo(dT) priming in reverse transcription results in the inclusion of various lengths of poly(dA/dT) sequences at the 3′ end of cDNA templates. The majority of genes in a given cell are expressed at lower levels and they constitute only a small portion of the total transcripts, whereas a small number of genes expressed at a high level constitute a large portion of the total transcripts. Direct screening of standard cDNA libraries will only identify highly expressed genes. Normalization and subtraction are required to reduce the high-abundance copies and to increase the representation of the low-abundance copies, thus allowing us to identify the genes expressed at a low level. Because of the presence of 3′ poly(dA/dT) sequences in the cDNA templates, random hybridization can occur anywhere along the poly(dA) and poly(dT) sequences during the normalization and subtraction process. This random hybridization results in the formation of tangled poly(dA)/poly(dT) double-stranded hybrids, independent of the sequence specificity. As double-stranded hybrids are removed, copies of many genes inadvertently annealed to the hybrids are lost. The genes expressed at low levels will be particularly affected. This phenomenon may contribute directly to the low efficiency of novel gene identification in efforts of genome-wide gene identification.

An experimental strategy was developed called screening poly(dA/dT)⁻ cDNAs for gene identification to overcome the above-described imbalances. The methodology experimentally increased the rate of novel gene identification in direct screening and SAGE tag collection.

Applying this strategy significantly enhances the efficiency of genome-wide gene identification and has an positive effect on gene identification in functional genomic studies for the identification of rare gene expression.[13]

The combination of microarrays and the studious application of programs to scan these resulting databases provide insight into complex phenomena like Human Leukocyte Antigen group DR (HLA-DR) in the immune response.

In the defense mechanisms of the immune system, helper T cell activation is essential for the initiation of a protective immune response to pathogens and tumors. HLA-DR, the predominant isotype of the human class II major histocompatibility complex (MHC), plays a central role in helper T cell selection and activation. HLA-DR proteins bind peptide fragments derived from protein antigens and display them on the surface of antigen-presenting cells (APC) for interaction with antigen-specific receptors of T lymphocytes.

The pockets in the HLA-DR groove are primarily shaped by clusters of polymorphic residues, and have a distinct chemical and specific size characteristics in different HLA-DR alleles. Each HLA-DR pocket can be characterized by pocket profiles – a quantitative representation of the molecular interaction of all natural amino acid residues with a given pocket. Pocket profiles have been shown to be nearly independent of the remaining HLA-DR cleft. A small sample database of profiles is sufficient to generate a large number of HLA-DR matrices, representing the majority of human HLA-DR peptide-binding specificity. These virtual matrices were incorporated in software (TEPITOPE) capable of predicting promiscuous HLA class II ligands. This software, in combination with DNA microarray technology, provides for the generation of comprehensive databases of candidate promiscuous T cell epitopes in human disease tissues. DNA microarrays are used to reveal genes that are specifically expressed or up-regulated in disease tissues. Subsequently, the prediction software enables the scanning of these genes for promiscuous HLA-DR binding sites. Starting from nearly 20,000 genes, a database of candidate colon cancer-specific and promiscuous T cell epitopes could be fully populated within a matter of days. The approach has provided directions for the development of epitope-based vaccines.[14]

DNA microarrays have the ability to analyze the expression of thousands of the same set of genes under at least two different experimental conditions. DNA microarrays require substantial amounts of RNA to generate the probes, especially when bacterial RNA is used for hybridization (50 μg of bacterial RNA contains approximately 2 μg of mRNA). A computer-based algorithm was developed for the prediction of the minimal number of primers to specifically anneal to all genes in a given genome. The algorithm predicts that 37 oligonucleotides should prime all genes in the *Mycobacterium tuberculosis* genome. The usefulness of the genome-directed primers (GDPs) was demonstrated in comparison to random primers for gene expression profiling using DNA microarrays. Both types of primers were used to generate fluorescent-labeled probes and to hybridize to an array of 960 mycobacterial genes. The GDP probes were more sensitive and more specific than the random-primer probes, especially when mammalian RNA samples were spiked with mycobacterial RNA. The GDPs were used for gene expression profiling of mycobacterial cultures grown to log or stationary growth phases. This approach is useful for accurate genome-wide expression analysis, in particular for *in vivo* gene expression profiling, as well as directed amplification of sequenced genomes.[15]

Interactions between protein complexes and DNA are at the core of essential cellular processes such as transcription, DNA replication, chromosome segregation and genome maintenance. Techniques are therefore needed to identify DNA loci that interact *in vivo* with specific proteins. A limited repertoire of techniques is presently available.[16,17]

One method involves *in situ* cross-linking followed by purification of protein–DNA complexes. This technique does have the inherent risk of artifacts induced by the cross-linking agent, but it requires specific antibodies against each protein of interest as well as a large number of cells. Another method employs *in vivo* targeting of a nuclease to mark binding sites of a specific protein. Induction of protein breaks is, however, likely to cause major changes in chromatin structure and activation of DNA damage checkpoint pathways – both being distinct disadvantages.

A novel technique was developed, named DamID, for the identification of DNA loci that interact *in vivo* with specific nuclear proteins in eukaryotes. By tethering *Escherichia coli* DNA adenine methyltransferase (Dam) to a chromatin protein, Dam can be targeted *in vivo* to native binding sites of this protein, resulting in local DNA methylation. Sites of methylation can subsequently be mapped using methylation-specific restriction enzymes or antibodies. The successful application of DamID both in *Drosophila* cell cultures and in whole flies was demonstrated. When Dam is tethered to the DNA-binding domain of GAL4, targeted methylation is limited to a region of a few kilobases surrounding a GAL4 binding sequence. By using DamID, a number of expected and unexpected target loci for *Drosophila* heterochromatin protein 1 were identified. DamID has usefulness for the genome-wide mapping of *in vivo* targets of chromatin proteins in various eukaryotes.[17]

The number of targets for therapeutic intervention is assessed by considering the number of genes, the different splicing of the RNAs, the resulting larger number of proteins, and the numerous processes involved in generating membranes, complexes and supramolecular structures.

Higher-order chromatin is essential for epigenetic gene control and for the functional organization of chromosomes. Differences in higher-ordered chromatin structure are linked with distinct covalent modifications of histone tails that regulate transcriptional 'on' or 'off' states, and influence chromosome condensation and segregation. Post-translational modifications of histone N-termini, particularly of H4 and H3, are well documented and have

functionally been characterized as changes in acetylation, phosphorylation and, most recently, methylation. In contrast to the large number of histone acetyltransferases (HATs) and histone deacetylases (HDACs) described, genes encoding enzymes that regulate phosphorylation or methylation of histone N-termini are only now being identified. The interdependence of the different histone tail modifications for the integration of transcriptional output or higher-order chromatin organization is as yet not fully understood.

Human SUV39H1 and murine Suv39h1 – mammalian homologs of *Drosophila Su(var)3-9* and of *Schizosaccharomyces pombe clr4* – encode histone H3-specific methyltransferases that selectively methylate Lys9 of the N-terminus of histone H3 *in vitro*. The catalytic motif was mapped to the evolutionarily conserved SET domain, which requires adjacent cysteine-rich regions to confer histone methyltransferase activity. Methylation of Lys9 interferes with phosphorylation of Ser10, but is also influenced by pre-existing modifications in the N-terminus of H3. *In vivo*, deregulated SUV39H1 or disrupted Suv39h1 activity modulate H3 Ser10 phosphorylation in native chromatin and induce aberrant mitotic divisions. The data demonstrate a functional interdependence of site-specific H3 tail modifications and propose a dynamic mechanism for the regulation of higher-order chromatin.[18]

Transcription is controlled in part by the dynamic acetylation and deacetylation of histone proteins. The latter process is mediated by HDACs. Analysis of the regulation of HDAC activity in transcription has focused primarily on the recruitment of HDAC proteins to specific promoters or chromosomal domains by association with DNA-binding proteins. To characterize the cellular function of the identified HDAC4 and HDAC5 proteins, complexes were isolated by immunoprecipitation. Both HDACs were found to interact with 14-3-3 proteins at three phosphorylation sites. The association of 14-3-3 with HDAC4 and HDAC5 results in the sequestration of these proteins in the cytoplasm. Loss of this interaction allows HDAC4 and HDAC5 to translocate to the nucleus, interact with HDAC3 and repress gene expression. Regulation of the cellular localization of HDAC4 and HDAC5 represents a mechanism for controlling the transcriptional activity of these class II HDAC proteins.[19]

In *Drosophila*, compensation for the reduced dosage of genes located on the single male X chromosome involves doubling their expression in relation to their counterparts on the female X chromosomes. Dosage compensation is an epigenetic process involving the specific acetylation of histone H4 at lysine 16 by the histone acetyltransferase MOF. Although MOF is expressed in both sexes, it only associates with the X chromosome in males. Its absence causes male-specific lethality. MOF is part of a chromosome-associated complex comprising male-specific lethal (MSL) proteins and at least one non-coding roX RNA. The integration of MOF into the dosage compensation complex is still not understood. The association of MOF with the male X chromosome depends on its interaction with RNA. MOF binds specifically through its chromodomain to roX2 RNA *in vivo*. *In vitro* analyses of the MOF and MSL-3 chromodomains indicate that these chromodomains may function as RNA interaction modules. Their interaction with non-coding RNA may target regulators to specific chromosomal sites.[20]

The structural and functional organization of chromatin needs to be considered in studies of gene function, gene expression and molecular interaction in pharmaceutical interventions.

The functional regulation of chromatin is closely related to its spatial organization within the nucleus. In yeast, perinuclear chromatin domains constitute areas of transcriptional repression. These silent domains are defined by the presence of perinuclear telomere clusters. The only protein found to be involved in the peripheral localization of telomeres is Yku70/Yku80. This conserved heterodimer can bind telomeres and functions in both repair of DNA double-strand breaks and telomere maintenance. These findings do not describe the underlying structural basis of perinuclear silent domains. Nuclear pore complex extensions formed by the conserved TPR homologs Mlp1 and Mlp2 are responsible for the structural and functional organization of perinuclear chromatin. Loss of MLP2 results in a severe deficiency in the repair of double-stranded breaks. Double deletions of MLP1 and MLP2 disrupt the clustering of perinuclear telomeres and releases telomeric gene expression. These effects are probably mediated through the interaction with Yku70. Mlp2 physically tethers Yku70 to the nuclear periphery, thus forming a link between chromatin and the nuclear envelope. This structural link is docked to nuclear pore complexes through a cleavable nucleoporin, Nup145. Through these interactions, nuclear pore complexes organize a nuclear subdomain that is intimately involved in the regulation of chromatin metabolism.[21]

The packaging of the eukaryotic genome in chromatin presents barriers that restrict the access of enzymes that process DNA. To overcome these barriers, cells possess a number of multi-protein, ATP-dependent chromatin remodeling complexes, each containing an ATPase subunit from the SNf2/SW12 superfamily. Chromatin remodeling complexes function by increasing nucleosome mobility and are clearly implicated in transcription. SNF2/SW12- and ISWI-related proteins were analyzed to identify remodeling complexes that potentially assist other DNA transactions. A complex from *S. cerevisiae* was purified that contains the Ino80 ATPase. The Ino80 complex contains about 12 polypeptides including two proteins related to the bacterial RuvB DNA helicase, which catalyzes branch migration of Holliday junctions. The purified complex remodels chromatin, facilitates transcription *in vitro* and displays 3′ to 5′ DNA helicase activity. Mutations of Ino80 show hypersensitivity to agents that cause DNA damage, in addition to defects in transcription. Chromatin remodeling driven by the Ino80 ATPase may be connected to transcription as well as DNA damage repair.[22]

SNPs are point mutations that constitute the most common type of genetic variation and are found at a rate of 0.5–10 per 1000 base pairs within the human genome. SNPs are stable mutations that can be contributory factors for human disease and can also serve as genetic markers. The complex interaction between multiple genes and the environment necessitates the tracking of SNPs in large populations in order to elucidate their contribution to disease development and progression. Several projects are intensively pursuing the identification of human SNPs through large-scale mapping projects with high-density arrays, mass spectrometry (MS), molecular beacons, peptide nucleic acids and the 5′ nuclease assay. A study has integrated microelectronics and molecular biology for the discrimination of SNPs, and a rapid assay for SNP detection was developed that utilizes electronic circuitry on silicon microchips. The method was validated by the accurate discrimination of blinded DNA samples for the complex quadra-allelic SNP of mannose-binding protein. The microchip directed the transport, concentration and attachment of amplified patient DNA to selected electrodes (test sites), creating an array of DNA samples. Through control of the electric field, the microchip enabled accurate genetic identification of these samples using fluorescent-labeled DNA reporter probes. The accuracy was established by internal controls of dual-labeled reporters and by using mismatched sequences in addition to the wild-type and variant reporter sequences to validate the SNP genotype. The ability to customize this assay for multiple genes offers advantages for bringing the assay to the clinical laboratory.[23]

Dynamic allele-specific hybridization, a method to detect SNPs, is based on dynamic heating and coincident monitoring of DNA denaturation and avoids the use of additional enzymes or reaction steps.[24]

The most common DNA sequence variations, SNPs, are stable and widely scattered across the chromosome. Once constructed, a high-density SNP map of several hundred thousand markers will be an indispensable tool for genome-wide association studies to identify genes that contribute to disease risk and individual differences in drug response. To facilitate large-scale SNP identification, new technologies are being developed to replace gel-based resequencing. Highly redundant, sequence-specific oligonucleotide arrays were hybridized against fluorescent-labeled DNA targets. The hybridization patterns are scanned for possible mismatches in sequences (references 2–5 in Tang *et al.*[25]).

A different experimental approach to SNP detection combines mass spectrometric detection with enzymatic extension of primers hybridized to immobilized DNA target arrays. The advantage of this combination is high specificity and high accuracy of allele identification.

Silicon chips with immobilized target DNAs were used for accurate genotyping by MS. Genomic DNAs were amplified with PCR and the amplified products were covalently attached to chip wells via *N*-succinimidyl(4-iodoacetyl)amino benzoate (SIAB) chemistry.

Primer annealing, extension and termination were performed on at the microliter scale directly in the chip wells in parallel. Diagnostic products thus generated were detected *in situ* by using matrix-assisted laser desorption ionization (MALDI)-MS. This miniaturized method has applicability for accurate, high-throughput, low-cost identification of genetic variations.[25] With the accumulation of large-scale sequence data, emphasis in genomics is shifting from determining gene structure to testing gene function, relying on reverse genetic methodology. The feasibility of screening for chemically induced mutations in target sequences in *Arabidopsis thaliana* was explored. The TILLING (Targeted Induced Local Lesions In Genomes) method combines the efficiency of ethyl methanesulfonate (EMS)-induced mutagenesis with the ability of denaturing high-performance liquid chromatography (DHPLC) to detect base pair changes by heteroduplex analysis. This method generates a wide range of mutant alleles, is fast and automatable, and is applicable to any organism that can be chemically mutagenized.[26]

Strategies to experimentally detect translocations are important because of the numerous cases of genes in leukemia-associated translocations. Such methods include Southern blot analysis, which is not as sensitive as PCR, karyotype analysis and fluorescence *in situ* hybridization (FISH) with specific probes. Reverse transcriptase (RT)-PCR with gene-specific primers detects only a fraction of translocations because there are no primers available for many of the genes involved.

Translocations of the *MLL* gene at chromosome band 11q23 occur in leukemias of infants and in leukemias associated with DNA topoisomerase II inhibitors. The ability to rapidly identify *MLL* translocations, whether by cytogenetic or molecular approaches, is relevant for diagnosis, prognosis, and treatment. *MLL* is an example of a gene involved in translocations with numerous different partner genes and the specific partner gene with which *MLL* is fused may have an impact on the clinical response.

Identifying translocations of the *MLL* gene at chromosome band 11q23 is important for the characterization and treatment of leukemia. Cytogenetic analysis does not always find the translocations and the many partner genes of *MLL* make molecular detection difficult. cDNA panhandle PCR was developed to identify der(11) transcripts regardless of the partner gene. By reverse transcribing first-strand cDNAs with oligonucleotides containing coding sequence from the 5' *MLL* breakpoint cluster region at the 5' ends and random hexamers at the 3' ends, the known *MLL* sequence was attached to the unknown partner sequence. This enabled the formation of stem–loop templates with the fusion point of the chimerical transcript in the loop and the use of *MLL* primers in two-sided PCR. The assay was validated by detection of the known fusion transcript and the transcript from the normal *MLL* allele in the cell line MV4-11. cDNA panhandle PCR then was used to identify the fusion transcripts in two cases of treatment-related acute myeloid leukemia where the karyotypes were normal and the partner genes unknown. cDNA panhandle PCR revealed a fusion of *MLL* with *AF-10* in one case and a fusion of *MLL* with *ELL* in the other. Spliced transcripts and exon scrambling were detectable by the method. Leukemias with normal karyotypes may contain cryptic translocations of *MLL* with a variety of partner genes. cDNA panhandle PCR is useful for identifying *MLL* translocations and determining unknown partner sequences in the fusion transcripts.[27]

An efficient and rapid subtraction hybridization technique (RaSH) allows the identification and cloning of differentially expressed genes[688].

1.3   Proteomics

Proteomics is the large-scale analysis of proteins and constitutes a valuable tool for understanding gene function. Proteomics deals mainly with protein microcharacterization for large-scale identification of proteins and their post-translational modifications, differential-display proteomics for comparison of protein levels with potential application in a wide range of diseases and studies of protein–protein interactions using techniques such as

MS or the yeast two-hybrid system. Due to the difficulty in predicting the function of a protein based on homology to other proteins or even their three-dimensional structure, the determination of components of a protein complex or of a cellular structure is central to functional analysis.

Proteomics provides a powerful set of tools for the large-scale study of gene function at the protein level. In particular, the MS studies of gel-separated proteins are leading to a re-emphasis of biochemical studies of protein function. Protein characterization continues to improve in terms of throughput, sensitivity and completeness. Post-translational modifications are increasingly being studied.[28]

Proteomics is the linguistic equivalent to genomics (from genome) and refers to the concept of the whole set of expressed proteins – the proteome. It involves research into the proteome using the technologies of protein separation (e.g., by two-dimensional electrophoresis) plus identification.[29]

Genome sequencing projects are only the starting point for understanding the structure and, in particular, the function of proteins. A major challenge is the study of the co-expression of thousands of genes under physiological and pathophysiological conditions, and the definition of an organism by this pattern of gene expression. To define protein-based gene expression analysis, the concept of the proteome and the field of proteomics (studies of the proteome) were defined as the proteome being the entire PROTEin complement expressed by a genOME.[30]

The field of proteomics is rapidly expanding towards increases in the number of proteins studied, automation of separation and subsequent structural analyses, studies of protein–protein interactions, applications of automated MS analyses, and development of software to process the resulting data.[31] Further to the structural identification of proteins, the protein interactions are crucial to understanding the cellular system. Protein interactions are analyzed by biochemical, physical, cellular and genetic means.

A substantial number of proteins involved in transcriptional regulation have been identified, but the majority are probably still unknown. Genetic strategies such as the one-hybrid assay and phage-display techniques suffer from the inability to detect proteins whose specific binding to a DNA element is dependent upon accessory proteins. An approach relying on MALDI time-of-flight (TOF) MS identifies DNA-binding proteins isolated from cell extracts by virtue of their interaction with double-stranded DNA probes immobilized onto small, paramagnetic particles.

This method enables the rapid identification of DNA-binding proteins. Immobilized DNA probes harboring a specific sequence motif are incubated with cell or nuclear extract. Proteins are analyzed directly off the solid support by MALDI-TOF. The determined molecular masses are often sufficient for identification. If not, the proteins are subject to MS peptide mapping followed by database searches. Apart from protein identification, the protocol also yields information on post-translational modifications. The protocol was validated by the identification of known prokaryotic and eukaryotic DNA-binding proteins, and is use provided evidence that poly(ADP-ribose) polymerase exhibits DNA sequence-specific binding to DNA.[32]

A method for solving the three-dimensional structures of protein–protein complexes in solution on the basis of experimental nuclear magnetic resonance (NMR) restraints provides requisite translational [i.e. intermolecular nuclear Overhauser enhancement (NOE) data] and orientational (i.e. backbone ^1H–^{15}N dipolar couplings and intermolecular NOEs) information. Providing high-resolution structures of the proteins in the unbound states are available and no significant backbone conformational changes occur upon complexation (which can readily be assessed by analysis of dipolar couplings measured on the complex), accurate and rapid docking of the two proteins can be achieved. The method, which is demonstrated for the 40 kDa complex of enzyme I and the histidine phosphocarrier protein, involves the application of rigid body minimization using a target function comprising only three terms, i.e. experimental NOE-derived intermolecular interproton distance and dipolar coupling restraints, and a simple intermolecular van der Waals' repulsion potential. This approach promises to dramatically reduce the amount of time and effort required to solve the structures of protein–protein complexes by NMR and to extend the capabilities of NMR to larger protein–protein complexes, possibly up to molecular masses of 100 kDa and more.[33]

The genomics revolution has changed the paradigm for the comprehensive analysis of biological processes and systems. Genetic, biochemical and physiological biological processes and systems may be described by comparison of global, quantitative gene expression patterns from cells or tissues representing different states. For these comparisons, applicable methods for the precise measurement of gene expression are being developed and applied.

Proteome analysis is most commonly accomplished by a combination of two-dimensional gel electrophoresis to separate and visualize proteins, and MS for protein identification. This technique is powerful, mature and sensitive, but challenges remain concerning the characterization all of the elements of a proteome. More than 1500 features were visualized by silver staining a narrow pH range (4.9–5.7) two-dimensional gel in which 0.5 mg of total soluble yeast protein was separated. Fifty spots migrating to a region of 4 cm^2 were subjected to MS protein identification. Despite the high sample load and extended electrophoretic separation, proteins from genes with codon bias values of <0.1 (lower abundance proteins) were not found, even though fully one-half of all yeast genes fall into that range. Proteins from genes with codon bias values of <0.1 were found, however, if protein amounts exceeding the capacity of two-dimensional gels were fractionated and analyzed. The large range of protein expression levels limits the ability of the two-dimensional gel/MS approach to analyze proteins of medium to low abundance, and thus the potential of this technique for total proteome analysis is limited.[34]

Table 1.2 points to another difficulty, co-migration, in identifying proteins from two-dimensional gels.

Table 1.3 lists the theoretical amounts of starting protein to visualize individual proteins of different abundances.

Protein–protein interactions are studied, for example, by the yeast two-hybrid-system, a genetic technique designed to identify novel protein–protein interactions that were previously detected by biochemical studies. All two-hybrid screening systems rely on the fact that transcriptional activation and DNA-binding domains of transcription factors are modular in nature. In these systems, the coding sequence for the DNA-binding domain of a transcription factor such as Gal4 or LexA is fused to the cDNA of a protein of interest, termed

Table 1.2. Proteins comigrating in a single silver-stained spot on a 2D gel [Refs in 34].

Gene name*	Peptide sequences identified†	pl*	Molecular mass, kDa
GDH1	(K)VIELGGTVVSLSDSK (K)FIAEGSNMGSTPEAIAVFETAR (R)EIGYLFGAYR (K)VLPIVSVPER	5.50	49.6
RPT3	(R)ENAPSIIFIDEVDSIATK	5.32	47.9
SUB2	(R)DVQEIFR (K)LTLHGLQQYYIK (R)INLAINYDLTNEADQYLHR (K)NKDTAPHIVVATPGR (R)FLQNPLEIFVDDEAK	5.36	50.3
TIF3	(R)GSNFQGDGREDAPDLDWGAAR (R)ADLVAVLK (K)ITIPIETANANTIPLSELAHAK (R)EREEVDIDWTAAR (R)EREEPDIDWSAAR	5.17	48.5
VMA 1	(K)VGHDNLVGEVIR	5.09	67.7
YFRO44C	(R)YPSLSIHGVEGAFSAQGAK (K)LVYGVDPDFTR (K)FISEQLSQSGFHDIK (R)TELIHDGAYWVSDPFNAQFT'AAK (K)ILIDGIDEMVAPLTEK	5.54	52.9

* Gene names, predicted pl, and molecular mass values are from the Yeast Proteome Database (YPD) (13).
† Peptide sequences were identified automatically and verified manually by using SEQUEST (12).

Table 1.3. Theoretical required total starting protein amounts for individual protein visualization by staining [Refs in 34].

Protein abundance, copies per cell	Silver staining*		Coomassie staining*	
	Protein amount mg†	Number of cells	Protein amount mg†	Number of cells
10	20.073	1.20×10^9	2007.3	1.20×10^{11}
100	2.007	1.20×10^8	200.7	1.20×10^{10}
1,000	0.201	1.20×10^7	20.1	1.20×10^9
10,000	0.020	1.20×10^6	2.0	1.20×10^8
100,000	0.002	1.20×10^5	0.2	1.20×10^7

* Protein detection limits for silver and Coomassie staining were 1 and 100 ng, respectively.
† Soluble yeast protein was calculated based on 1 mg of yeast protein being derived from harvesting 6×10^7 cells. Calculations are based on a protein molecular mass of 50 kDa and 100% efficiencies of the procedures used.

the bait. The fusion protein thus encoded tethers the bait to the promoter region of a reporter gene. A second fusion of a cDNA library with the coding sequence of a transcriptional activation domain is termed the prey. Functional reconstitution of transcription factor activity occurs upon association of the bait and prey protein domains. This interaction is detected by expression of reporter genes that are dependent upon the bait's DNA-binding domain. The two-hybrid system is a powerful tool for screening libraries for novel protein–protein interactions and for the isolation of factors that promote or disrupt protein interactions. A differential two-hybrid yeast system can screen for interactions between prey proteins and two different bait proteins through the activation of bait-specific reporters. It allows the identification of proteins that interact differentially with one bait tethered to the Gal4 DNA-binding domain and another bait tethered to the LexA DNA-binding domain.[35]

To detect interactions between proteins of vaccinia virus, a two-hybrid analysis was carried out to assay every pair wise combination. An array of yeast transformants that contained each of the 266 predicted viral ORFs as Gal4 activation domain hybrid proteins was constructed. The array was individually mated to transformants containing each ORF as a Gal4 DNA-binding domain hybrid and diploids expressing the two-hybrid reporter gene were identified. Of the 70,000 combinations, 37 protein–protein interactions were found, including 28 that were previously unknown. In some cases, e.g., late transcription factors, both proteins were known to have related roles although there was no prior evidence of physical associations. For some other interactions, neither protein had a known role. In the majority of cases, one of the interacting proteins was known to be involved in DNA replication, transcription, virion structure or host evasion, thereby providing a clue to the role of the other uncharacterized protein in a specific process.[36]

Direct interaction between proteins is an important means of relaying information in a network or chain of signaling molecules.

Of the numerous classes of cell-surface-receptor signaling molecules, synaptic transmission between individual neurons is mediated largely by two major structurally and functionally distinct neurotransmitter receptor families, i.e. ligand-gated channels and G protein-coupled receptors (GPCRs). Although both are integral membrane proteins, ligand-gated receptors modulate synaptic neurotransmission directly through the formation and opening of an inherent ion channel, whereas GPCRs are single-polypeptide proteins containing seven hydrophobic transmembrane domains that transduce extracellular neurotransmitter signals into the cell interior by interacting with heterotrimeric G proteins. These in turn modulate a diverse array of cellular effectors to produce changes in cellular second-messenger systems and/or ionic conductance, and ultimately physiological responsiveness.

GABA(A) (γ-aminobutyric-acid A) and dopamine D_1 and D_5 receptors represent two structurally and functionally divergent families of neurotransmitter receptors. The former comprises a class of multi-subunit ligand-gated channels mediating fast interneuronal synaptic transmission, whereas the latter belongs to the seven-transmembrane-domain single-polypeptide receptor super family that exerts its biological effects, including the modulation of GABA(A) receptor function, through the activation of second-messenger signaling cascades by G proteins. GABA(A)-ligand-gated channels complex selectively with D_5 receptors through the direct binding of the D_5 C-terminal domain with the second intracellular loop of the GABA(A) γ2(short) receptor subunit. This physical association enables mutually inhibitory functional interactions between these receptor systems. The data highlight a new signal transduction mechanism whereby subtype-selective GPCRs dynamically regulate synaptic strength independently of classically defined second-messenger systems and provide a heuristic framework in which to view these receptor systems in the maintenance of psychomotor disease states.[37]

With the completion of an increasing number of genomic sequences, attention is focusing on the interpretation of the data contained in sequence databases in terms of structure, function and control of biological systems. Approaches for global profiling of gene expression analysis at the mRNA level are identifying clusters of genes for which the expression is idiotypic for a specific state. These sensitive methods of profiling do not indicate changes in protein expression. Quantitative proteome analysis, the global analysis of protein expression, is a complementary method to study steady-state gene expression and perturbation-induced changes. In a further step towards functional studies, proteome analysis provides more accurate information about biological systems and pathways because the measurements directly focus on the actual biological effector molecules.

Quantitative protein analysis is accomplished by combining protein separation, most commonly by high-resolution two-dimensional polyacrylamide gel electrophoresis (PAGE), with MS-(mass spectrometry) based or tandem MS (MS/MS)-based sequence identification of selected, separated protein species. This method is sequential, labor intensive and difficult to automate. It selects against specific classes of proteins, such as membrane proteins, very large and small proteins, and extremely acidic or basic proteins. The technique has a bias toward highly abundant proteins, as lower abundance regulatory proteins (e.g., transcription factors and protein kinases) are rarely detected when total-cell lysates are analyzed.

A method has been developed for the accurate quantification and concurrent sequence identification of the individual proteins within complex mixtures. The method is based on a class of new chemical reagents termed isotope-coded affinity tags (ICATs) and tandem MS. With this strategy, protein expression in the yeast *S. cerevisiae* was compared, using either ethanol or galactose as a carbon source. The measured differences in protein expression correlated with known yeast metabolic function under glucose-repressed conditions. The method is redundant if multiple cysteinyl residues are present and the relative quantification is highly accurate because it is based on stable isotope dilution techniques. The ICAT approach provides a broadly applicable means to compare quantitatively global protein expression in cells and tissues.[38]

Phage antibody libraries provide a source of binders to almost any antigen, including many that were previously considered difficult targets, such as self-antigens or cell-surface proteins. Phage selection involves repeated rounds of growth, panning and infection, which selects both for binding and for antibody fragments that are well expressed on phages. When selecting against highly complex targets, there is often a strong bias for antibodies directed against immunodominant epitopes and abundant proteins.

A technique for high-throughput screening of recombinant antibodies, based on the creation of antibody arrays, uses robotic picking and high-density gridding of bacteria containing antibody genes followed by filter-based enzyme-linked immunosorbent assay (ELISA) screening to identify clones that express binding antibody fragments. By eliminating the need for liquid handling, up to 18,342 different antibody clones can be screened at a time, and, because the clones are arrayed from master stocks, the same antibodies can be double-spotted and screened simultaneously against 15 different antigens. The technique was applied in several different applications, including the isolation of antibodies against impure proteins and complex antigens, where several rounds of phage display often fail. The results indicate that antibody arrays can be used to identify differentially expressed proteins.[39]

The array format for analyzing peptide and protein function offers an attractive experimental alternative to traditional library screens. Approaches range from synthetic peptide arrays to whole proteins expressed in living cells. Comprehensive sets of purified peptides and proteins permit high-throughput screening for discrete biochemical properties, whereas formats involving living cells facilitate large-scale genetic screening for novel biological activities. Three major genome-scale studies using yeast as a model organism have investigated different aspects of protein function, including biochemical activities, gene disruption phenotypes and protein–protein interactions. Such studies show that protein arrays can be used to examine in parallel the functions of thousands of proteins previously known only by their DNA sequence.[40]

The systematic approach of arrays towards the simultaneous studies of structure and function is applied to proteins, promising advances in the research of protein interactions on a large scale.

Systematic efforts are currently under way to construct defined sets of cloned genes for high-throughput expression and purification of recombinant proteins. To facilitate subsequent studies of protein function, miniaturized assays were developed that accommodate extremely low sample volumes and enable the rapid, simultaneous processing of thousands of proteins. A high-precision robot designed to manufacture complementary DNA arrays was used to spot proteins onto chemically derivatized glass slides at extremely high spatial densities. The proteins attached covalently to the slide surface yet retained their ability to interact specifically with other proteins, or with small molecules, in solution. Three applications for protein microarrays were demonstrated: screening for protein–protein interactions, identifying the substrates of protein kinases and identifying protein targets of small molecules.[41]

Protein–DNA interactions are crucial processes in gene regulation and gene function. Study of complexes, receptors and assemblies in cells, organs and organisms is necessary to understand and manipulate the functions.

The study of components of transcriptional complexes and the effects of ligands in altering the pattern of association has been studied with μESI-MS. This allows the screening of transcriptionally active libraries and enables rapid data acquisition, especially data on the molecular masses of complexes. These methods and data have potential for automation in drug discovery efforts by the identification of non-covalent ligand interactions with associating molecules and the identification of specific ligands that effect changes in transcription. The technique is useful for studying functional and regulatory protein–DNA and protein–DNA complexes.[42]

The new sciences (genomics and proteomics) and the new technologies (combinatorial chemistry, bioinformatics, biochips and biosensors) are increasing the speed of drug discovery, and provide large numbers of target molecules, but modest numbers of new lead structures.[43]

An assessment of the contributions of proteomics to the biotechnological industrial value creation outlines key factors for successful technologies.[44]

Protein–protein interactions play pivotal roles in various aspects of the structural and functional organization of the cell, and their complete description is indispensable for the thorough understanding of the cell. A comprehensive system for addressing this understanding is to examine two-hybrid interactions in all possible combinations between proteins of *S. cerevisiae*. All of the yeast ORFs were cloned individually as a DNA-binding domain fusion (bait) in a MATα strain and as an activation domain fusion (prey) in a MATα strain, and subsequently divided into pools, each containing 96 clones. These bait and prey clone pools were systematically mated with each other and the transformants were subjected to strict selection for the activation of three reporter genes followed by sequence tagging. Initial examination of about 4×10^6 different combinations, constituting around 10% of the total to be tested, revealed 183 independent two-hybrid interactions, more than half of which were entirely novel. The obtained binary data allow us to describe more complex interaction networks, including one that may explain a mechanism for the connection between distinct steps of vesicular transport. The approach described provides many leads for the integration of various cellular functions and serves as a major technology for the completion of the protein–protein interaction map.[45]

Table 1.4 summarizes the results of two-hybrid screening. Figure 1.1 describes complex two-hybrid interaction networks.

Two large-scale yeast two-hybrid screens were carried out to identify protein–protein interactions between full-length ORFs predicted from the *S. cerevisiae* genome sequence. In one approach, a protein array of about 6000 yeast transformants was constructed, with each transformant expressing one of the ORFs as a fusion to an activation domain. This array was screened by a simple and automated procedure for 192 yeast proteins, with positive responses identified by their positions in the array. A second approach consisted of studying cells expressing one of about 6000 activation domain fusions pooled to generate a library. A high-throughput screening procedure was used to screen nearly all of the 6000

Table 1.4. Two-hybrid screening summary [Refs in 45].

Mating reactions	430
Combinations to be examined	~4×10^6
Positive colonies	866
Sequence tag pairs obtained	750
Independent two-hybrid interactions	183
Bidirectionally detected interactions	16
Total independent interactions	175
Known interactions	12
Previously unreported interactions	163
Highly likely	26
Homotypic	32
Novel	105

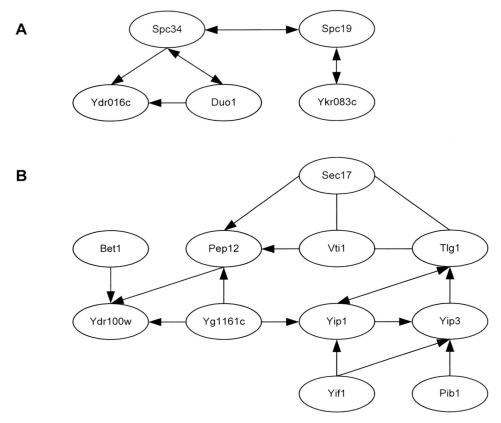

Figure 1.1. Complex two-hybrid interaction networks. Two-hybrid interaction networks for proteins related to spindle pole body (A) and vesicular transport (B) are shown. Arrows indicate two-hybrid interactions, beginning from the bait and ending at the prey. Double-headed arrows mean that the interactions
were detected bidirectionally. Note that arrows indicate the direction of two-hybrid interactions but not any
biological orientation. Solid lines indicate known interactions recorded in the Yeast Proteome Database
(14) but not yet detected by our two-hybrid screening [Refs in 45].

predicted yeast proteins, expressed as Gal4 DNA-binding domain fusion proteins, against
the library and characterized positive by sequence analysis. These experiments yielded the
detection of 957 putative interactions involving 1004 *S. cerevisiae* proteins. The data reveal
interactions that place functionally unclassified proteins in a biological context, interactions between proteins involved in the same biological function and interactions that link
biological functions into larger cellular processes.[46]

Selection and screening methods are effective tools for studying macromolecular interactions. Valuable methods are the yeast-based one-hybrid and two-hybrid systems (for studying protein–DNA and protein–protein interactions, respectively) and bacterial-based phage
display methods (for studying either type of interaction). These systems have been used to

identify interaction partners for particular DNA or protein targets and they were used in combination with mutagenesis or randomization strategies to study the details of biologically important interactions (reviewed in 1–5 in Joung *et al.*[47]).

A bacterial 'two-hybrid' system that readily allows selection from libraries larger than 10^8 in size was developed. The bacterial system may be used to study either protein–DNA or protein–protein interactions, and it offers a number of potentially significant advantages over existing yeast-based one-hybrid and two-hybrid methods. The system was tested by selecting zinc finger variants from a large randomized library that bind tightly and specifically to desired DNA target sites. The method allows sequence-specific zinc fingers to be isolated in a single selection step. Thus it is faster than phage display strategies that typically require multiple enrichment/amplification cycles. Given the large library sizes that can be handled with the bacterial-based selection system, this method is an efficient tool for the identification and optimization of protein–DNA and protein–protein interactions.[47]

Many intracellular processes are mediated through inducible protein–protein interactions. Methods that allow such processes to be manipulated are powerful tools for understanding and controlling cellular activities. The use of chemical inducers of dimerization (dimerizers) is a versatile approach. Cells are engineered to express chimeric proteins comprising a signaling domain fused to a drug-binding domain; treatment with bivalent ligands cross-links the proteins and initiates signaling. This strategy has been used to create inducible alleles of numerous signaling proteins that are activated by dimerization. A strategy for achieving the opposite mode of control, wherein proteins are constitutively associated until addition of a drug, is valuable for probing the consequences of rapidly abolishing oligomerization events inside cells. A point mutant was discovered and characterized that has the unusual property of forming discrete dimers that can be dissociated by ligand. This is the basis of a reverse dimerization system that is applicable as a disaggregation switch for intracellular processes.

Chemically induced dimerization provides a general way to gain control over intracellular processes. Typically, FK506-binding protein (FKBP) domains are fused to a signaling domain of interest, allowing cross-linking to be initiated by addition of a bivalent FKBP ligand. In the course of protein engineering studies on human FKBP, a single point mutation was discovered in the ligand-binding site (Phe36 → Met) that converts the normally monomeric protein into a ligand-reversible dimer. Two-hybrid, gel filtration, analytical ultra-centrifugation and X-ray crystallographic studies show that the mutant (F_M9) forms discrete homodimers with micromolar affinity that can be completely dissociated within minutes by addition of monomeric synthetic ligands. These unexpected properties form the basis for a 'reverse dimerization' regulatory system involving F_M fusion proteins, in which association is the ground state and addition of ligand abolishes interactions. This strategy was used to rapidly and reversibly aggregate fusion proteins in different cellular compartments, and to provide an off switch for transcription. Reiterated F_M domains should be generally useful as conditional aggregation domains (CADs) to control intracellular events where rapid, reversible dissolution of interactions is required. Dimerization is apparently a latent property of the FKBP fold. The crystal structure reveals a remarkably complementary interaction between the monomer binding sites, with only subtle changes in side chain disposition accounting for the dramatic change in quaternary structure.[48]

Table 1.5A. Application of MALDI-MS profiling of mass-limited tissue and single cell samples [Refs in 49].

Specimen	Sample	MALDI	Sample preparation	Refs
Microorganisms *Haemophilus.*	Proteins from intact and disrupted bacterial cells allow rapid species determination.	Molecular-weight profiling.	Whole-cell lysate followed by on-plate matrix dried-droplet deposition.	23
Plants *Dahlia variabilis L.* Onion bulbs *Allium cepa* L.	Fructans have been characterized from onion bulb extracts. Epidermal or parenchymal cell layers of the onion bulb were subjected to direct MALDI-MS. Both proteins and oligosaccharides were mass profiled	The size distribution of fructans was determined by MALDI-MS. Metastable ion scanning was performed for identification	Tissue homogenate and extract without analyte purification. The supernatant of the extract was mixed with MALDI matrix on-plate and allowed to air dry. Single cell was covered with matrix.	29
Pulmonate mollusk *Lymnaea stagnalis*	Distribution and processing of cardioactive, egg laying, myomodulin, and insulin-related peptides in cells and nerves.	First demonstration of cell profiling, use of different mass analyzers, direct peptide sequencing.	Matrix deposition on fresh tissue or cells by the dried-droplet method.	27, 30– 37
Pulmonate mollusk *Helix aspersa*	Myomodulin-related peptides in neuropil.	MALDI on ganglion extracts used to determine the ratio of peptide concentrations.	Tissues were homogenized and extracted in MALDI matrix, the mixture was then co-deposited with matrix on-plate.	38
Opistobranch mollusk *Aplysia californica*	Distribution and processing of egg-laying peptides, neuropeptide Y, cardioactive peptides, new peptides in F-cluster including *Aplysia* insulin, post-translational modifications of R3-14 peptides in cells, peptide distribution and transport in connectives, *Aplasia mytilus* inhibitory-peptide-related peptides.	Matrix-rinsing for marine samples, direct peptide sequencing, sampling releaseates and single secretory granules, imaging of cultural neurons and dried-droplet samples, ligation experiments to determine direction of peptide transport, combination of single-cell MALDI and whole amount *in situ* hybridization and immunocytochemistry.	Following tissue dissection or single-cell isolation, MALDI matrix was deposited on single cells or nerve sections using dried-droplet approach. Salt removal by matrix-rinsing protocol. Peptide sequencing using dual-matrix approach.	28, 39– 47

Table 1.5B. Application of MALDI-MS profiling of mass-limited tissue and single cell samples [Refs in 49].

Specimen	Sample	MALDI	Sample preparation	Refs
Crayfish *Orconectes limosus*	Hyperglycemic hormone and related peptides in neurosecretory cells and glands.	Combination of MALDI-MS profiling and immunocytochemistry on the same sample.	Different matrices were used with dried-droplet, thin-layer and sandwich methods for gland and organ extracts. Cells were recovered for MALDI-MS after fixation and immunodetection by rinsing with matrix.	48
Insects *Schistocerca gregaria; Locusta migratoria; Periplaneta americana.*	Adipokinetic hormone I and II in the principal insect neurohaemal organs and neurosecretory glands, corpus cardiacum.	Direct MALDI-MS profiling of tissues in both linear and reflectron mode, matrix solution is 50 mM α-cyan-4-dioxy-cinnamic acid in acetonitrile : ethanol (1:1).	Matrix solution was directly deposited on freshly dissected tissues to allow air dry.	40
Insect *Helocovoptera zea*	Peptides encoded by the pheromone biosynthesis activating neuropeptide gene in neuronal clusters from the sub esophageal ganglion.	Direct comparison of MS and immunocytochemistry results.	Matrix deposition using dried droplet method.	50
Frog *Xenopus laevis*	POMC gene expression in single melanotrope cells, light- and dark-adapted cells.	Direct peptide sequencing of unfractionated tissue extract.	Micro scale extraction (100 cells) in matrix solution.	51, 52
Frog *Rana ridibunda*	Different POMC prohormone processing in two melanotrope cell subsets.	MALDI-MS profiling for semi quantitative comparison of peak intensities.	Matrix deposition using dried-droplet method.	53
Rat	Peptides in pancreas, pituitary sections and blots, effects of salt loading.	Semi quantitative MS profiling, MALDI imaging of cellular samples.	Matrix application via electro spray; tissue blot on organic membranes.	25, 26, 54, 55

Abbreviation: MALDI, matrix-assisted laser desorption-ionization; MS, mass spectroscopy; LC, liquid chromatography; POMC, proopiomelanocortin

The quest for large-scale automated proteomic studies is based on the desire to increase the sensitivity of detection, in combination with powerful two-dimensional resolution.

MALDI MS is an analytical approach suitable for obtaining molecular weights of peptides and proteins from complex samples. MALDI MS can profile the peptides and proteins from single-cell and small tissue samples without the need for extensive sample preparation, except for cell isolation and matrix preparation. Strategies for peptide identification and characterization of post-translational modifications are versatile and broadly applicable.[49]

Mass spectrometry is also a useful technique for the analysis of nucleic acids. Matrix-assisted laser adsorption/ionization time of flight (MALDI-TOF) mass spectrometry serves to analyze genetic variations such as micro satellites, insertions/deletions, and especially single-nucleotide polymorphisms (SNPs). The ability to resolve oligonucleotides varying in mass by less than a single nucleotide makes MALDI-TOF mass spectrometry an applicable technology for SNP and mutant analysis. The technique was used for quantitative analysis of mutations in attenuated mumps virus vaccines and the results showed excellent correlation with data from mutant analysis by PCR and restriction enzyme cleavage (MAPREC)[49].

The protein-folding enigma, the shaping of linear protein-sequences into three-dimensional structures, is addressed both experimentally and by computational methods.

Membrane proteins acquire their unique functions through specific folding of their polypeptide chains stabilized by specific interactions in the membrane. Classical methods to study their stability or resistance to unfolding are usually investigated by chemical and thermal denaturation. Atomic force microscopy and single-molecule force spectroscopy were combined to image and manipulate purple membrane patches from *Halobacterium salinarum*. Individual bacteriorhodopsin molecules were first localized and then extracted from the membrane; the remaining vacancies were imaged again. Anchoring forces between 100 and 200 pN for the different helices were found. Upon extraction, the helices were found to unfold. The force spectra revealed the individuality of the unfolding pathways. Helices G and F as well as helices E and D unfolded pair wise, whereas helices B and C occasionally unfolded one after the other. Experiments with cleaved loops revealed the origin of the individuality: stabilization of helix B by neighboring helices.[50]

The studies of protein folding show that the mechanisms and principles of protein folding may be simpler than the complexity of the process and proteins indicate, and may be determined by the topology of the native state.[51]

Cellular processes as different as growth factor signaling and transcription depend on interactions between proteins. Large-scale protein–protein interaction screens in model systems together with global gene expression profiling after receptor activation provide the identification of many unexpected interactions between GPCRs and other proteins.[52]

Somatostatin and dopamine are two major neurotransmitter systems that share a number of structural and functional characteristics. Somatostatin receptors and dopamine receptors are co localized in neuronal subgroups, and somatostatin is involved in modulating dopamine-mediated control of motor activity. The molecular basis for such interaction between the two systems is unclear. It was shown that dopamine receptor D2R and somatostatin receptor SSTR5 interact physically through hetero-oligomerization to create a novel receptor with enhanced functional activity. The results provide evidence that receptors from different GPCR families interact through oligomerization. Such direct intramembrane association defines a new level of molecular cross talk between related GPCR families.[53]

The identification and characterization of all proteins expressed by a genome in biological samples is the major challenge in proteomics. The high-throughput approaches combine two-dimensional electrophoresis with peptide mass finger-printing (PMF) analysis. Automation is often possible but a number of limitations still adversely affect the rate of protein identification and annotation in two-dimensional gel electrophoresis databases: the sequential excision process of pieces of gel containing protein; the enzymatic digestion step; the interpretation of mass spectra (reliability of identifications); and the manual updating of two-dimensional gel electrophoresis databases. A highly automated method has been developed that generates a fully annotated two-dimensional gel electrophoresis map. Using a parallel process, all proteins of a two-dimensional gel electrophoresis are first simultaneously digested proteolytically and electro-transferred onto a poly(vinylidene difluoride) membrane. The membrane is then directly scanned by MALDI-TOF MS. After automated protein identification from the obtained peptide mass fingerprints using PeptIdent software (from www.expasy.ch), a fully annotated two-dimensional map is created on-line. It is a multidimensional representation of a proteome that contains interpreted PMF data in addition to protein identification results. This MS imaging method represents a major step toward the development of a clinical molecular scanner.[54]

The analysis of multi-protein complexes is essential for analyzing the biological activities of proteins, since most cellular functions are performed by protein assemblies or multi-protein complexes rather than by individual proteins.

The identity of the members of such complexes can now be determined by mass spectrometry (MS). MS can also be used to define the spatial organization of these complexes. Thus, components of a protein complex are purified via molecular interactions using an affinity tagged member and the purified complex is then partially cross-linked. The products are separated by gel electrophoresis and their constituent components identified by MS, yielding nearest-neighbor relationships. A member of the yeast nuclear pore complex (Nup85p) was tagged and a six-member subcomplex of the pore was cross-linked and analyzed by one-dimensional sodium dodecylsulfate (SDS)–PAGE. Cross-linking reactions were optimized for yield and number of products. Analysis by MALDI MS resulted in the identification of protein constituents in the cross-linked bands even at a level of a few hundred femtomoles. Based on those results, a model of the spatial organization of the complex was derived that was supported by biological experiments. The use of MS is the method of choice for analyzing cross-linking experiments aimed at nearest neighbor relationships.[55]

A large-scale protein-protein interaction map of the human gastric pathogen *Helicobacter pylori* was constructed by using a high-throughput strategy of the yeast two-hybrid assay to screen 261 H. pylori proteins against a highly complex library of genome-encoded polypeptides. More than 1,200 interactions were identified between *H. pylori* proteins, connecting 46.6% of the proteome. This technology is applicable to higher eukaryotic organisms as well and enables the generation of proteomes for studies of diseaes, for the construction of mutants, and the development of screening assays for drug discovery[507].

Table 1.5C. Proteins identified containing three or more predicted transmembrane domains[a] [Refs in 563].

Number of predicted transmembrane domains	Number of proteins in class by MudPIT	Number of proteins in class identified	Percentage of total predicted
3	185	31	17
4	101	16	16
5	57	12	21
6	58	14	24
7	56	7	13
8	54	13	24
9	71	12	17
10	53	14	26
11	30	4	13
12	15	4	27
13	8	3	38
14	3	0	0
15	4	1	25
16	1	0	0
20	1	0	0
Totals	697	131	19

[a]   The Munich Information Center for Protein Sequences Website was used to obtain this information[24]. The prediction of transmembrane domains at this site is based on Klein *et al.*[34] and Goffeau *et al.*[35]

A largely unbiased method for rapid and large-scale proteome analysis by a combination of multidimensional liquid chromatography, tandem mass spectrometry, and database searching by the SEQUEST algorithm, named multidimensional protein identification technology (MudPIT) was apllied to the proteome of the Saccharomyces cerevisiae strain BJ5460 grown to mid-log phase. A total of 1,484 proteins were detected, identified, and characterized with demonstration of a particular advantage of this methodology, namely the identification of low-abundance proteins like transcription factors and protein kinases as well as three or more predicted transmembrane domains. This allowed the mapping of the soluble domains of several of the integral membrane proteins. MudPIT is thus useful for proteome analysis of rare and difficult-to-access membrane proteins.[563] Table 1.5C lists identified proteins with three or more predicted transmembrane domains.

The strategy for building a *Helicobacter pylori* proteome-wide interaction map is shown in Figure 1.2.

A production database (the PIMBuilder) was built that contains information related to the genomic sequence of *H. pylori,* which codes for 1,590 putative proteins or ORFs It was populated with raw data from screening experiments The PIMBuilder tracks all biotechnological or bioinformatics operations performed during the production processes, stores information about all biological objects produced during experiments, and interfaces with robots and bioinformatics modules. It also implements the procedure used to construct PIMs from raw experimental data. After identification of almost all positive clones, over-

Data collection: PIMBuilder

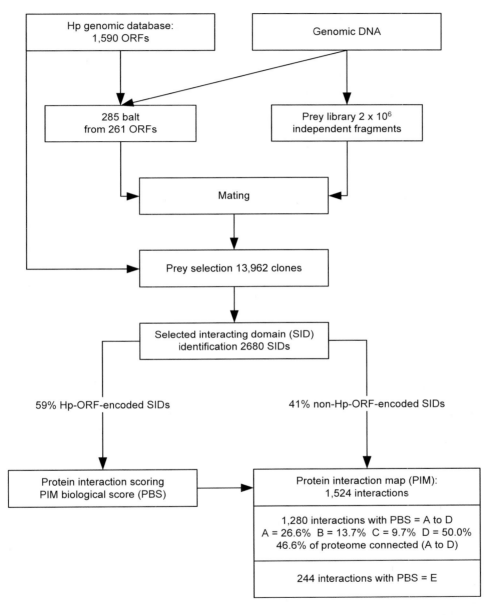

Figure 1.2. Outline of the strategy for building an *H. pylori* (Hp) proteome-wide interaction map.

lapping prey fragments were clustered into families to define SIDs. Those families that had no biological coding capability (antisepses or intergenic region, out of frame fragments occurring in a single frame) were discarded. A PIM biological score (PBS; see Methods) was then calculated for *H. pyloriORF*-encoded SIDs. Interactions were grouped into categories A to D (from high to 1ow heuristic values). The global connectivity of the PIM was also analysed to detect highly connected prey polypeptides. Those interactions were grouped in the E category. Processing of data and visualization of interactions were performed by an in-house bioinformatic platform (PIMRider) [Refs in 507].

The structures, the interactions, and the dynamics of proteins are fundamental to their function. The dynamic properties of proteins in the nucleus are essential for their function in nuclear architecture and gene expression. High mobility of proteins ensures their availability throughout the nucleus and the dynamic interaction creates the architectural framework for nuclear processes. Thus, nuclear morphology is shaped by the functional interactions of nuclear protein components. Studies with fluorescently tagged protein molecules reveal high mobility, compartmentation, and repeated transient interactions.[564]

A Human Transcriptome Map was generated by integration of the mapping data about the chromosomal position of human genes with genome-wide messenger RNA expression profiles from serial analysis of gene expression (SAGE). Over 2.45 million SAGE transcript tages, including 160,000 tags of neuroblastomas, known for 12 tissue types were assigned by algorithms to UniGene clusters and their chromosomal position. The resulting expression profiles for chromosomal regions in 12 normal and pathologic tissue types reveal a clustering of highly expressed genes to specific chromosomal regions and provide the tool for identifying genes that are overexpressed or silenced in cancer.[636]

The technologies and applications in proteomic studies are expanding and include the generation of novel proteins from distantly related sequences by sequence homology-independent protein recomination (SHIPREC),[669] the directed evolution of proteins by exon shuffling to generate libraries of therapeutic proteins,[673] the creation of genome-wide protein expression libraries by random activation of gene expression,[665] the development of multi-channel devices with electrospray tips for mass-spectrometry,[694] and make the study of the protein complement of the genome, the proteome, an essential basis for functional biology which requires the integration with DNA sequence data, mRNA profiles, metabolite con-centrations (the metabolome), structural studies, and biological models and systems.[667]

1.4   Cytomics

The study of biochemical processes at the cellular level is yielding results integrating the lower level mechanisms into cellular and ultimately organismal processes.

Chemoselective ligation reactions designed to modify only one cellular component among all others have provided insight into cellular processes. The goal in developing such transformations is to relate the selectivity of non-covalent recognition events, such as antibody–antigen binding, to biological processes.

Selective chemical reactions enacted within a cellular environment can be powerful tools for elucidating biological processes or engineering novel interactions. A chemical transformation that permits the selective formation of covalent adducts among richly functionalized biopolymers within a cellular context has been developed. A ligation modeled after the Staudinger reaction forms an amide bond by coupling of an azide and a specifically engineered triarylphosphine. Both reactive partners are abiotic and chemically orthogonal to native cellular components. Azides installed within cell-surface glycoconjugates by metabolism of a synthetic azidosugar were reacted with a biotinylated triarylphosphine to produce stable cell-surface adducts. The selectivity of the transformation permits its execution within a cell's interior, offering new possibilities for probing intracellular interactions.[56]

The level of complexity of biological studies and pharmaceutical intervention is increased by the necessity to take structures and modifications at the chromatin level into account.

The manipulation of eukaryotic genomes within a chromatin environment is a crucial and fundamental issue in biology. The chromatin structure is composed of highly conserved histone proteins (H3, H4, H2A, H2B and H1) that function as building blocks to package eukaryotic DNA into repeating nucleosomal units that are folded into higher-order chromatin fibers. The histones are integral and dynamic components of the machinery responsible for gene transcription. Distinct histone modifications, on one or more tails, act sequentially or in combination to form a 'histone code' that is read by other proteins to bring about distinct downstream events.[57]

The overall efficiency of protein biogenesis in cells has been the target of an investigation focussing on proteasomes. MHC class I molecules function to present peptides eight to 10 residues long to the immune system. These peptides originate primarily from a cytosolic pool of proteins through the actions of proteasomes and are transported into the endoplasmic reticulum, where they assemble with nascent class I molecules. Most peptides are generated from proteins that are apparently metabolically stable. To explain this, it was proposed that peptides arise from proteasomal degradation of defective ribosomal products (DRiPs). DRiPs are polypeptides that never attain native structures owing to errors in translation or post-translational processes necessary for proper protein folding. DRiPs constitute upwards of 30% of newly synthesized proteins as determined in a variety of cell types. Some DRiPs represent ubiquinated proteins und ubiquinated proteins are formed from human immunodeficiency virus Gag polyprotein, a long-lived viral protein that serves as a source of antigenic peptides.[59]

It is noteworthy that the 70% efficiency for protein synthesis is in line with the estimated 50% efficiency of producing functional mRNA from heterogeneous nuclear RNA.

The cell being the basic unit of functional tissues, organs, and organisms, the understanding and reconstruction of the whole-cell biology including genome, proteome, metabolome, and the interacting networks requires the description and simulation of cells. *In silico* analysis, description, and simulation complements the *in vitro* and *in vivo* processes and the construction of minimal organisms and cells.[663]

The analysis of the intracellular metabolites, the metabolome, by means of comparison of mutants' metabolic profiles (comparative metabolomics) aids in understanding the networks of proteins and the presence of silent genes.[698]

1.5   Micro- and Nanotechnology

Oligonucleotide microarrays, also called DNA chips, are currently made by a light-directed chemistry that requires a large number of photolithographic masks for each chip. A maskless array synthesizer (MAS) was described that replaces the chrome masks with virtual masks generated on a computer, which are relayed to a digital micro-mirror array. A 1:1 reflective imaging system forms an ultraviolet image of the virtual mask on the active surface of the glass substrate, which is mounted in a flow cell reaction chamber connected to a DNA synthesizer. Programmed chemical coupling cycles follow light exposure and these steps are repeated with different virtual masks to grow desired oligonucleotides in a selected pattern. This instrument has been used to synthesize oligonucleotide microarrays containing more than 76,000 features measuring 16 μm^2. The oligonucleotides were synthesized at high repetitive yield and, after hybridization, could readily discriminate single-base pair mismatches. The MAS is adaptable to the fabrication of DNA chips containing probes for thousands of genes, as well as any other solid-phase combinatorial chemistry to be performed in high-density microarrays.[60]

Nanotechnology involves the ability to arrange molecules and atoms into molecular structures. This will have major impacts on computation (higher density of computational power), materials (properties of catalysts and strengths of structures), and the manufacture of tools and devices (surgical tools and instruments of molecular dimensions), capable of intervening at the fundamental level of disease.[61]

The application of micro-machining techniques is growing rapidly and has applications in microfluidics (for labs-on-a-chip), in sensors as well as in fiber optics and displays. The two most widespread methods for the production of microelectromechanical structures (MEMS) are bulk micromachining and surface micromachining. Bulk micromachining is a subtractive fabrication method whereby single-crystal silicon is lithographically patterned and then etched to form three-dimensional structures. Surface micromachining is an additive method where layers of semiconductor-type materials (polysilicon, metals, silicon nitride, silicon dioxide, etc.) are sequentially added and patterned to make three-dimensional structures.

An alternative microfabrication technique is based on replication moulding. An elastomer is patterned by curing on a micromechanical mold. This technique is termed 'soft lithography', and is used to make blazed grating optics, stamps for chemical patterning and microfluidics devices. Soft lithography is essentially a subtractive method.

A technique was developed called 'multilayer soft lithography' that combines soft lithography with the capability to bond multiple patterned layers of elastomer.

Soft lithography is an alternative to silicon-based micromachining that uses replica moulding of non-traditional elastomeric materials to fabricate stamps and microfluidic channels. An extension to the soft lithography paradigm, multilayer soft lithography, is described, with which devices consisting of multiple layers may be fabricated from soft materials. This technique was used to build active microfluidic systems containing on/off valves, switching valves and pumps entirely out of elastomer. The softness of these materials allows the device areas to be reduced by more than two orders of magnitude compared with silicon-based devices. The other advantages of soft lithography, such as rapid prototyping, ease of fabrication and biocompatibility, are retained.[62]

Microscopic fluidic devices, ranging from surgical endoscopes and MEMS to the commercial 'lab-on-a-chip', allow chemical analysis and synthesis on scales unimaginable a decade ago. These devices transport miniscule quantities of liquid along networked channels. Several techniques have been developed to control small-scale flow, including micromechanical and electrohydrodynamic pumping, electro-osmotic flow, electrowetting and thermocapillary pumping. Most of these schemes require micromachining of interior channels and kilovolt sources to drive electrokinetic flow. Recently the use of temperature instead of electric fields to derive droplet movement was suggested. A simple, alternative technique utilizing temperature gradients to direct microscopic flow on a selectively patterned surface (consisting of alternating stripes of bare and coated SiO_2) was demonstrated. The liquid is manipulated by simultaneously applying a shear stress at the air–liquid interface and a variable surface energy pattern at the liquid–solid interface. To further advance the technology, a theoretical estimate of the smallest feature size attainable with this technique.[63]

Direct-write technologies are of increasing importance in materials processing. Structures are built directly without the use of masks, allowing rapid prototyping. The techniques comprise plasma spray, laser particle guidance, matrix-assisted pulsed-laser evaporation, laser chemical vapor deposition, micro-pen, ink jet, e-beam, focused ion beam and several droplet micro-dispensing methods. Micrometer-scale patterns of viable cells are required for the next generation of tissue engineering, fabrication of cell-based microfluidic biosensor arrays, and selective separation and culturing of microorganisms.

Patterns of viable *E. coli* bacteria have been transferred onto various substrates with laser-based forward transfer technique. These tools can be used to create three-dimensional mesoscopically engineered structures of living cells, proteins, DNA strands and antibodies and to co fabricate electronic devices on the same substrate to generate cell-based biosensors and bioelectronic interfaces and implantates.[64]

To allow researchers the design and construction of nanodevices with scale-up for large product runs, a special serial scanning probe lithography method, called dip-pen nanolithography (DPN), an eight-pen nanoplotter capable of doing parallel DPN, for the deposition of organic molecules, has been developed.

Because line width and patterning speed in DPN are independent of contact force, only one of the tips in the parallel writing mode (the imaging tip) has a feedback system to monitor tip position and to write the pattern. All other tips reproduce what occurs at the imaging tip in a passive fashion. Proof-of-concept experiments that demonstrate eight-pen parallel writing, ink and rinsing wells, and 'molecular corralling' via a nanoplotter-generated structure have been reported. This device enables research to do high-resolution and aligned patterning of nanostructures on a large scale in a fast and automated way.[65]

Work on supramolecular structures has opened access to useful nanocomposites by influencing phase-separating fluids. Simulations show that when low-volume fractions of nanoscale rods are immersed in a binary, phase-separating blend, the rods self-assemble into needle-like, percolating networks. The interconnected network arises through the dynamic interplay of phase-separation between the fluids, through preferential adsorption of the minority component onto the mobile rods, and through rod-rod repulsion. Such cooperative effects provide a means of manipulating the motion of nanoscopic objects and directing their association into supramolecular structures. Increasing the rod concentration beyond the effective percolation threshold drives the system to self-assemble into a lamellar morphology, with layers of wetted rods alternating with layers of the majority-component fluid. This approach can potentially yield organic/inorganic composites that are ordered on a nanometer scale and exhibit electrical or structural integrity.[66]

Whereas most fabrication of microelectronic devices is carried out by photolithography and thus is intrinsically two-dimensional, increasingly modern devices and micro- (respectively, nano-) structures for biomedicine require three-dimensional structures and manufacturing methods.

Self-assembly is a demonstrated strategy to achieve interconnections and to construct three-dimensional circuits and structures. Self-assembly of millimeter-scale polyhedra, with surfaces patterned with solder dots, and light-emitting diodes, generated electrically functional, three-dimensional networks. The patterns of dots and wires controlled the structure of the networks formed. Both parallel and serial connections were formed. This self-assembly mechanism may be suitable for generating other structures for implantates and biosensors.[67]

Self-assembly and controlled organization are fabrication methods of increasing importance for nanodevices in order to achieve precision and economies-of-scale.

Colloidal inorganic nanoparticles have size-dependent optical, optoelectronic and material properties that are expected to lead to superstructures with a range of practical applications. Discrete nanoparticles with controlled chemical composition and size distribution are readily synthesized using reverse micelles and microemulsions as confined reaction media, but their assembly into well-defined superstructures amenable to practical use remains a difficult and demanding task. This usually requires the initial synthesis of spherical nanoparticles, followed by further processing such as solvent evaporation, molecular cross-linking or template patterning. The interfacial activity of reverse micelles and microemulsions can be exploited to couple nanoparticle synthesis and self-assembly over a range of length scales to produce materials with complex organization arising from the interdigitation of surfactant molecules attached to specific nanoparticle crystal faces. This principle is demonstrated by producing three different barium chromate nanostructures – linear chains, rectangular super-lattices and long filaments – as a function of reactant molar ration, which in turn is controlled by fusing reverse micelles and microemulsion droplets containing fixed concentrations of barium and chromate ions, respectively. If suitable soluble precursors and amphiphiles with head groups complementary to the crystal surface of the Nan particle target are available, it should be possible to extend this approach to the facile production of one-dimensional wires and higher-order colloidal architectures made of metals and semi-conductors.[68]

DNA microarrays can be manufactured by either synthesizing DNA on a substrate, or delivering and covalently attaching presynthesized DNA to the microchips. Technologies such as photolithographic DNA synthesis allow the manufacture of high-density oligonucleotide microarrays. These methods, however, are costly, time consuming and not easily amenable to custom-manufacture.

A method for fabricating DNA microarrays was developed which uses a bubble jet ink jet device to eject 5′-terminal-thiolated oligonucleotides to a glass surface. The oligonucleotides are covalently attached to the glass surface by heterobifunctional cross-linkers that react with the amino group on the substrate and a thiol group on the oligonucleotide probe. Using this method, DNA microarrays were fabricated that carried 64 groups of 18mer oligonucleotides encoding all possible three-base mutations in the mutational 'hot spot' of the p53 tumor suppressor gene. These were screened with a fluorescent labeled synthetic 18mer oligonucleotide derived from the p53 gene, or segments of the p53 gene that had been PCR amplified from genomic DNA of two cell lines of human oral squamous cell carcinoma (SCC). This allowed to discriminate between matched hybrids and 1-bp mismatched hybrids.[69]

Microrobots for fluid media have been developed that show potential for acting as tiny mobile robots in the body fluids. Conducting polymers are excellent materials for actuators that are operated in aqueous media. Microactuators based on polypyrrole-gold bilayers enable large movement of structures attached to these actuators and are of particular interest for the manipulation of biological objects, such as single cells. A fabrication method for creating individually addressable and controllable polypyrrole-gold microactuators was developed. With these individually controlled microactuators, a micrometer-size manipulator (microrobotic arm) was fabricated. This microrobotic arm can pick up, lift, move and place micrometer-size objects within an area of about 250 μm by 100 μm, making the microrobot an excellent tool for single-cell manipulation.[70]

The application of micro- and nanorobots in the human body requires the development of co-operative mechanism for targeted action of the devices on desired phenomena, e.g., removal of obstacles, construction of scaffolds, measurements of cellular parameters. One of the greatest challenges in robotics is to create machines that are able to interact with unpredictable environments in real time. A possible solution may be to use swarms of robots behaving in a self-organized manner, similar to workers in an ant colony. Efficient mechanisms of division of labor, in particular series–parallel operation and transfer of information among group members, are key components of the tremendous ecological success of ants. The general principles regulating division of labor in ant colonies allow the design of flexible, robust and effective robotic systems. Groups of robots using ant-inspired algorithms of decentralized control techniques foraged more efficiently and maintained higher levels of group energy than single robots. However, the benefits of group living decreased in larger groups, most probably because of interference during foraging. A similar relationship between group size and efficiency has been documented in social insects. When food items were clustered, groups where robots could recruit other robots in an ant-like manner were more efficient than groups without information transfer, suggesting that group dynamics of swarms of robots may follow rules similar to those governing social insects.[71]

The construction and the evolutionary improvement of micro- and nanodevices may be carried out using evolving populations of artificial creatures. Biological life is in control of its own means of reproduction, which generally involves complex, auto-catalyzing chemical reactions. But this autonomy of design and manufacture has not yet been realized artificially. Robots are still laboriously designed and constructed by teams of human engineers, usually at considerable expense. Few robots are available because these costs must be absorbed through mass production, which is justified only for toys, weapons and industrial systems such as automatic teller machines. A combined computational and experimental approach in which simple electromechanical systems are evolved through simulations from basic building blocks (bars, actuators, and artificial neurons) were reported; the 'fittest' machines (defined by their locomotive ability) are then fabricated robotically using rapid manufacturing technology. Autonomy of design and construction are achieved using evolution in a 'limited universe' physical simulation coupled to automatic fabrication.[72]

Microinjection of fluorochromes, antibodies, proteins and genetic material is widely practiced by biologists despite negative effects on the target and disadvantages associated with insertion of microcapillaries.

A galinstan expansion femto-syringe enables femtoliters to attoliters to be introduced into prokaryotes and subellular compartments of eukaryotes. The method uses heat-induced expansion of galinstan (a liquid metal alloy of gallium, indium and tin) within a glass syringe to expel samples through a tip diameter of about 0.1 µm. The narrow tip inflicts less damage than conventional capillaries and the heat-induced expansion of the galinstan allows fine control over the rate of injection. Injection of Lucifer Yellow and Lucifer Yellow–dextrin conjugates into cyanobacteria, and into nuclei and chloroplasts of higher organisms was demonstrated. Injection of a plasmid containing the *bla* gene into the cyanobacterium *Phormidium laminosum* resulted in transformed ampicillin-resistant cultures. Green fluorescent protein (GFP) was expressed in attached leaves of tobacco and *Vicia faba* following injection of DNA containing its gene into individual chloroplasts.[73]

The study of individual cells or even compartments within a cell requires appropriately sized nanosensors. Optical nanosensors are usually grouped into chemical and biological nanosensors, depending on the probe being used. Both types of sensors have been used to measure chemicals in microscopic environments and to detect different entities within single cells.

The construction of nanosensors is crucially dependent on the construction of nanometer-sized tips on optical fibers. Near-field optical microscopy has spawned nanoscale optical fibers produced by either pulling with micropipette pullers or by chemical etching.[74]

The synthesis of materials that are ordered on all length scales, from molecular via nano to meso ('hierarchical materials') is a major goal of materials science. In this way, the properties of the large-scale structures and devices can be controlled by fabricating the desired molecular structures and characteristics. The methods deployed include biomimetic methods with polypeptides as building blocks as well as amphiphile and colloidal templating with appropriate mesophases as templates for inorganic mesophorous materials.[75]

The technologies and materials for creating nanodevices are increasingly applicable both in the realm of microelectronic manufacturing and chemical as well as biotechnological synthesis.

Carbon nanotubes find increasing application in fields as diverse as microelectronics, molecular electronics, nanodevices and actuators as well as sensors and chemical processes.

For constructing mobile mechanical devices, molecular surfaces need to be engineering. The controlled and reversible telescopic extension of multiwall nanotubes was demonstrated, realizing ultralow-friction nanoscale linear bearings and constant-force nanosprings. Measurements performed *in situ* on individual custom-engineered nanotubes inside a high-resolution transmission electron microscope demonstrated the anticipated van der Waals energy-based retraction force and enabled the placement of quantitative limits on the static and dynamic interwall frictional forces between nested nanotubes. Repeated extension and retraction of telescoping nanotube segments revealed no wear or fatigue on the atomic scale. These nanotubes may be almost perfect wear-free surfaces for nanodevices and nanomechanical apparatus.[76]

The versatility of carbon nanotubes in the field of molecular electronics was evidenced by a concept exploiting carbon nanotubes as both molecular device elements and molecular wires for reading and writing information. Each device element is based on a suspended, crossed nanotube geometry that leads to bistable, electrostatically switchable ON/OFF states. The device elements are naturally addressable in large arrays by the carbon nanotube molecular wires making up the devices. These reversible, bistable device elements could be used to construct non-volatile random access memory and logic function tables at an integration level approaching 10^{12} elements/cm^2 and an element operation frequency in excess of 100 GHz. The viability of this concept was demonstrated by detailed calculations and by experimental realization of a reversible, bistable nanotube-based bit.[77]

The construction of complicated three-dimensional structures on a sub-micrometer scale requires increasingly sensitive and sophisticated methods amenable to both custom and mass production. Electrochemical machining is increasingly important (in addition to lithography, direct writing, molecular assembly) to achieve the construction of micro- and nanodevices.

The application of ultra short voltage pulses between a tool electrode and a workpiece in an electrochemical environment allows the three-dimensional machining of conducting materials with sub-micrometer precision. The principle is based on the finite time constant for double-layer charging, which varies linearly with the local separation between the electrodes. During nanosecond pulses, the electrochemical reactions are confined to electrode regions in close proximity. The technique was used for local etching of copper and silicon as well as for local copper deposition.[78]

A molecular machine was constructed which uses DNA both as structural material and as fuel. Molecular recognition between complementary strands of DNA allows construction on a nanometer length scale. DNA tags may be used to organize the assembly of colloidal particles, and DNA templates can direct the growth of semiconductor nanocrystals and metal wires. As a structural material in its own right, DNA can be used to make ordered static arrays of tiles, linked rings and polyhedra. The construction of active devices is also possible, such as a nanomechanical switch, whose conformation is changed by inducing a transition in the chirality of the DNA double helix. Melting of chemically modified DNA has been induced by optical absorption, and conformational changes caused by the binding of small groups have been shown to change the enzymatic activity of ribozymes. The construction of a DNA machine was described in which the DNA is used not only as a structural material, but also as fuel. The machine, made from three strands of DNA, has the form of a pair of tweezers. It may be closed and opened by addition of auxiliary strands of fuel DNA; each cycle produces a duplex DNA waste product.[79]

The methods to produce nanoscale devices are developing rapidly and involve numerous physical and chemical methodologies.

The production of materials with micrometer- and sub-micrometer-scale patterns is of importance in a range of applications, such as photonic materials, high-density magnetic storage devices, microchip reactors and biosensors. One method of preparing such structures is through the assembly of colloidal particles. Micropatterned colloidal assemblies have been produced with lithographically patterned electrodes or micro-molds. Another method combines the well-known photochemical sensitivity of semiconductors with electric-field-induced assembly to create ordered arrays of micrometer-sized colloidal particles with tunable patterns. Light affects the assembly process and patterns are produced using electrophoretic deposition in the presence of an ultraviolet illumination motif. The distribution of current across an indium tin oxide (ITO) electrode can be altered by varying the illumination intensity; during the deposition process, this causes colloidal particles to be swept from darkened areas into lighted regions. Illumination also assists in immobilizing the particles on the electrode surface. The patterning effects of the ultraviolet light may result from alterations in the current density that affects particle assembly on an ITO electrode.[80]

A rapid-prototyping method that uses readily available equipment has been described which will facilitate research and experimental construction of nanostructured devices.

Living systems exhibit form and function on multiple length scales and at multiple locations.

In order to mimic such natural structures, it is necessary to develop efficient strategies for assembling hierarchical materials. Conventional photolithography, although ubiquitous in the fabrication of microelectronics and MEMS, is impractical for defining features below 0.1 µm and poorly suited to pattern chemical functionality. 'Soft' lithographic approaches have been combined with surfactant and particulate templating procedures to create materials with multiple levels of structural order. The materials thus formed have been limited primarily to oxides with no specific functionality and the associated processing times have ranged from hours to days. Using a self-assembling 'ink', silica surfactant self-assembly was combined with three rapid printing procedures – pen lithography, ink-jet printing and dip-coating of patterned self-assembled monolayers – to form functional, hierarchically organized structures in seconds. The described rapid-prototyping procedures are simple, employ readily available equipment, and provide a link between computer-aided design and self-assembled nanostructures. The ability to form arbitrary functional designs on arbitrary surfaces will be of practical importance for directly writing sensor arrays and fluidic or photonic systems.[81]

Another production technology combines biomolecules and their molecular recognition properties with semiconductor materials.

In biological systems, organic molecules exert significant control over the nucleation and mineral phase of inorganic materials such as calcium carbonate and silica, and over the assembly of crystallites and other nanoscale building blocks into complex structures required for biological function. This ability to direct the assembly of nanoscale components into controlled and sophisticated structures has motivated intense efforts to develop assembly methods that mimic or exploit the recognition capabilities and interactions found in biological systems. Of particular value would be methods that could be applied to materials with interesting electronic or optical properties, but natural evolution has not selected for interactions between biomolecules and such materials. Peptides with limited selectivity for binding to metal surfaces and metal oxide surfaces have been selected. This approach was extended to demonstrate that combinatorial phage display libraries can be used to evolve peptides that bind to a range of semiconductor surfaces with high specificity, depending on the crystallographic orientation and composition of the structurally similar materials used. As electronic devices contain structurally related materials in close proximity, such peptides may find use for the controlled placement and assembly of a variety of practically important materials, thus broadening the scope for bottom-up fabrication approaches.[82]

An approach which may lend itself to the mass production of organic–silicon structures is based on self-directed chemical growth processes. Advances in techniques for the nanoscale manipulation of matter are important for the realization of molecule-based miniature devices with new or advanced functions. A promising approach involves the construction of hybrid organic molecule–silicon devices. However, challenges exist both in the formation of nanostructures that will constitute the active parts of future devices and in the construction of commensurately small connecting wires. Atom-by-atom crafting of structures with scanning tunneling microscopes, although essential to fundamental advances, is too slow for any practical fabrication process; self-assembly approaches may permit rapid fabrication, but lack the ability to control growth location and shape. Furthermore, molecular diffusion on silicon is greatly inhibited, thereby presenting a problem for self-assembly techniques. An approach is described for fabricating nanoscale organic structures on silicon surfaces, employing minimal intervention by the tip of a scanning tunneling microscope and a spontaneous self-directed chemical growth process. Growth of straight molecular styrene lines is demonstrated – each composed of many organic molecules – and the crystalline silicon substrate determines both the orientation of the lines and the spacing within these lines. The described process should in principle allow parallel fabrication of identical complex functional structures.[83]

Sol–gel synthesis is widely used for making transition metal oxide solids with fine-scaled microstructures. Pore and particle sizes no greater than a couple of nanometers can easily be achieved in the freshly derived gels. Maintaining such microstructural dimensions when the fresh gels are subsequently crystallized at elevated temperatures is difficult. Freshly derived metal oxide gels have a hydrous solid skeleton that contains many hydroxyl groups and is either amorphous or paracrystalline. A post-firing step is therefore indispensable for dehydroxylation and for achieving a sufficient degree of crystallinity in order to give the desired combination of mechanical, catalytic or optoelectronic properties for their applications. Upon firing of gels, condensation among surface hydroxyl groups, nucleation of new oxide crystals and growth of existing crystals occur concurrently over a fairly wide temperature range. Accordingly, some crystals grow extensively before the gel is fully dehydroxylated and crystallized, which reduces surface area, enlarges pores and increases the difficulty of removing intracrystal defects.

Crystal growth upon firing of hydrous metal oxide gels can be effectively inhibited by replacing the surface hydroxyl groups before firing with another functional group that does not condense and that can produce small, secondary-phase particles that restrict advancing of grain boundaries at elevated temperatures. Fully crystallized SnO_2, TiO_2 and ZrO_2 materials with mean crystallite sizes of around 20, 50 and 15 Å, respectively, were synthesized by replacing the hydroxyl group with methyl siloxyl before firing at 500 °C. An ultrasensitive SnO_2-based chemical sensor resulting from the microstructural miniaturization was demonstrated.[84]

Micro-devices need power for operation and several approaches to supplying this power requirement exist, e.g., external power supply. An intriguing method for generating power by stimuli-responsive hydrogels has been developed. Hydrogels have been developed to respond to a variety of stimuli, but their use in macroscopic systems has been hindered by slow response times (diffusion being the rate-limiting factor governing the swelling process). However, there are many natural examples of chemically driven actuation that rely on short diffusion paths to produce a rapid response. It is therefore expected that scaling down hydrogel objects to the micrometer scale should greatly improve response times. At these scales, stimuli-responsive hydrogels could enhance the capabilities of microfluidic systems by allowing self-regulated flow control. The fabrication of active hydrogel components inside microchannels via direct photo-patterning of a liquid phase was reported. The approach greatly simplifies system construction and assembly as the functional components are fabricated *in situ*, and the stimuli-responsive hydrogel components perform both sensing and actuating functions. Significantly improved response times were demonstrated (less than 10 s) in hydrogel valves capable of autonomous control of local flow.[85]

Another phenomenon linking nano-sized structures and biochemical interaction has been reported in nanomechanics. The specific transduction, via surface stress changes, of DNA hybridization and receptor–ligand binding into a direct nanomechanical response of micro-fabricated cantilevers was reported. Cantilevers in an array were functionalized with a selection of biomolecules. The differential deflection of the cantilevers was found to provide a true molecular recognition signal despite large non-specific responses of individual cantilevers. Hybridization of complementary oligonucleotides showed that a single base mismatch between two 12mer oligonucleotides is clearly detectable. Similar experiments on Protein A–immuno-globulin interactions demonstrate the wide-ranging applicability of nanomechanical trans-duction to detect biomolecular recognition.[86]

Techniques have been described which allow interaction between energetic fields and matter on a nanoscale, and may thus support motion and operate sensors in micro- or nanodevices. Analogs of mechanical devices that operate on the molecular level, such as shuttles, brakes, ratchets, turnstiles and unidirectional spinning motors, are current targets of both synthetic chemistry and nanotechnology. These structures are designed to restrict the degrees of freedom of submolecular components such that they can only move with respect to each other in a predetermined manner, ideally under the influence of some external stimuli. Alternating current (a.c.) electric fields are commonly used to probe electronic structure, but can also change the orientation of molecules (a phenomenon exploited in liquid crystal displays), or interact with large-scale molecular motions, such as the backbone fluctuations of semi-rigid polymers. It was shown that modest a.c. fields can be used to monitor and influence the relative motion within certain rotaxanes, molecules comprising a ring that rotates around a linear 'thread' carrying bulky 'stoppers' at each end. Strong birefringence was observed at frequencies that correspond to the rate at which molecular ring pirouettes about the thread, with the frequency of maximum birefringence, and by inference also the rate of pirouetting giving rise to it, changing as the electric field strength is varied. Computer simulations and NMR spectroscopy show the ring rotation to be the only dynamic process occurring on a timescale corresponding to the frequency of maximum birefringence, thus confirming that mechanical motion within the rotaxanes can be addressed, and to some extent controlled, by oscillating electric fields.[87]

Another observation concerning magnetic fields may contribute to sensors in micro- and nanodevices that allow directional motion in operation.

Animals can detect small changes in the Earth's magnetic field by two distinct mechanisms, one using the mineral magnetite as the primary sensor and one using magnetically sensitive chemical reactions. Magnetite responds by physically twisting, or even reorienting the whole organism in the case of some bacteria, but the magnetic dipoles of individual molecules are too small to respond in the same way. In order to asses whether reactions whose rates are affected by the orientation of reactants in magnetic fields could form the basis of a biological compass, a general model was used, incorporating biological components and design criteria, to calculate realistic constraints for such a compass. This model compares a chemical signal produced due to magnetic field effects with stochastic noise and with changes due to physiological temperature variation. The analysis shows that a chemically based biological compass is feasible, with its size, for any give detection limit, being dependent on the magnetic sensitivity of the rate constant of the chemical reaction.[88]

The combination of branched DNA molecules and sticky ends creates a powerful molecular assembly kit for structural DNA nanotechnology. Polyhedra, complex topological objects, a nanomechanical device and two-dimensional arrays with programmable surface features have already been produced in this way. Future applications range from macromolecular crystallography and new materials to molecular electronics and DNA-based computation.[89]

The field of DNA computers has been extended to RNA and presents a general approach for the solution of satisfiability problems. A variant of the 'Knight problem' was considered, which asks generally what configurations of knights can one place on an $n \times n$ chess board such that no knight is attacking any other knight on the board. Using specific ribonuclease digestion to manipulate strands of a 10-bit binary RNA library, a molecular algorithm was developed and applied to a 3×3 chessboard as a 9-bit instance of this problem. The nine spaces on the board correspond to nine 'bits' or placeholders in a combinatorial RNA library. A set of winning molecules was recovered that describes solutions to this problem.[90]

Molecular computation (DNA computing) as a massively parallel type of computing can complement rather than replace silicon computing as a computational medium in the future.[91]

Another mode of DNA computation is aiming at autonomous computing. Hairpin formation by single-stranded DNA molecules was exploited in DNA-based computation in order to explore the feasibility of autonomous molecular computing. An instance of the satisfiability problem, a famous hard combinatorial problem, was solved by using molecular biology techniques. The satisfiability of a given Boolean formula was examined autonomously, on the basis of hairpin formation by the molecules that represent the formula. This computation algorithm can test several clauses in the given formula simultaneously, which could reduce the number of laboratory steps required for computation.[92]

DNA computing was proposed as a means of solving a class of intractable computational problems in which the computing time can grow exponentially with problem size (the 'NP-complete' or non-deterministic polynomial time complete problems). The principle of the technique has been demonstrated experimentally for a simple example of the Hamiltonian path problem (in this case, finding an airline flight path between several cities, such that each city is visited only once). DNA computational approaches to the solution of other problems have also been investigated. One technique involves the immobilization and manipulation of combinatorial mixtures of DNA on a support. A set of DNA molecules encoding all candidate solutions to the computational problem of interest is synthesized and attached to the surface. Successive cycles of hybridization operations and exonuclease digestion are used to identify and eliminate those members of the set that are not solutions. Upon completion of all the multistep cycles, the solution to the computational problem is identified using PCR to amplify the remaining molecules, which are then hybridized to an addressed array. The advantages of this approach are its scalability and potential to be automated (the use of solid-phase formats simplifies the complex repetitive chemical processes, as has been demonstrated in DNA and protein synthesis). The use of this method to solve a NP-complete problem is reported. A small example of the satisfiability problem (SAT) is considered, in which the values of a set of Boolean variables satisfying certain logical constraints are determined.[93]

The future of such devices may lie in fully organic or organic–silicon computing devices to be implanted in living bodies that can integrate signals from several sources and compute a response in terms of an organic molecule delivery device for a drug or a signal.[94]

1.6 Cellular Cloning

Cellular cloning offers major advances in the breeding of higher organisms with specific traits, in the field of therapeutic cloning, the generation of organs, tissues and cells with genetic identity and immunological selfness,[95] and even primate cloning.

Significant efforts are necessary to turn the various techniques of generating transgenic animals into a routine method.[607]

The automated sorting of living transgenic embryos, demonstrated with *Drosophila* mutants, facilitates research in generating and characterizing transgenics.[600]

Cloning of pigs is attracting major efforts because the pig is physiologically very close to humans and there is interest in using pigs as donors for transplantation (xenotransplantation). There are two major problems facing pig cloners: activating development of the oocyte after nuclear transfer and the need for at least four pig embryos in the uterus of a surrogate mother sow for embryonic development to proceed normally.

One of the techniques used is piezo-actuated microinjection, whereby a vibrating needle pierces the donor cell and oocyte plasma membrane. This technique was used to directly inject porcine fetal fibroblast donor nuclei into enucleated oocytes. The injection was followed by activation by means of an electric pulse after nuclear transfer by microinjection.

Another method involves the fusion of porcine granulosa-derived donor cells with enucleated mature oocytes. After 18 h, the donor nucleus is removed from the first oocyte and transferred to the cytoplasm of a fertilized egg. This double nuclear transfer strategy was used to overcome the putatively insufficient activation stimulus provided after nuclear transfer in the prior one-step method.

A key feature of cloning by nuclear transfer is that the donor nucleus must be reprogrammed by oocyte-specific factors so that it can direct the development of the embryo. The components within a somatic cell that are responsible for directing its differentiation, such as transcription factors, histones and nuclear lamins, are associated with chromatin and their composition and abundance changes with the differentiation status of the cell. If the entire donor cell is fused with an enucleated oocyte, those cell-specific factors are also transferred into the cytoplasm of the recipient oocyte and can block the ability of oocyte-specific factors to reprogram the nucleus. If the nucleus alone is injected, then only those factors associated with chromatin are transferred, increasing the likelihood that oocyte-specific factors will be able to reprogram the donor nucleus. Another approach would be the microinjection of donor cell metaphase chromosomes alone into the recipient enucleated oocyte. This would prevent the transfer of most donor cell-specific factors, and thus allow for a greater degree of nuclear remodeling and reprogramming.

In addition to retention of oocyte-specific factors to restructure the donor nucleus, other factors are important for efficient nuclear transfer and cloning nuclear reprogramming, such as retaining the correct pattern of histone acetylation and DNA methylation during *in vitro* culture of the donor cell line. For successful reprogramming of the donor nucleus, the methylation pattern must be faithfully maintained during manipulations *in vitro*. Another important area of cloning concerns the development of defined culture media which should allow successful maturation of oocytes *in vitro* as well as early development of embryos before transfer to the uterus. A highly efficient, well-defined culture system has been developed which enabled the successful maturation and fertilization of immature pig oocytes.[96]

ES cells, derived from the inner cell mass of developing blastocysts, can grow continuously in an undifferentiated state in culture, but have the potential to develop into all adult tissues and cell types. ES cells are thus widely regarded as important vehicles for tissue transplantation and gene therapy. One approach is to generate lineage-committed progenitor cells from ES cells *in vitro* and then to transplant such cells so they can undergo further differentiation *in vitro*. Another approach is to generate from ES cells a differentiated cell type that can express cell lineage-specific function *in vitro* and then to transplant such differentiated cells to restore such functional activity *in vivo*.

An important goal of tissue engineering is to achieve reconstitution of specific functionally active cell types by transplantation of differentiated cell populations derived from normal or genetically altered ES cells *in vitro*. Mast cells derived *in vitro* from wild-type or genetically manipulated ES cells can survive and orchestrate immunologically specific IgE-dependent reactions after transplantation into mast cell-deficient Kit^W/Kit^{W-v} mice. These findings define a unique approach for analyzing the effects of mutations of any genes that are expressed in mast cells, including embryonic lethal mutations, *in vitro* or *in vivo*.[97]

Stem cells, one source of cloning tissues and organs, hold a promise for contributing to a vast array of ailments, since they have the potential to produce various cell types and could provide replacements for tissues damaged by age, trauma or disease. Stem cells demonstrate a remarkable variety and plasticity. They have been discovered in the adult CNS, epithelial tissues (requiring matrix structures for differentiation) and in bone marrow, whereby hematopoietic stem cells (HSC) may provide transplants for blood and immune cells. The other source of cells for generating tissues and organs, ES cells, may have a wider range of applications and potential. The distinct advantage of ES cells lies in their ability to differentiate into any cell type.

Mouse ES cells have been shown to become glial cells producing the protective myelin sheath *in vitro* and when injected into the brains of myelin-deficient mice.[98] Spinal cord function could be partially restored by implanting immature nerve cells in rats.[99]

Purging of tumor cells and selection of stem cells are key technologies for enabling cell transplantation and stem cell gene therapy. A strategy was reported for cell selection based on physical properties of the cells. Exposing cells to an external pulsed electric field (PEF) increases the natural potential difference across the cell membrane until a critical threshold is reached and pore formation occurs, resulting in fatal perturbation of cell physiology. Attaining this threshold is a function of the applied field intensity and cell size, with larger cells porated at lower field intensities than smaller cells. Since HSC are smaller than other hematopoietic cells and tumor cells, the exposure of peripheral blood mononuclear cells (PBMCs) to PEFs caused stepwise elimination of monocytes without affecting the function of smaller lymphocyte populations. Mobilized peripheral blood exposed to PEFs was enriched for $CD34^+/CD38^+$ cells and stem cell function was preserved. Furthermore, PEF treatment was able to selectively purge blood preparations of tumor cells and eradicate transplantable tumor.[100]

The vast possibilities of organs and tissues derived from stem cells or ES cells are matched by the tremendous demand building up in the population. From the standpoint of patients or patients-to-be, stem cell research holds the promise of delivering health to millions of sufferers. In the USA alone, out of a population of 265 million, an estimated 128 million patients may be helped by human pluripotent stem cell research for afflictions such as cardiovascular diseases, autoimmune diseases, diabetes, osteoporosis, cancer, Alzheimer's disease, Parkinson's disease, severe burns, spinal cord injuries and birth defects.[101]

A key regulatory protein called Oct-3/4, a transcription factor protein, is involved in the determination of the development of ES cells.

Pluripotential cells in the embryo have the capacity to give rise to differentiated progeny representative of all embryonic germ layers, as well as the extraembryonic tissues that support development. In mammals the property of pluripotentiality is restricted to the oocyte, the zygote, early embryonic cells, primordial germ cells and the stem cells of tumors derived from pluripotential cells (embryonic carcinomas). Under certain conditions, pluripotential stem cells can be propagated indefinitely *in vitro* and still maintain the capacity for differentiation into a wide variety of somatic and extraembryonic tissues. Pioneering work on mouse embryonic carcinoma (EC) cells led to the derivation of pluripotent diploid ES cells directly from the mouse blastocysts. Since the first description of mouse ES cells, it has been recognized that the derivation of human ES cells could provide a unique resource for the functional analysis of early human development. Human ES cells could be used to identify polypeptide factors involved in differentiation and proliferation of committed embryonic progenitor cells. Since ES cells can in principle serve as an unlimited source of any cell type in the body, human ES cells could yield highly effective *in vitro* models for use in drug discovery programs and provide a renewable source of cells for use in transplantation therapy.

The derivation of pluripotent ES cells from human blastocysts was described. Two diploid ES cell lines have been cultivated *in vitro* for extended periods while maintaining expression of markers characteristic of pluripotent primate cells. Human ES cells express the transcription factor Oct-4, essential for the development of pluripotential cells in the mouse. When grafted into SCID mice, both lines give rise to teratomas containing derivatives of all three embryonic germ layers. Both cell lines differentiate *in vitro* into extraembryonic and somatic cell lineages. Neural progenitor cells may be isolated from differentiating ES cell cultures and induced to form mature neurons. ES cells provide a model to study early human embryology, an investigational tool for discovery of novel growth factors and medicines, and a potential source of cells for use in transplantation therapy.[102]

The utilization of this regulatory protein or its gene under desired regulatory control will allow us to influence the direction of stem cells.

The primary role of cytokines in hemato-lymphopoiesis is assumed to be the regulation of cell growth and survival. A clonogenic common lymphoid progenitor, a bone marrow-resident cell that gives rise exclusively to lymphocytes (T, B and natural killer cells), can be redirected to the myeloid lineage by stimulation through exogenously expressed interleukin (IL-2) and granulocyte/macrophage colony-stimulating factor (GM-CSF) receptors. Analysis of mutants of the β-chain of the IL-2 receptor revealed that the granulocyte- and monocyte-differentiation signals are triggered by different cytoplasmic domains, showing that the signaling pathway(s) responsible for these unique developmental outcomes are separable. The endogenous myelo-monocytic cytokine receptors for GM-CSF and macrophage colony-stimulating factor (M-CSF) are expressed at low to moderate levels on the more primitive HSC, are absent on common lymphoid progenitors and are up-regulated after myeloid lineage induction by IL-2. Cytokine signaling can thus regulate cell-fate decisions. A critical step in lymphoid commitment is down regulation of cytokine receptors that drive myeloid cell development.[103]

The ability of cells to migrate is critical during developmental events in embryogenesis, in the function of mature cells, particularly those of the immune and hematopoietic systems, in vascular remodeling during angiogenesis, and in most immune and infectious diseases. Signals affecting cell movements can be mediated by a number of gene families and are a defining characteristic of the chemokine family of cytokines. Significant advances in understanding these events can be gained by investigating cell migration *in vivo*. In particular, the goal of non-invasive imaging of cell migration, and specifically of genetically engineered stem and progenitor cells has been sought. Developments in magnetic resonance imaging (MRI) have enabled *in vivo* imaging at near microscopic resolution.

The ability to track the distribution and differentiation of progenitor and stem cells by high-resolution *in vivo* imaging has significant clinical and research applications. A cell-labeling approach was developed using short HIV-Tat peptides to derivatize superparamagnetic nanoparticles. The particles are efficiently internalized into hematopoietic and neural progenitor cells in quantities up to 10–30 pg of superparamagnetic iron per cell.

Following intravenous injection into immunodeficient mice, 4% of magnetically CD34$^+$ cells homed to bone marrow per gram of tissue, and single cells could be detected by MRI in tissue samples. In addition, magnetically labeled cells that had homed to bone marrow could be recovered by magnetic separation columns. Localization and retrieval of cell populations *in vivo* enable detailed analysis of specific stem cell and organ interactions critical for advancing the therapeutic use of stem cells.[104]

Understanding of the regulation of the quiescent state of stem cells and their release into differentiation and proliferation is crucial for controlled use of stem cells in therapy and regenerative medicine. Relative quiescence is a defining characteristic of HSC, while their progeny have dramatic proliferative ability and inexorably move toward terminal differentiation. The quiescence of stem cells is considered to be of critical biological importance in protecting the stem cell compartment, which was directly assessed using mice engineered to be deficient in the G$_1$ checkpoint regulator, cyclin-dependent kinase inhibitor, p21$^{cip1/waf1}$ (p21). In the absence of p21, hematopoietic stem cell proliferation and absolute number were increased under normal homeostatic conditions. Exposing the animals to cell cycle-specific myelotoxic injury resulted in premature death due to hematopoietic cell depletion. Self-renewal of primitive cells was impaired in serially transplanted bone marrow from p21$^{-/-}$ mice, leading to hematopoietic failure. It is thus demonstrated that p21 is the molecular switch governing the entry of stem cells into the cell cycle, and, in its absence, increased cell cycling is crucial to prevent premature stem cell depletion and hematopoietic death.[105]

An intriguing prospect of the merger of tissue engineering and gene therapy is glimpsed from research on treating liver diseases. Liver diseases affect approximately 20 million people in the USA alone and many of the 40,000 deaths attributable to liver failures could be prevented by treatments of liver damage.[106]

The restoration of liver function was achieved in rodents by using gene therapy that prevents the withering of telomeres.[107]

In a step towards demonstrating the effectiveness of generating liver tissue for transplantation, researchers grew enough human liver cells in culture to help rats through an acute liver failure. They used a transformation of human liver cells with a cancer gene to overcome previously poor growth characteristics of the cultured cells and, having achieved growth, cut the transforming gene out again.[108]

The first demonstration that mouse ES cells could be used to transfer a pre-determined genetic modification to a whole animal was published in 1989. The extension of this technique to other mammalian species, particularly livestock, might bring numerous biomedical benefits, e.g., ablation of xenoreactive transplantation antigens, inactivation of genes responsible for neuropathogenic disease and precise placement of transgenes designed to produce proteins for human therapy. Gene targeting has not yet been achieved in mammals other than mice, because functional ES cells have not been derived. Nuclear transfer from cultured somatic cells provides an alternative means of cell-mediated transgenesis. An efficient and reproducible gene targeting in fetal fibroblasts to place a therapeutic transgene at the ovine α1(I) procollagen (*COL1A1*) locus and the production of live sheep by nuclear transfer was reported.[109]

Live births of cloned sheep, cattle and goats have been achieved by somatic cell transfer, in which a nucleus donor cell is fused with an enucleated oocyte. Clonal propagation of selected porcine phenotypes is especially important in meat production. In addition, genetic modification could be combined with cloning in the provision of potential donors for xenotransplantation to humans.

Pig cloning will thus have a marked impact on the optimization of meat production and xenotransplantation. To clone pigs from differentiated cells, the nuclei of porcine (*Sus scrofa*) fetal fibroblasts were injected into enucleated oocytes, and the development was induced by electro activation. The transfer of 110 cloned embryos to four surrogate mothers produced an apparently normal female piglet. The clonal provenance of the piglet was indicated by her coat color and confirmed by DNA microsatellite analysis.[110]

Cloning has been extended to animals while using adult cells as nucleus donors. Cloning of whole animals with somatic cells as parents offers the possibility of targeted genetic manipulations *in vitro* such as 'gene knock-out' by homologous recombination. Such manipulation requires prolonged culture of nuclear donor cells. Previous successes in cloning were limited to the use of cells collected either fresh or after short-term culture. Demonstration of genetic totipotency of cells after prolonged culture is pivotal to combining site-specific genetic manipulations and cloning. The birth of six clones of an aged (17-year-old) Japanese Black Beef bull using ear skin fibroblast cells as nuclear donor cells after up to 3 months of *in vitro* culture (10–15 passages) was reported. Higher developmental rates were observed for embryos derived from later passages (10 and 15) as compared with those embryos from an early passage. The four surviving clones are at the time of publication 10–12 months of age and appear normal, similar to their naturally reproduced peers. The data show that fibroblasts of aged animals remain competent for cloning and prolonged culture does not affect the cloning competence of somatic adult donor cells.[111]

Eight calves had been derived from differentiated cells of a single adult cow, five from cumulus cells and three from oviductal cells out of 10 embryos transferred to surrogate cows (80% success). All calves were visibly normal, but four died at or soon after birth from environmental causes and post-mortem analysis revealed no abnormality. These results show that bovine cumulus and oviductal epithelial cells of the adult have the genetic content to direct the development of newborn calves.[112]

Since the first report of live mammals produced by nuclear transfer from a cultured differentiated population in 1995, successful development has been obtained in sheep, cattle, mice and goats using a variety of somatic cell types as nuclear donors. The methodology used for embryo reconstruction in each of these species is essentially similar: diploid donor nuclei have been transplanted into enucleated MII oocytes that are activated on, or after, transfer. Preactivated oocytes have also proved successful as cytoplast recipients in sheep and goats. The reconstructed embryos are then cultured and selected embryos transferred to surrogate recipients for development to term. In pigs, nuclear transfer has been significantly less successful; a single piglet was reported after transfer of a blastomere nucleus from a four-cell embryo to an enucleated oocyte; however, no live offspring were obtained in studies using somatic cells such as diploid or mitotic fetal fibroblasts as nuclear donors. The development of embryos reconstructed by nuclear transfer is dependent upon a range of factors. Some of these factors were investigated and the successful production of cloned piglets from a cultured adult somatic cell population using a new nuclear transfer procedure was reported.[113]

Unprecedented ethical issues need to be addressed and resolved, especially where ES cells are concerned (see Chapter 11 on 'Ethics').

A procedure for cloning pigs by the use of *in vitro* culture systems after somatic cell nuclear transfer enables the use of genetic modification procedures to produce tissues and organs from cloned pigs with reduced immunogenicity for use in xenoptransplantation and the modification of existing as well as the introduction of new traits in additional models of human disease. The method involves extended *in vitro* culture of fetal cells preceding nuclear transfer, *in vitro* maturation and activation of oocytes, and *in vitro* embryo culture.[606] Several different procedures for pig cloning have been described.[608] Table 1.5D lists several successfulcloning procedures in pigs.

In cloning experiments of mouse embryos, X inactivation was monitored to study epigenetic differences between the two X chromosomes of a somatic female nucleus. In the trophectoderm (TE), X inactivation was nonrandom with the inactivated X of the somatic donor being inactivated. When using female embryonic stem cells with two active X chromosomes, random X inactivation is observed in TE and embryo. Thus, epigenetic markers imposed on the X chromosome during gametogenesis, responsible for normal imprinted X inactivation in the TE, are functionally equivalent to the marks imparted to the chromosomes during somatic X inactivation.[601]

Table 1.5D. Comparison of successful pig cloning procedures [Refs in 608].

Research group	Oocyte maturation	Method of nuclei transfer	Method of oocyte activation	Donor cell type	Number of embryos transferred	Time of embryo culture prior to transfer	Donor cell culture conditions
Betthouser et al.[3]	In vitro	Fusion	Ionomycin and 6-di-methyl-amino-purine	Fetal fibroblast (3 passages)[a] genital ridge (0 passage)[b]	143 164	8 hours (one cell stage)[b] 72 hours (4 cell stage)[b]	Confluent no serum starvation
Onishi et al.[2]	In vivo	Piezo-activated micro-injection[5]	Electrical	Fetal fibroblast (2 passages)	36	40 hours (4-8-cells)	Confluent 16 days without fresh sera
Polejaeva et al.[1]	In vivo, first and second round	Fusion (2 rounds) round	Electrical (for both rounds)	Granulosa cells	72	One day for first round; no culture for second round	Confluent no serum starvation

[a] first litter, [b] second litter

Stem cell research is pivotal for understanding the development and function of organs in mammals and for the generation of tissues for regenerative medicine,[614] such as CNS stem cells, hepatocytes, and other differentiated cells of other organs, including muscle, blood, liver, and heart. Adult human liver cells can be derived from stem cells originating in the bone marrow or circulating outside the liver, implying that blood-system stem cells might be used clinically to generate hepatocytes for replacement therapy.[613]

Human embryonic stem (ES) cells derived from the inner cell mass of *in vitro* fertilized human blastocysts were subjected to eight growth factors (basic fibroblast growth factor, transforming growth factor β1, activin-A, bone morphogenetic protein 4, hepatocyte growth factor, epidermal growth factor, β nerve growth factor, retinoic acid) to direct the differentiation of human ES-derived cells *in vitro*. Each factor has a unique effect and the results indicate the potential for directing differentiation and for the enrichment of human cell types in vitro.[611]

Clonogenic proliferative cells were isolated from adult human dental pulp. These dental pulp stem cells (DPSCs) produced sporadic densely calcified nodules. In immuno-compomised mice, DPSCs generated a dentin-like structure lined with human odontoblast-like cells surrounding a pulp-like interstitial tissue. The DPSCs may be used for dentinal repair and, after proliferation ex vivo, for the fabrication of a viable dental implant.[602]

Epidermal stem cells offer the potential for research on and treatment of various skin diseases.[603] Bone-marrow-derived cells offer an alternative source of neurons in patients with neurodegenerative disease.[605] A genome-wide expression analysis was performed to define regulatory pathways in hematopoietic stem cells and their global genetic program. Subtracted complementary DNA libraries studied with array hybridization techniques enabled the characterization of several thousand previously undescribed molecules with apparently regulatory properties, thus defing a molecular-genetic phenotype of the hematopoietic stem cell.[616] A clonogenic common myeloid progenitor giving rise to all myeloid lineages was identified providing the basis for identifying determinants of lineage commitment.[612]

The remodeling of chromosomal architecture following transplantation of somatic nuclei into unfertilized eggs is a key feature of cloning. The chromatin-remodeling nucleosomal adenosine triphosphatase (ATPase) ISWI erases the TATA binding protein from association with the nuclear matrix. Dissecting the biochemical processes of nuclear remodeling will improve the efficiency of cloning and the control of differention in stem cell development.[617] Murine GATA-2 and GATA-3 are specifically expressed during adipogenesis in white adipocyte precursors. Constitutive GATA- and GATA-3 expression suppresses adipocyte differentiation and traps cells at the preadipocyte stage. GATA-3-deficient embryonic stem cells show an enhanced capacity for differentiation into adipocytes, and defective GATA-2 and GATA-3 expression is associated with obesity. GATA-2 and GATA-3 regulate adipocyte differentiation and provide targets for therapeutic intervention in differentiation-related diseases.[618] Stem cell factor (SCF) is influencing hematopiesis and the survival, proliferation, and differentiation of mast cells, melanocytes, and germ cells. The elucidation of the crystal structure provides the bassis for SCF-complex formation and function.[620]

Neural stem cells display a broad developmental capacity and may be used to generate a number of cell types for transplantation in different diseases.[615] Neural stem cells (NSCs) may even be used as therapeutic agents to destroy migratory brain tumor cells plus providing repair of tumor-associate damage.[609] NSCs implanted into experimental intracranial gliomas in vivo in adult rodents, distributed quickly and extensively throughout the tumor bed, actively targeting the tumor cells (including human glioblastomas). NSCs can deliver a therapeutically active molecule – cytosine deaminase – causing a quantifiable reduction in tumor burden. Inherently migratory NSCs may function as a delivery vehicle for therapeutic genes and vectors to refractory, migratory, invasive brain tumors.[610]

Identifying the niches of stem cells *in vivo* and the requirements for stem cell viability and repopulation is essential for future therapies in regenerative medicine, possibly by activating stem cells *in situ*. In *Drosophila*, three cell types act as germ line stem cell niche.[619] The creation of complex biological structures such as limbs requires detailed knowledge of the genetic developmental program for successful tissue regeneration.[604]

1.7   Tissue Engineering (Organ Cultivation)

The 'Holy Grail' of tissue engineering, the generation of specific tissues and organs with immunologic identity to the recipient and the conveyance of desired attributes, is getting closer with the possibilities of cloning without embryos, by using stem cells which directly develop into cells or tissues for the patient (without the need for embryos and the related ethical obstacles).[114] The regeneration of tissues and organs – regenerative medicine – offers vastly expanded teatment opportunities in injury and disease.[566] Table 1.5E lists a selection of programs and companies in regenerative medicine, covering human organs, xenoptransplantation, growth factors, stem cells, and biomaterials.

The usefulness of genetic modification of cells which are to be used in engineering tissue is evidenced by non-malignant transformation with human telomerase reverse transcriptase of adrenocortic bovine cells.[115]

The expanding list of tissues that use differentiation from stem to progenitor to mature cells includes blood cells, immune system cells, central and peripheral nervous tissues, skeletal muscles, and epithelial cells. The resulting clinical stem cell transplantation could greatly increase the physician's armamentarium against degenerative diseases and the consequences of acute failures.

Due to the tremendous research effects, the large numbers of afflicted individuals and the central role of the respective tissues, particularly pertinent areas are cells of the immune system, the pancreas, and the liver.

Fine-tuning of the immune rejection by co-transplantation of human stem cells and stem cells for other organs might be used for tolerance induction and cell- and organ-specific regeneration of diseased or damaged organs.

A crucial technique in therapeutic cloning is the mastery of nuclear transfer to endow the recipient cell, destined to become an organ, with a cell nucleus from the donor, i.e. the patient.

The successful application of nuclear transfer techniques to a range of mammalian species has brought the possibility of human therapeutic cloning significantly closer. The objective of therapeutic cloning is to produce pluripotent stem cells that carry the nuclear

Table 1.5E. A selection of programs and of companies working in regenerative medicine [Refs in 566].

Company	Program	Product/Stage
Human organs, tissues, and cells		
Advanced Tissue Sciences (La Jolla, CA)	Skin and vascular tissue	TransCyte, a temporary skin substitute; Dermagraft, a skin graft (pending FDA approval)
Organogenesis (Canton, MA)	Skin and vascular tissue	Apligraf, a skin substitute
LifeCell (Branchburg, NJ)	Skin and vascular tissue; orthopedics	Alloderm, an injectable tissue matrix for plastic surgery
Ortec International (New York, NY)	Skin substitute	Composite Cultured Skin
Genzyme BioSurgery (Cambridge, MA)	Cartilage repair, wound-healing products	Carticel (approved) for cartilage repair
Curis (Cambridge, MA)	Cartilage for urological condition	Chondrogel (phase 3)
Modex (Lausanne, Switzerland)	Encapsulated human cells for therapeutic protein delivery	Preclinical
Xenotransplantation		
PPL Therapeutics (Edinburgh, UK and Blacksburg, VA)	Transgenic pigs	
BioTransplant (Charlestown, MA)	PERV-free miniature swine; immunesuppression system (with Novartis)	
Alexion Pharmaceuticals (New Haven, CT)	Transgenic pigs	UniGraft for spinal injury and Parkinson's disease (preclinical)
Growth factors		
Human Genome Sciences (Rockville, MD)	Wound healing; blood and immune system	Repifermin for wound healing (phase 2)
The Genetics Institute (Cambridge, MA)	Bone morphogenic	rhBMP-2 (phase 1)
Curis (Cambridge, MA)	Bone regeneration	OP-1 (pending approval)
Regeneron (Tarrytown, NY)	Growth factors involved in blood cell, vessels, cartilage, and nerve regeneration	Preclinical
Guilford Pharmaceuticals (Baltimore, MD)	Nerve-regenerating factors	NIL-1 for Parkinson's disease (phase 2)

Table 1.5E. A selection of programs and of companies working in regenerative medicine (Cont'd)

Company	Program	Product/Stage
Stem Cells		
Geron (Menlo Park, CA)	Human stern cells for range, of disorders, telomerase inhibition; cloning	Telomerase inhibitors for cancer; telomerized cells for research (preclinical)
Layton Bioscience (Sunnyvale, CA)	Human neuronal stem cells	Preclinical
Cell Based Delivery (Providence, RI)	Genetically modified human muscle stem cells for application in cardiovascular disease	Preclinical
Aegera Therapeutics (Montreal, PQ)	Skin stem cells	Preclinical
MorphoGen Pharmaceuticals (San Diego, CA)	Pluripotent stern cells and matrix protein delivery systems	Preclinical
Ixion Biotechnologies (Alachua, FL)	Pancreatic islet stem cell technology	Preclinical
Osiris Therapeutics (Baltimore, MD)	Mesenchymal stem cells-progenitors for connective tissues such as bone, cartilage, muscle and tendons	Allogen (phase 2) and Osteocel (phase 1) for bone regeneration
Incara Pharmaceuticals (Research Triangle Park, NC)	Liver stem cells	Preclinical
Biomaterials and matrices		
Advanced Materials Design (New York, NY)	Development of resorbable polymers for bone and musculoskeletal uses	Preclinical
Selective Genetics (San Diego, CA)	Matrix-based delivery of DNA for various applications	Preclinical
Interpore International (Irvine, CA)	Coral bone graft	ProOsteon (preclinical)
Collagenesis (Beverly, MA)	Tissue matrix from cadaver	Dermalogen; Dermaplant skin (for damaged skin, wrinkles); Urogen and Duraderm (urological and gynecological reconstruction and support)
Protein Polymer Technologies (San Diego, CA)	Recombinant protein polymer hydrogels	Pilot studies

genome of the patient and then induce them to differentiate into replacement cells, such as cardiomyocytes to replace damaged heart tissue or insulin-producing β cells for patients with diabetes. Although cloning would eliminate the critical problem of immune incompatibility, there is also the task of reconstituting the cells into more complex tissues and organs *in vitro*.[116]

Liver transplants are the desired therapy for a number of conditions wherein the liver tissue has been damaged by disease, infections, toxins, medications, drugs or metabolic (inborn or induced) defects.

Islet cell transplantation or extraneously cultivated organs would be preferable to multiple daily insulin treatments. These cells or organs will sense the circulating levels of glucose and respond appropriately and immediately by releasing the right dose of human correct insulin (glycosylation patterns) at the right dose and speed. The complications of diabetes, due to imperfect matching of the body's demands with the substituting therapy, are frequent, life shortening and difficult to manage.[117]

The achievement of growing human corneas in the laboratory will impact the transplantation of damaged or impaired corneas, the studies of corneal wound healing and eye diseases, as well as the testing of household products for toxicity.[118]

Human corneal equivalents comprising the three main layers of the cornea (epithelium, stroma and endothelium) were constructed. Each cellular layer was fabricated from immortalized human corneal cells that were screened for use on the basis of morphological, biochemical and electrophysiological similarity to their natural counterparts. The resulting corneal equivalents mimicked human corneas in key physical and physiological functions, including morphology, biochemical marker expression, transparency, ion and fluid transport, and gene expression. Morphological and functional equivalents to human corneas that can be produced *in vitro* have immediate applications in toxicity and drug efficacy testing, and form the basis for future development of implantable tissues.[119]

Atherosclerotic vascular disease, in the form of coronary artery and peripheral vascular disease, is the largest cause of mortality in the USA. A tissue engineering approach was developed to produce arbitrary lengths of vascular graft material from smooth muscle and endothelial cells that were derived from a biopsy of vascular tissue. Bovine vessels cultured under pulsatile conditions had rupture strengths greater than 2000 mmHg, suture retention strengths of up to 90 g and collagen contents of up to 50%. Cultured vessels also showed contractile responses to pharmacological agents and contained smooth muscle cells that displayed markers of differentiation such as calponin and myosin heavy chains. Tissue-engineered arteries were implanted in miniature swine, with patency documented up to 24 days by digital angiography.[120]

Angiogenesis and vascular remodeling play significant roles in wound healing, tumor growth, cardiovascular disease and tissue transplantation. Significant effort is directed at developing models to study and manipulate these processes. Angiogenesis is typified by the elongation of thin-walled capillary tubes from existing vascular structures, followed by growth of mesenchymal cells such as pericytes and smooth-muscle cells during the formation of mature stable vessels. Mechanisms were studied of early vascular remodeling in experiments of suspending EC in three-dimensional culture, where they form tubular structures resembling immature capillaries.

Conditions for forming cultured human umbilical vein EC (HUVEC) into tubes within a three-dimensional gel that on implantation into immuno-incompetent mice undergo re-modeling into complex microvessels lined by human endothelium were identified. HUVEC suspended in mixed collagen/fibronectin gels organize into cords with early lumena by 24 h and then apoptose. Twenty-hour constructs, s.c. implanted in immunodeficient mice, display HUVEC-lined thin-walled microvessels within the gel 31 days after implantation. Retroviral-mediated overexpression of a caspase-resistant Bcl-2 protein delays HUVEC apoptosis *in vitro* for over 7 days. Bcl-2 transduced HUVEC produce an increased density of HUVEC-lined perfused microvessels *in vivo* compared with untransduced or control-transduced HUVEC. Bcl-2, but not control-transduced HUVEC recruit an ingrowth of perivascular smooth muscle α-actin-expressing mouse cells at 31 days, which organize by 60 days into HUVEC-lined multilayered structures resembling true microvessels. This sys-tem provides an *in vivo* model for dissecting mechanisms of microvascular remodeling by using genetically modified endothelium. Incorporation of such human endothelial-lined microvessels into engineered synthetic skin may improve graft viability, especially in re-cipients with impaired angiogenesis.[121]

The problems of organ replacements have been addressed by so far five approaches. Transplantation, which suffers from lack of donors and entails immune rejection and im-munosuppression problems; autografting, which uses parts of organs to restore organ func-tion in other body parts; replacement with an implantate or prothesis, often lasting insuffi-cient time before another surgery or treatment becomes necessary; *in vitro* synthesis, in which an organ is synthesized in culture and then implanted; and *in vivo* synthesis, where only a minimum of tissue is implanted to induce organ regeneration *in situ*. This combina-tion of *in vitro* and *in vivo* methodologies may advance some hitherto intractable problems of transplantation, especially in older patients, which suffer from chronically open wounds resulting from lack of vascular supply to the skin. The reported enrichment of vascular structures in host skin by implanting a skin substitute addresses these needs.

In vitro conditions appear to allow the synthesis of the avascular, epithelium-like tissue on one side of the basement membrane but not the synthesis of the vascularized supporting connective tissue on the other side. Synthesis of supporting tissue requires the presence of non-diffusible regulators. The synthesis of a near-physiological peripheral nerve connect-ing two transected nerve stumps across a gap and synthesis of a physiological conjunctiva in a conjunctiva-free bed, which had been prepared by excision to bare sclera, required for *in vivo* synthesis the presence of non-diffusible regulator. Studies of conditions required for organ synthesis, in particular the requirements for extracellular matrix-like regulators, will advance the field of organ replacement.[122]

The search for stem cells has revealed such cells in tissues considered previously un-likely to harbor cells capable of regenerative capacities.

The mature mammalian retina is thought to lack regenerative capacity. The identifica-tion of a stem cell in the adult mouse eye has been reported, which represents a possible substrate for retinal regeneration. Single pigmented ciliary margin cells clonally proliferate *in vitro* to form sphere colonies of cells that can differentiate into retinal-specific cell types, including rod photoreceptors, bipolar neurons and Müller glia. Adult retinal stem cells are localized to the pigmented ciliary margin and not to the central and peripheral retinal pigmented epithelium, indicating that these cells may be homologous to those found in the eye germinal zone of other non-mammalian vertebrates.[123]

ES cells are clonal cell lines derived from the inner cell mass of the developing blastocyst that can proliferate extensively *in vitro* and are capable of adopting all the cell fates in a developing embryo. Clinical interest in the use of ES cells has been stimulated by studies showing that isolated human cells with ES properties from the inner cell mass or developing germ cells can provide a source of somatic precursors. Previous studies have defined *in vitro* conditions for promoting the development of specific somatic fates, specifically, hematopoietic, mesodermal and neurectodermal. A method was presented for obtaining dopaminergic (DA) and serotonergic neurons in high yield from mouse ES cells *in vitro*. Furthermore, the ES cells can be obtained in unlimited numbers and these neuron types are generated efficiently. CNS progenitor populations were generated from ES cells, these cells were expanded and their differentiation promoted into DA and serotonergic neurons in the presence of mitogen and specific signaling molecules. The differentiation and maturation of neuronal cells was completed after mitogen withdrawal from the growth medium. This experimental system provides a powerful tool for analyzing the molecular mechanisms controlling the functions of these neurons *in vitro* and *in vivo*, and potentially for understanding and treating neurodegenerative and psychiatric diseases.[124]

The differentiation potential of stem cells in tissues of the adult has been thought to be limited to cell lineages present in the organ from which they were derived, but there is evidence that some stem cells may have a broader differentiation repertoire. It was shown that neural stem cells from the adult mouse brain can contribute to the formation of chimerical chick and mouse embryos, and give rise to cells of all germ layers. This demonstrates that an adult neural stem cell has a very broad developmental capacity and may potentially be used to generate a variety of cell types for transplantation in different diseases.[125]

Tissue engineering of human bone is a complex process, as the functional development of bone cells requires that regulatory signals be temporally and spatially ordered. The role of three-dimensional cellular interactions is well understood in embryonic osteogenesis, but *in vitro* correlates are lacking. *In vitro* serum-free transforming growth factor (TGF)-$\beta 1$ stimulation of osteogenic cells results immediately after passage in the formation of three-dimensional cellular condensations (bone cell spheroids) within 24–48 h. In turn, bone cell spheroid formation results in the up-regulation of several bone-related proteins (e.g., alkaline phosphatase, type I collagen and osteonectin) during days 3–7, and the concomitant formation of microcrystalline bone. This system of *ex vivo* bone formation should provide important information on the physiological, biological and molecular basis of osteogenesis.[126]

Massive bone defects are a great challenge to reconstructive surgery. The preferred treatment is an autologous bone graft. Bone lesions above a critical size become scarred rather than re-generated, leading to non-union. A greater degree of regeneration was obtained by using a resorbable scaffold with regeneration-competent cells to recreate an embryonic environment in injured adult tissues, and thus improve clinical outcome. A combination of a coral scaffold with *in vitro*-expanded marrow stromal cells (MSC) was used to increase osteogenesis more than that obtained with the scaffold alone or the scaffold plus fresh bone marrow. The efficiency of the various combinations was assessed in a large segmental defect model in sheep. The tissue-engineered artificial bone underwent morphogenesis leading to complete recorticalization and the formation of a medullary canal with mature lamellar cortical bone in the most favorable cases. Clinical union never occurred when the defects were left empty of filled with the scaffolding alone. In contrast, clinical union was obtained in three out of seven operated limbs when the defects were filled with the tissue-engineered bone.[127]

Approximately 400 million persons worldwide suffer from bladder disease. Individuals with end-stage bladder disease often require bladder replacement or repair. The bladder is responsible for being an elastic reservoir for the storage of urine, which maintains a low intraluminal pressure as it fills. In general, human organ replacement is limited by a donor shortage, problems with tissue compatibility and rejection. Creation of an organ with autologous tissue would be advantageous. Transplantable urinary bladder neo-organs were reproducibly created *in vitro* from urothelial and smooth muscle cells grown in culture from canine native bladder biopsies and seeded onto preformed bladder-shaped polymers. The native bladders were subsequently excised from canine donors and replaced with the tissue-engineered neo-organs. In functional evaluations for up to 11 months, the bladder neo-organs demonstrated a normal capacity to retain urine, normal elastic properties, and histological architecture. Successful re-constitution of an autologous hollow organ is possible using tissue-engineering methods.[128]

One of the main obstacles to successful islet transplantation for both type 1 and 2 diabetes is the limitation of available insulin-producing tissue. Only about 3000 cadaver pancreases become available in the USA each year, while about 35,000 new cases of type 1 diabetes are diagnosed each year. This lack of tissue has given a high priority to efforts to stimulate the growth of new pancreatic islet tissue. The need for transplantable human islets has stimulated efforts to expand existing pancreatic islets and to grow new ones. In order to demonstrate that human adult duct tissue could be expanded and differentiated *in vitro* to form islet cells, digested pancreatic tissue that is normally discarded from eight human islet isolations was cultured under conditions that allowed the expansion of the ductal cells as a monolayer where-upon the cells were overlaid with a thin layer of Matrigel. With this manipulation, the mono-layer of epithelial cells formed three-dimensional structures of ductal cysts from which 50–150 µm diameter islet-like clusters of pancreatic endocrine cells budded. Over 3–4 weeks culture the insulin content per flask increased 10- to 15-fold as the DNA content increased up to 7-fold. The cultivated human islet buds were shown by immunofluorescence to consist of cytokeratin 19-positive duct cells and hormone-positive islet cells. Double staining of insulin and non-β cell hormones in occasional cells indicated immature cells still in the process of differentiation. Insulin secretion studies were done over 24 h in culture. Compared with their basal secretion at 5 mM glucose, cysts/cultivated human islet buds exposed to stimulatory 20 mM glucose had a 2.3-fold increase in secreted insulin. Duct tissue from human pancreas can be expanded in culture and then be directed to differentiate into glucose responsive islet tissue *in vitro*. This approach may provide a potential new source of pancreatic islet cells for transplantation.[129]

Expression profiling was studied in pancreatic β cells that will contribute to under standing regulation of development, function and regeneration of islet tissue. The β cells of the pancreatic islets of Langerhans are a primary component in vertebrate glucose hemostasis. Pancreatic β cells respond to changes in blood glucose by secreting insulin and increasing insulin synthesis. To identify genes used in thee responses, expression profiling was carried out of β cells exposed to high (25 mM) or low (5.5 mM) glucose by using oligonucleotide microarrays. Functional clustering of genes that averaged a 2.2-fold or greater change revealed large groups of secretory pathway components, enzymes of intermediate metabolism, cell-signaling components and transcription factors. Many secretory pathway genes were up-regulated in high glucose, including seven members of the endoplasmic reticulum translocon. In agreement with array analysis, protein levels of translocon components were increased by high glucose. Most dramatically, the α subunit of the signal recognition particle receptor was increased over 20-fold. These data indicate that the translocon and ribosome docking are major regulatory targets of glucose in the β cell. Analysis of genes encoding enzymes of intermediary metabolism indicated that low glucose brought about greater utilization of amino acids as an energy source. This is supported by observations of increased urea production under low-glucose conditions. Genome-wide integration of β cell functions at the level of transcript abundance validates the efficacy of expression profiling in identifying genes in the β cell glucose response.[130]

An experiment in xenotransplantation and on porcine retrovirus is included in this chapter, in order to show the potential of grafts in axonal regeneration and in highlighting viral dangers lurking in xenografts.

Transplantation of olfactory ensheathing cells (OECs) or Schwann cells derived from transgenic pigs expressing the human complement inhibitory protein, CD59 (hCD59), into transected dorsal column lesions of the spinal cord of the immunosuppressed rat to induce axonal regeneration. Non-transplanted lesion-controlled rats exhibited no impulse conduction across the transection site, whereas in animals receiving transgenic pig OECs or Schwann cells impulse conduction was restored across and beyond the lesion site for more than a centimeter. Cell labeling indicated that the donor cells migrated into the denervated host tract. Conduction velocity measurements showed that the regenerated axons conducted impulses faster than normal axons. By morphological analysis, the axons seemed thickly myelinated with a peripheral pattern of myelin expected from the donor cell type. These results indicate that xenotransplantation of myelin-forming cells from pigs genetically altered to reduce the hyperacute response in humans are able to induce elongated axonal regeneration and re-myelination and restore impulse conduction across the transected spinal cord.[131]

Oligodendrocyte precursor cells (OPCs) are the best-characterized precursors in the CNS of mammals. They arise from multi-potential cells in spatially restricted germinal zones and then migrate through the developing CNS.

During animal development, cells become progressively more restricted in the cell types to which they can give rise. In the CNS, multipotential stem cells produce various kinds of specified precursors that divide a limited number of times before they terminally differentiate into either neurons or glial cells. Certain extracellular signals can induce OPCs to revert to multipotential neural stem cells, which can self-renew and give rise to neurons and astrocytes, as well as to oligodendrocytes, thus demonstrating a great developmental potential.[132]

Animal donors such as pigs could provide an alternative source of organs for transplantation. The promise of xenotransplantation is offset by the possible public health risk of cross-species infection. All pigs contain several copies of porcine endogenous retroviruses (PERV), and at least three variants of PERV can infect human cell lines *in vitro* in coculture, infectivity and pseudotyping experiments. If xenotransplantation of pig tissues results in PERV viral replication, there is a risk of spreading and adaptation of this retrovirus to the human host. C-type retroviruses related to PERV are associated with malignancies of hematopoietic lineage cells in their natural hosts. Pig pancreatic islets were shown to produce PERV and can infect human cells in culture. After transplantation into NOD/SCID (non-obese diabetic/severe combined immunodeficiency) mice, ongoing viral replication was detected and several tissue compartments became infected. PERV is thus demonstrated to be transcriptionally active and infectious cross-species *in vivo* after transplantation of pig tissues. The concern for PERV infection risk associated with pig islet xenotransplantation in immunosuppressed human patients appears justified.[133]

The progress toward human therapeutic cloning requires the successful application of nuclear transfer techniques to a range of mammalian species and the subsequent transfer of technologies to human cells. The objective of therapeutic cloning is to produce pluripotent stem cells that carry the nuclear genome of the patient and then induce them to differentiate into replacement cells, such as cardiomyocytes to replace damaged heart tissue or insulin-producing β cells for patients with diabetes. Although cloning would eliminate the critical problem of immune incompatibility, there is also the task of reconstituting the cells into more complex tissues and organs *in vitro*. The range of disorders where human therapeutic cloning may find application is impressive.[134]

The isolation of pluripotent ES cells opens the way to generate many different tissues and organs for gene and tissue therapies.

Pluripotent human stem cells isolated from early embryos thus represent a potentially un-limited source of many different cell types for cell-based gene and tissue therapies. If the full potential of cell lines derived from donor embryos is to be realized, the problem of donor-recipient tissue matching needs to be overcome. One approach, which avoids the problem of transplant rejection, would be to establish stem cell lines from the patient's own cells through therapeutic cloning. Recent studies have shown that it is possible to transfer the nucleus from an adult somatic cell to an unfertilized oocyte that is devoid of maternal chromosomes, and achieve embryonic development under the control of the transferred nucleus. Stem cells isolated from such a cloned embryo would be genetically identical to the patient and pose no risk of immune rejection. The isolation of pluripotent murine stem cells was reported from reprogrammed adult somatic cell nuclei. Embryos were generated by direct injection of mechanically isolated cumulus cell nuclei into mature oocytes. ES cells isolated from cumulus-cell-derived blastocysts displayed the characteristic morphology and marker expression of conventional ES cells and underwent extensive differentiation into all three embryonic germ layers (endoderm, mesoderm, and ectoderm) in tumors and in chimeric fetuses and pups. The ES cells were also shown to differentiate readily into neurons and muscle in culture. Thus, pluripotent stem cells can be derived from nuclei of terminally differentiated adult somatic cells and offer a model system for the development of therapies that rely on autologous, human pluripotent stem cells.[135]

The senescence of normal cells poses a barrier against their use in organ generation and transplantation.

Most normal somatic cells from humans and other mammals cannot divide indefinitely because of the progressive loss of DNA from telomeres, the specialized ends of the linear chromosomes. The gradual erosion of telomeres during repeated cell division results from the combined effects of the inability of normal DNA replication to completely extend telomeric DNA and the lack of expression of the reverse transcriptase component of the telomerase ribonucleoprotein complex (TERT). Stem cells and germ line cells, which normally express TERT and have telomerase activity, are able to fully replicate telomeric DNA and thereby are able to divide indefinitely. Cells that do not have telomerase activity are able to divide only a limited number of times before they senesce. In cells with forced expression of TERT, the progressive shortening of telomeres is prevented and these cells continue to proliferate rapidly at a cumulative population doubling level (PDL) greatly exceeding the PDL at which they would otherwise senesce.

These results indicate that expression of TERT is sufficient to immortalize at least some cell types that normally undergo cellular senescence. Moreover, immortalization resulting from TERT expression does not appear to be accompanied by the acquisition of characteristics of cancer cells, such as chromosomal abnormalities, anchorage-independent growth in culture, loss of normal cell cycle checkpoints, or tumorigenicity in immunodeficient mice. Such findings encouraged speculation that cells expressing TERT could be used in cell-based therapies, such as *ex vivo* gene therapy, tissue engineering, and transplantation of cells to correct metabolic or endocrine defects. The fulfillment of this promise depends on ectopic expression of TERT not affecting cellular regulatory systems such as those that control differentiation.

Human telomerase reverse transcriptase (hTERT) expression in experimental xenotransplantation was demonstrated. It had been shown that bovine adrenocortical cells can be transplanted into SCID mice and that these cells form functional tissue that replaces the animals' own adrenal glands. Primary bovine adrenocortical cells were co-transfected with plasmids encoding hTERT, SV40 T antigen, neo and GFP. These clones do not undergo loss of telomeric DNA and appear to be immortalized. Two clones were transplanted beneath the kidney capsule of SCID mice. Animals that received cell transplants survived indefinitely despite adrenalectomy. The mouse glucocorticoid, corticosterone, was replaced by the bovine glucocorticoid, cortisol, in the plasma of these animals. The tissue formed from the transplanted cells resembled that formed by transplantation of cells that were not genetically modified and was similar to normal bovine adrenal cortex. The proliferation rate in tissues formed from these clones was low and there was no indication of malignant transformation.[136]

Retroviral-mediated transduction of hTERT (human telomerase reverse transcriptase) in human dermal microvascular endothelial cells (EC) (HDMEC) results in cell lines that form microvascular structures upon subcutaneous implantation in severe combined immunodeficiency mice (SCID) mice. Anti-human type IV collagen basement membrane immunoreactivity and visualization of enhanced green fluorescent protein (eGFP)-labeled microvessels confirmed the human origin of these vessels. Primary HDMEC-derived vessel density decreased with time, while telomerized HDMEC maintained durable vessels six weeks after xenografting. The modulation of implant vessel density by exposure to different angiogenic and angiostatic factors confirms the utility of the system for human microvascular remodeling and pertinent in vivostudies.[565]

Slowing the rate of telomere shortening could have benefits in specific tissues where telomere-based growth arrest, i.e. senescence, occurs. The medical implications of this technology are profound. Some promising areas of cell engineering include rejuvenation of HSC for improving bone marrow transplants or enhancing general immunity for older patients. Other possibilities include an unlimited supply of skin cells for grafts in burn patients and for treating ulcerated lesions that do not heal. The immortalization of chondrocytes or osteoprogenitor cells to treat osteoarthritis or for bone grafts, and endothelial cells for the generation of tissue-engineered blood vessels are other areas of interest. In the future we could try to grow cells in the laboratory for disorders for which there are currently no cures, e.g., muscle satellite cells for muscular dystrophy, retinal cells for the treatment of macular degeneration (a leading cause of age-related blindness) and immune cells for HIV patients. This could avoid immune rejections and could slow down or reverse some of the problems associated with these disorders. Proteins could also be produced from normal human cell cultures to treat the donor patient. Engineered cells may also be useful for *in vitro* markets, as alternatives to animal testing or to produce products for cosmetic applications. Of course, safety and efficacy standards, quality and control assurances will need to be developed as well as preclinical and clinical evaluations.[137]

Gene therapy and tissue engineering are new concepts in the treatment of human disease. Gene therapy uses DNA as a 'super pharmaceutical', an agent with the potential to alter cell function for an extended period of time in relation to more established therapeutic agents. Tissue engineering aims to regrow tissue structure lost as a result of trauma or disease through the application of engineered materials. Both fields can demonstrate initial success, such as the replacement of adenosine deaminase genes in children with severe immunodeficiency and the use of synthetic materials to accelerate healing of burns and skin ulcers. Major limitations are the low level of expression in gene therapy and the tissue engineers have not yet learned to reproduce complex architecture, such as vascular networks, which are essential for normal tissue function. A combination of both methods has been used as a new strategy to overcome problems remaining.

DNA delivery has been described *in situ* from polymer coatings, microspheres and synthetic matrices. DNA–material hybrid systems promise to enable forms of gene therapy not possible with other gene delivery systems, e.g., polymeric microspheres facilitate the expression of orally administered genes.[138]

Loss or deficient tissue function leads to millions of surgical procedures each year and to a loss of, for example, the US economy alone of hundreds of billions of dollars. Tissue engineering has emerged as a potential means of growing new tissues and organs to treat these patients, and several approaches are currently under investigation to engineer structural tissues. One approach involves transplanting cells on biodegradable polymer matrices. Matrices serve to deliver cells to specific anatomical sites, create and maintain a space for tissue development, and guide tissue formation before being degraded. Limitations of this approach include the need to isolate and expand cells *in vitro*, and poor survival of many cell types following tissue transplantation. A separate strategy involves the delivery of tissue-inductive proteins [(e.g., bone morphogenetic proteins and platelet-derived growth factor (PDGF)]. One important drawback of this approach is decreased protein stability in the delivery system. Delivery of plasmid DNA encoding for inductive factors has been proposed as a replacement to direct delivery of the protein. The delivery of plasmid DNA *in vivo* is typically associated with low levels of gene transfer and cellular expression, perhaps due to a limited exposure of cells to the plasmid.

The engineering of tissues is proposed by the incorporation and sustained release of plasmids encoding tissue-inductive proteins from polymer matrices. Matrices of poly(lactide-co-glycolide) (PLG) were loaded with plasmid, which was subsequently released over a period ranging from days to a month *in vitro*. Sustained delivery of plasmid DNA from matrices led to the transfection of large numbers of cells. Furthermore, *in vivo* delivery of a plasmid encoding PDGF enhanced matrix deposition and blood vessel formation in the developing tissue. This contrasts with direct injection of the plasmids, which did not significantly affect tissue formation. This method of DNA delivery may find utility in tissue engineering and gene therapy applications.[139]

Biocompatible materials are commonly used in reconstructive medicine. One such material, CellFoam, has been used to repair bone defects *in vivo*; the open pore structure permits infiltration and integration of cells into the surrounding engraftment site. CellFoam is fabricated by high-temperature precipitation of tantalum, a highly biocompatible metal with an established history of *in vivo* medical use, as a thin layer onto a highly regular reticulated carbon skeleton. The resulting material is highly porous and has a heavily contoured surface with nanostructure ridges that facilitate cell attachment. Three-dimensionality is advantageous for bone reconstruction and is critical for organ tissue function.

Biocompatible inorganic matrices have been used to enhance bone repair by integrating with endogenous bone structure. Hypothesizing that a three-dimensional framework might support reconstruction of other tissues as well, the capacity of tantalum-coated carbon matrix to support reconstitution of functioning thymic tissue was assessed. A thymic organoid was an engineered by seeding matrices with murine thymic stoma. Co-culture of human bone marrow-derived hematopoietic progenitor cells within this xenogenic environment generated mature functional T cells within 14 days. The proportionate T cell yield from this system was highly reproducible, generating over 70% CD3$^+$ T cells from either AC133$^+$ or CD34$^+$ progenitor cells. Cultured T cells expressed a high level of T cell receptor excision circles (TREC), demonstrating *de novo* T lymphopoiesis, and function of fully mature T cells. This system not only facilitates analysis of the T lymphopoietic potential of progenitor cell populations; it also permits *ex vivo* genesis of T cells for possible applications in treatment of immunodeficiency.[140]

Cell lineage studies are essential basis for elucidating the genetic and biochemical pathways and switches in stem cells and differentiation. Hematopoietic stem cells give rise to all blood cell lineages in mammals. Molecular decisions that differentiate one lineage from another are regulated by unique protein complexes constituted of general as well as lineage-specific transcription factors. Gene inactivation studies in mice have identified an important role for a number of hematopoietic transcription factors, such as GATA-1 for erythroid development, GATA-2 for definitive hematopoiesis and GATA-3 for T cell development. The *Drosophila* GATA homolog *serpent* (*srp*) is required for the embryonic blood cell development. Another mammalian gene encoding the AML1 protein is the most frequent target of chromosomal translocations in acute myeloid leukemia.

Two major classes of cells observed within the *Drosophila* hematopoietic repertoire are plasmocytes/macrophages and crystal cells. The transcription factor Lz (Lozenge), which resembles human AML1 (acute myeloid leukemia-1) protein, is necessary for the development of crystal cells during embryonic and larval hematopoiesis. Another transcription factor, Gcm (glial cells missing), has previously been shown to be required for plasmocytes development. Misexpression of Gcm causes crystal cells to be transformed into plasmocytes. The *Drosophila* GATA protein Srp (Serpent) is required for both Lz and Gcm expression and is necessary for the development of both classes of hemocytes, whereas Lz and Gcm are required in a lineage-specific manner. Given the similarities of Srp and Lz to mammalian GATA and AML1 proteins, observations in *Drosophila* are likely to have broad implications for understanding mammalian hematopoiesis and leukemia.[141]

Therapeutic cloning describes the potential use of nuclear transfer to provide cells, tissues and organs for patients requiring the replacement or supplementation of diseased or damaged tissues. The concept of using ES cells as a source of multiple cell types for use in tissue repair is being developed towards biomedical application based on breakthroughs in somatic nuclear transfer and human ES cell derivation. ES cell lines that are customized and genetically identical to those of the patient are becoming feasible.[142] Locally delivered bone marrow cells can generate *de novo* myocardium, thus ameliorating the outcome of coronary artery disease.[659] Cells amenable to regenerative tissue studies include neurons derived from human neurospheres[671] and insulin-secreting structures similar to pancreatic islets.[672]

1.8   Micro- and Nanotechnologies for Medicine

The most profound effects of microtechnologies are evident in the development of microarrays (biochips) for drug discovery, testing and genotyping.

Biological microchips (biochips) are revolutionizing gene expression analysis and classical genotyping as well as diagnostics and testing. The worldwide markets for biochips are primarily composed of DNA chips, protein chips and laboratory chips. Generally, arrays are differentiated from microfluidic systems, which can actively operate analysis, separation and synthetic processes by means of microscopic capillary systems, mini-pumps and mini-valves. These are often called a 'lab-on-a-chip'.

The total market volume for biochips was estimated for 1999 to be approximately US$ 180 million, whereby the percentage of DNA chips is above 90%. Annual growth of more than 30% is expected for the years until 2005. In 2005, the total market volume is expected to amount to US$ 950 million (US$ 725 million DNA chips, US$ 157 million for lab chips and US$ 68 million for protein chips).

The market is segmented (pharmaceutical, biotechnology and academia), and out of a total of anticipated US$ 660 million for 2003, the pharmaceutical industry will comprise

US$ 260 million, the biotechnology industry US$ 185 million and the academic institutions US$ 212 million.

DNA chips are provided by more than 20 companies (e.g., Affymetrix, A Gene, Amersham Pharmacia Biotech, Genometrix, Genomic Solutions, Hyseq Incyte Pharmaceuticals, Nanogen, etc.), protein chips by one company (Ciphergen) and microfluidic chips by several companies [Aclara Biosciences, Caliper (Cellomics), Cepheid and Orchid Biocomputer].[143]

The requirements for ever more discriminating methods of separating synthesized molecules, especially during later stages of drug development, is facilitated by improved methods in the design of separating phases. Biomimetic synthesis offers the opportunity to design and synthesize specific inorganic structures to fit hand-in-glove with target molecules. Self-assembled block co-polypeptide architecture provides the matrix for silica shapes. The synthetic capability to control directly silica shape, hydrolysis and condensation rate by adjustment of block co-polypeptide composition offers a route to the environmentally benign, biomimetic synthesis of inorganic materials.

In biological systems such as diatoms and sponges, the formation of solid silica structures with precisely controlled morphologies is directed by proteins and polysaccharides, and occurs in water at neutral pH and ambient temperature. Laboratory methods, in contrast, have to rely on extreme pH conditions and/or surfactants to induce the condensation of silica precursors into specific morphologies or patterned structures. This contrast in processing conditions and the growing demand for benign synthesis methods that minimize adverse environmental effects have spurred much interest in biomimetic approaches in materials science. The recent demonstration that silicatein – a protein found in the silica spicules of the sponge *Thethya aurantia* – can hydrolyze and condense the precursor molecule tetraethoxysilane to form silica structures with controlled shapes at ambient conditions seems particularly promising in this context. Synthetic cysteine–lysine block copolypeptides were described that mimic the properties of silicatein: the copolypeptides self-assemble into structured aggregates that hydrolyze tetraethoxysilane while simultaneously directing the formation of ordered silica morphologies. Oxidation of the cystein sulfydryl groups, which is known to affect the assembly of the block co-polypeptide, allows to produce different structures: hard silica spheres and well-defined columns of amorphous silica are produced using the fully reduced and the oxidized forms of the copolymer, respectively.[144]

Another example of molecular imprinting has been demonstrated by a generally applicable imprinting strategy which is used to synthesize bulk, amorphous, microporous silicas with binding sites comprising either one, two or three propylamine groups that are covalently linked to the silicon atoms of the silicate.

Molecular imprinting aims to create solid materials containing chemical functionalities that are spatially organized by covalent or non-covalent interactions with imprint (or template) molecules during the synthesis process. Subsequent removal of the imprint molecules leaves behind designed sites for the recognition of small molecules, making the material ideally suited for applications such as separations, chemical sensing and catalysis. The molecular imprinting of bulk polymers and polymer and silica surfaces has been reported, but the extension of these methods to a wider range of materials remains problematic. For example, the formation of substrate-specific cavities within bulk silica, while conceptually straightforward, has been difficult to accomplish experimentally. The imprinting of bulk amorphous silicas with single aromatic rings carrying up to three 3-amino-propyl-triethoxysilane side groups has been described. This generates and occupies microporosity and attaches functional groups to the pore walls in a controlled fashion. The triethoxysilane part of the molecules' side groups is incorporated into the silica framework during sol–gel synthesis and subsequent removal of the aromatic core creates a cavity with spatially organized aminopropyl groups covalently anchored to the pore walls. The imprinted silicas act as shape-selective base catalysts. The strategy can be extended to imprint other functional groups, which should give access to a wide range of functionalized materials.[145]

Progress towards analytical precision required in nanotechnology has been reported in the search for optical reporters for diagnostic detection and labeling to be used in biomedical research, testing and application.

The use of colloidal silver plasmon-resonant particles (PRPs) as optical reporters in biological assays has been demonstrated. PRPs are ultrabright, nano-sized optical scatterers, which scatter light elastically and can be prepared with a scattering peak at any color in the visible spectrum. PRPs are readily observed individually with a microscope configured for dark-field microscopy, with white-light illumination of typical power. The use of PRPs, surface coated with standard ligands, was demonstrated as target-specific labels in an *in situ* hybridization and an immunocytology assay. PRPs can replace or complement established labels, such as those based on radioactivity, fluorescence, chemiluminescence or enzymatic/colorimetric detection that are used routinely in biochemistry, cell biology and medical diagnostic applications. Moreover, because PRP labels are non-bleaching and bright enough to be rapidly identified and counted, an ultrasensitive assay format based on single-target molecule detection is now practical. The results of a model sandwich immunoassay for goat anti-biotin antibody used the number of PRP labels counted in an image to constitute the measured signal.[146]

Fiber optic chemical sensors and biosensors offer important advantages for *in situ* monitoring applications because of the optical nature of the detection signal. Recent advances in nanotechnology leading to the development of optical fibers with submicron-sized dimensions have opened up new horizons for intracellular measurements.

The application of an antibody-based nanoprobe for *in situ* measurements of a single cell was reported. The nanoprobe employs antibody-based receptors targeted to a fluorescent analyte, benzopyrene tetrol (BPT), a metabolite of the carcinogen benzo[*a*]pyrene (BaP) and of the BaP–DNA adduct. Detection of BPT is of great biomedical interest, since this species can serve as a biomarker for monitoring DNA damage due to BaP exposure and for possible precancer diagnosis. The measurements were performed on the rat liver epithelial clone 9 cell line, which was used as the model cell system. Before making measurements, the cells were treated with BPT. Nanoprobes were inserted into individual cells, incubated 5 min to allow antigen-antibody binding, and then removed for fluorescence detection. A concentration of $96 \pm 0.2 \times 10^{11}$ M for BPT in the individual cells investigated was determined. The results demonstrate the possibility of *in situ* measurements inside a single cell using an antibody-based nanoprobe.[147]

Sequence-selective DNA detection is becoming increasingly important to unravel the genetic basis of disease and for the development of diagnostics and therapies. A DNA array probing technique with increased sensitivity has been developed which uses oligonucleotide-modified gold nanoparticle probes and a conventional flatbed scanner for analyzing combinatorial DNA arrays. Labeling oligonucleotide targets with nanoparticle rather than fluorophore probes substantially alters the melting profiles of the targets from an array substrate. This difference permits the discrimination of an oligonucleotide sequence from targets with single nucleotide mismatches with a selectivity that is over three times that observed for fluorophore-labeled targets. In addition, when coupled with a signal amplification method based on nanoparticle-promoted reduction of silver (I), the sensitivity of this scanometric array detection system exceeds that of the analogous fluorophore system by two orders of magnitude.[148]

Functional nanoelectromechanical systems (NEMS) require locomotive forces for power to act as biomolecular motors. The integration of enzymes as biomolecular motors with nano-scale engineered systems generates hybrid organic-inorganic devices capable of utilizing ATP as an energy source. Such a biomolecular motor consisting of an endineered substrate, an F1-ATPase biomolecular motor and fabricated nanopropellers was constructed.[459] The molecular dynamics of the F1-ATP-ase at the molecular level were described[652] and the suitability of biomolecular devices as nanomolecular machines comprehensively reviewed.[653]

The technological requirements for designing and constructing microfluidic devices for micro- and nano-technology are expanding. A microfluidic platform for the construction of microscale components and autonomous systems consists of a combination of liquid-phase photopolymerization, lithography, and laminar flow to generate channels, valves, actuators, sensors, and systems.[460]

Bioelectronics, the integration of biomaterials, cells and tissues or organs, electronics, optronics, electronic transducers, enables the electronic transduction of biorecognition events, biotransaformations and biosensing events.[666] Electrogenic cells can be coupled to electrodes and the resulting bio-electronic devices, such as cell-transistor hybrids, can be used for drug screening, the generation of biosensors, and implantates.[708] The capabilities to manipulate single biomolecules will lead to molecular motors and devices.[668]

Nanotechnologies provide increasing potential for modifications of surfaces and the construction of biocompatible devices.[656] There is an expanding range of tools available for developing, engineering, and constructing micro- and nano-devices, such as the submicrometer patterning of thin-film electrets with trapped charge by using a flexible, electriclly conductive electrode,[562] by replica-molding techniques.[557,558] Robust chemical methods allow the fabrication of ultrahigh-density arrays of nanopores with high aspect ratios using the equi-librium self-assembled morphology of asymmetric diblock copolymers.[556] The direct printing of functional electronic materials provides possibilities for the construction of polymer devices and integrated transistor circuits.[555] Sensor arrays, nanoreactors, photonic and fluidic devices can be constructed by optically defined multifunctional patterning of photo-sensitive thin-film silica mesophases.[554]

The required energy conversion in nanodevices may be achieved by constructing molecular motors[552,553] and providing molecular switches[551,554] such as those consisting of a metal cluster and redox-addressable groups.[550]. The integration of sensors, measurement and control will be faciliated by the development of mono-molecular electronics where single molecules integrate elementary functions and interconnections.[549]. The rules and methods as well as the design tools for nano-technology are rapidly spreading and developing.[529–543]

The potential of electronics with organic molecules and the technologies employed in their production, such as soft lithography and self-assembly of monolayers,[709,710] increases the feasibility of micro- and nano-devices for biomedical applications.

1.9 Structural Genomics

The rational design of drugs, therapies, agonists and antagonists requires knowledge of the structure and function of target molecules such as receptors, enzymes, RNA, complexes and regulatory proteins. Following the elucidation of gene sequences and the analysis of proteins and their interactions, the knowledge of the structure of target molecules become essential for further research into and development of the desired interactions. The structure is expected to provide useful information about biological and biochemical functions in order to facilitate the understanding of normal and disease phenomena, and the development of therapeutically active molecules.

The structure reveals the organization of the protein, RNA or complex in three dimensions, and thus allows the identification of the partial structures, domains, sequences and interactions that determine the function(s) of the molecule. Protein–ligand complexes are especially useful providers of functional information as they define the location of the ligand, the active site, the conformational effects and the mechanism of action of the molecule.[428]

The choice of targets from the numerous biological molecules is determined by the feasibility of isolating and purifying the molecules or complexes, as well as by the interest in the putative or known biological function or context. Structural genomics aims to determine the structure of all proteins in order to provide targets for R & D, and to construct a true structural and functional model of cells and organisms. The selected target molecules are studied primarily by crystallography and NMR in order to elucidate their structure, followed by biochemical and biological studies and validation.[429]

The production of biological molecules, in particular proteins, for structural analysis is a major challenge primarily met by cloning and by purification.[430]

Automation will be a major key for the success of structural genomics. The development of effective macromolecular structure determination facilities using intense X-rays from synchroton sources and advances in protein expression, purification and crystallization are accelerating the process of determining the structures of target macromolecules.[431, 432]

Figure 1.3 shows the procedure for structure determination by X-ray crystallography and Table 1.6 lists software used in structural analysis (data processing, molecular replacement, heavy atom site identification, model building, refinement and validation).

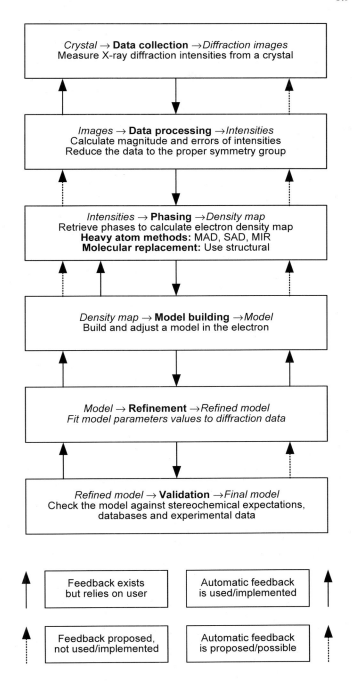

Figure 1.3. Procedure for structure determination by X-ray crystallography [Refs in 432].

Table 1.6. Software [Refs in 432].

General packages	
CCP4	http://www.dl.ac.uk/CCP/CCP4/
CNS	http://cns.csb.yale.edu/vLO/
Data processing	
D*TREK	http://www.msc.com/brochures/dTREK/
DPS	http://ultdev.chess.corneil.edu/MacCHESS/DPS/
HKL2000/DENZO	http://www.hkl-xray.com/
MOSFLM	http://www.mrc-lmb.cam.ac.uk/harry/mosflm/
XDS	not available
STRATEGY	http://www.crystal.chem.uu.nl/distr/strategy.html
PREDICT	http://biop.ox.ac.uk/www/distrib/predict.html
Phasing	
• **Molecular replacement**	
AMORE (and in CCP4)	ftp://b3sgi3.cep.u-psud.fr/pub/
CNS	http://cns.csb.yale.edu/vLO/
MOLREP (CCP4)	http://www.dl.ac.uk/CCP/CCP4/dist/html/molrep.html
• **Heavy atom sites identification**	
SHELXD	http://shelx.uni-ac.gwdg.de/SHELX/
SNB	http://www.hwi.buffalo.edu/SnB/
RANTAN, RSPS (CCP4)	http://www.dl.ac.uk/CCP/CCP4/dist/htmi/rantan.htmi;
	http://www.dl.ac.uk/CCP/CCP4/dist/html/rsps.html
• **Heavy atom phasing**	
CHART	http://crick.chem.gla.ac.uk/-pauie/chart/
MLPHARE (CCP4)	http://www.dl.ac.uk/CCP/CCP4/dist/html/miphare.htmi
PHASES	not available
SHARP	http://lagrange.mrc-lmb.cam.ac.uk/
SOLVE	http://www.solve.ianl.gov/
Model Building	
• **Pattern searching**	
ESSENS	http:7/alpha2.bmc.uu.se/-gerard/manuals/gerard_manuals.html
FFFEAR	http://www.ysbl.york.ac.uk/-cowtan/fffear/fffear.html
• **Interactive graphics for model building**	
MAIN	http://www-bmb.ijs.si/
0	http://origo.imsb.au.dk/-mok/o/
QUANTA	http://www.msi.com/life/products/quanta/index.html
TURBO-FRODO	http://afmb.cnrs-mrs.fr/TURBO_FRODO/
XTALVIEW	http://www.scripps.edu/pub/dem-web/
• **Automated model building**	
ARP/wARP	http://www.embl-hamburg.de/ARP/
Refinement	
BUSTER	http://lagrange.mrc-lmb.cam.ac.uk/
CNS	http://cns.csb.yale.edU/v1.0/
REFMAC (CCP4)	http://www.dl.ac.uk/CCP/CCP4/dist/html/refmac.htrnl
SHELXL	http://shelx.uni-ac.gwdg.de/SHELX/
TNT	http://www.uoxray.uoregon.edu/tnt/welcome.html

Table 1.6. Software (Cont'd).

Validation

PROCHECK (and in CCP4)	http://www.biochem.ucl.ac.uk/~roman/procheck/procheck.htmi
SFCHECK (CCP4)	http://www.dl.ac.uk/CCP/CCP4/dist/html/sfcheck.html
WHATCHECK	http://swift.embl-heidelberg.de/whatcheck/

NMR spectroscopy yields structural information that aids in the determination of three-dimensional protein structures and complements, or is increasingly substituting for, X-ray crystallographic data.[433]

The simulation and modeling of protein structures assists in refining the structural determinations and speeds up the structure analysis, especially when using comparative modeling, based on the fact that protein families share similar architectures and sequences.[434]

Table 1.7 lists websites related to the use of NMR in structural genomics.

Table 1.7. Web sites related to the use of NMR in structural genomics [Refs in 433].

Center or Consortium	URL
BioMagResBank-	www.bmrb.wisc.edu
Harvard Structural Genomics of Cancer Initiative, USA	sbweb.med.harvard.edu/-sgc/
New Jersey Commission on Science and Technology Initiative in Structural Genomics, USA	www-nmr.cabm.rutgers.edu/structuralgenomics
Northeast Structural Genomics Consortium, USA	www.nesg.org
Protein Structure Factory, Germany	userpage.chemie.fu-berlin.de/-psf/
Riken Genome Sciences Center, Tokyo, Japan; Toronto	www.gsc.riken.go.jp
Structural Proteomics Project, Canada	nmr.oci.utoronto.ca/arrowsmith/proteomics

In addition to the historically based focus on protein structure, RNA is also an important target of structural analysis. Due to the variety of important roles which RNA molecules play in cells, e.g., protein synthesis and targeting, the processing, splicing, editing and modification of RNA, and involvement in chromosome end maintenance, it will be necessary to elucidate the roles of all RNAs, their interactions and their complexes.[435]

Structural genomics is receiving increasing resources in North America, Europe and Japan,[436, 437, 438] and is seen as being of crucial importance for the biotechnology industry as it will provide a competitive advantage in drug discovery and development.[439]

Table 1.8 lists the websites for structural genomics in North America and Table 1.9 shows the European research initiatives related to structural genomics.

Table 1.8. Web sites for North American structural genomics [Refs in 436].

Organization	Web site
Northeast Structural Genomics Consortium	http://www.nesg-org/
	http://www-nmr.cabm.rutgers.edu/structuralgenomics
New York Structural Genomics Research Consortium	http://www.nysgrc.org/
Joint Center for Structural Genomics	http://ww.jcsg.org/
Berkeley Structural Genomics Center	http://www-kimgrp.lbl.gov/genomics/proteinlist.html
TB Structural Genomics Consortium	http://www.doe-mbi.ucla.edu/TB
Ontario Cancer Institute/University of Toronto/Pacific Northwest National Laboratory/University of British Columbia/McGill University	http://www.nmr.oci.utoronto.ca/arrowsmith/ proteomics
	http//:s2f.carb.nist.gov/
Center for Advanced Research in Biotechnology/University of Maryland	http://www.nigms.nih.gov/funding/psi.html
NIH structural genomics initiatives	htpp://grants.nih.gov/grants(guide/rfa-files/ RFAGM-006.html
Targeting	http://www.structuralgenomics.org/main.html
	http://presage.Berkeley.edu
International cooperation	http://www.nigms.nih.gov/news/meetings/hinxton.html
International cooperation	http://www.nigms.nih.gov/news/meetings/hinxton.html

The requirements for competitive, cost-effective and fast structure analysis call for a high-throughput structural genomics facility.[440] A screening technique which combines lead identification, structural assessment and compound optimization is based on X-ray crystallography, and provides detailed crystallographic information in a high-throughput fashion. A new orally available class of urokinase inhibitors for use in cancer therapy has been identified by such a technique.[441]

Structural and dynamic transient structural studies are feasible using X-ray diffraction and absorption as well as electron diffraction. Improvements in the precision of diffraction images of complex chemical reactions have been achieved with advances in pulsed electron flux, repetition rate, detection sensitivity and experimental stability. Ultrafast electron diffraction (UED) methodology can be used to study transient structures in complex chemical reactions initiated with femtosecond laser pulses. Direct imaging of reactions has been achieved with an advanced apparatus equipped with an electron pulse (1.07 ± 0.27 ps) source, a charge-coupled device camera and a mass spectrometer. Two prototypical gas-phase reactions were studied: (i) the non-concerted elimination reaction of a haloethane with structure determination of the intermediate and (ii) the ring opening of a cyclic hydrocarbon containing no heavy atoms. UED is suitable for studying ultrafast structural dynamics in complex molecular systems due to its improved sensitivity, resolution and versatility, thus enabling the determination of transition states for studies of biocatalytic and chemical process development, and providing structural information of biomolecules for drug discovery.[500]

Table 1.9. European research initiatives related to structural genomics [Refs in 437].

Initiative	Focus of Initiative	Coordinating Institution[1]
European initiatives		
NMR software	Methods for fast interpretation of NMR data for protein structure analysis	University of Utrecht[2]
X-ray diffraction software	Crystallographic software interfacing and development for structural genomics	University of York[3]
France		
Structural genomics of pathogens	Proteins from *M. tuberculosis* and other pathogens, new folds	Institut Pasteur (Paris)[4]
Structural genomics of eukaryotes	Protein families from humans and other eukaryotes, new folds	IGBMC Ilkirch[5]
Structural genomics of yeast	Yeast proteins, new folds	Université Paris-Sud[6]
Germany		
Protein Structure Factory	High-throughput technology for protein structure analysis, human proteins	Max-Delbrück-Centrum für Molekulare Medizin, Berlin[7]

[1] The other institutions in each initiative are listed in the respective footnotes.
[2] Universities of Florence, Frankfurt/Main and Norwich, EMBL Heidelberg, Aventis, Organon, and Smith-Kline-Beecham Pharmaceuticals.
[3] Universities of Göttingen and Nijmegen, MRC/LMB Cambridge, Daresbury Laboratory, and the EMBL Grenoble and Hamburg outstations.
[4] Universities of Marseille and Toulouse.
[5] EMBL Grenoble, and the University of Montpellier.
[6] LURE.
[7] Berlin area: Free, Humboldt and Technical University, Research Institute for Molecular Pharmacology (FMP), MPI for Molecular Genetics, Alpha Bioverfahrenstechnik, Larova, WTA, DKFZ and EMBL Heidelberg, Bayer.

Table 1.10 lists the current players in commercial structural genomics.

The examination of molecular motions is facilitated by the advancing technquies of physical structural measurement. Sub-femtosecond resolution was achieved over a wide range of x-ray wavelengths, enabling experimental attosecond investigations.[528]

Structural determinations are important pre-requisites for functional studies which require molecular interactions and motions. A strong correlation was determined between phosphorylation-driven activation of the signaling protein NtrC and microsecond time-scale backbone dynamics. With nuclear magnetic resonance relaxation, the motions of NtrC were characterized in three functional states: unphosphorylated (inactive), phosphorylated (active), and a partially active mutant. The dynamics indicate an exchange between inactive and active conformations. Both states are populated in unphosphorylated NtrC, and phosphorylation shifts the equilibrium toward the active species. A dynamic population shift between two preexisting conformations appears to be the underlying mechanism of activation.[634] Multilevel regulation involving conformational transitions and rapid dissociations may provide specificity in signal transduction.[635]

Table 1.10. Current players in commercial structural genomics [Refs in 439].

Company name	Year founded	Location	Technology	URL
Experimental companies				
Astex	1998	Cambridge, UK	High throughput X-ray crystallography/focus on co-complexes	www.astex-technology.com
Integrative Proteomics	2000	Toronto, Canada	Automation for protein expression	www.integrative proteomics.com
Structure-Function Genomics	1999	Piscataway, NJ	NMR, protein domain analysis and expression	www.Monmouth. com/~spidersigns/ SFG/index.html
Structural GenomiX	1999	San Diego, CA	High throughput X-ray crystallography and compound design	www.stromix.com
Syrrx	1999	La Jolla, CA	High throughput X-ray crystallography	www.syrrx.com
Modeling companies				
IBM (Blue Gene project)	2000		Computational protein folding	www.ibm.com/ news/1999/12/06. phtml
Inpharmatica	1998	London, UK	Biopendium database	www.inpharma tica. com
Geneformatics	1999	San Diego, CA	'Fuzzy functional form' modeling for identifying active sites	www.genefor matics.com
Prospect Genomics	1999	San Francisco, CA	Homology modeling	not available
Protein Pathways	1999	Los Angeles, CA	Phylogenetic profiling, domain analysis, expression profiling	www.proteinpath ways.com
Structural Bioinformatics	1996	San Diego, CA, and Copenhagen, Denmark	Homology modeling, docking	www.strubix.com

2 Organizational Structures

2.1 Virtual and Real Enterprises

The realm of healthcare covers prediction, prevention, treatment and follow-up; the pharmaceutical areas involved include diagnostics, pharmaceuticals, therapies and medical devices. The recognition of improved benefits for patients by combining several fields of healthcare services together, i.e. the 'healthcare service provider', leads to the formation of integrated organizations. Such integrated organizations aim to provide (usually focused on a therapy or disease area) diagnostics, treatment and follow-up.

The necessity to deliver state-of-the-art comprehensive care requires either the formation of large vertically integrated health providers, e.g., in the dialysis field, or the creation of a federation of complementary companies (respectively, organizations) which together command the range of competencies required.

The formation of such a federation of companies by contractual means can be described as a virtual enterprise, i.e. an enterprise formed temporarily (although possibly for a long term). Such a virtual enterprise consists of real components and real enterprises, which are combined by contractual, organizational and technological means. It has a high flexibility to adapt to changing circumstances and technology or market conditions, e.g., new therapies developed elsewhere may easily be integrated or component companies may be exchanged for better performers.

The possibility of 'telemedicine', i.e. the provision of diagnosis and therapy at a distance by use of telecommunication systems, and the access to databases for medical knowledge plus the existence of, for example, automated analytical devices and robot surgeons (obviating the physical presence of the surgeon in the operating theater) will facilitate the provision of the best expertise anywhere, anytime.

In recent years (decades), the pharmaceutical industry has been (and still is) changing dramatically, primarily due to the developments in molecular biology and the surrounding disciplines, grouped together as biotechnology. Observers have arranged the pharmaceutical and biotechnology companies into categories such as FIPCOs (fully integrated pharmaceutical companies), which span the business from research (discovery) to market. Others were categorized as VIPCOs (virtually integrated pharmaceutical companies) which specialize in one area such as drug discovery and outsource all other activities. There are platform technology companies evolving around a particular technology or data stream, and integrated platform technology companies which have assembled several hopefully synergistic technologies. In the future, the main characterizing function between a pharmaceutical or biotechnological company and the other participants in that industry, the CROs (contract research organizations), is having R & D as a main function.[149]

Mastering the technologies is the task of the organizations having the core competence of drug discovery and development as well as marketing.

It is crucial to start with the right approach, i.e. by defining the objectives to be achieved and deriving which technologies, resources, alliances, outsourcing, and financial means are required to achieve this objective. In the course of implementing the business strategy in the technology-determined modes, it is necessary to weigh the risks and benefits in a quantitative or semi-quantitative way.[150]

The complexity of technology management highlights the need for technology acquisition professionals. Enabling technologies fuel the innovation engine, and generate opportunities and information that improve decision making. Choosing the appropriate technology for pharmaceutical R & D demands full-time, professional attention. Sophisticated organizations accept this tenet and employ dedicated technology managers. Although most organizations recognize the importance of the aforementioned management practices, few organizations employ all of them. The pharmaceutical industry lies within a complex maze of social, economic, political and technical forces. Successfully managing this complexity maze demands technology leadership.[151]

2.2   R & D Networks

The expanding knowledge base and the increasing multitude of sophisticated technologies as well as the cost of acquiring and maintaining technological expertise and human resources are the major reasons for companies to enter into alliances with specialized partners, thus forming R & D networks.

The chain of required activities and competencies which generates a therapy, a pharmaceutical, a diagnostic procedure, a medical device or an implantate or an organ ranges from basic or fundamental research via discovery to (preclinical and clinical) development on to manufacturing, marketing and sales. The discovery portion offers a high reward and leverage in delivering new and better treatments to known diseases, treatments for previously intractable illnesses, and possibilities for diagnostics and prevention or remediation otherwise unattainable.

Over the last approximately 15 years, the biotechnology industry has delivered, together with bio-oriented academia, the major biopharmaceutical discoveries and the technologies which shape the discovery process.[461]

This shift in the source areas for discovery and technologies away from the 'big pharma' corporations (integrated global pharmaceutical companies) necessitates the creation of R & D networks and alliances to capture those innovations and bring them through the development phase to the patient and market.[462]

Table 2.1 illustrates the number of transactions in pharmaceutical deal making between established pharmaceutical companies and biotech corporations plus the cumulative deal values.[462] The distribution of biotech corporations focus between disease research, drug delivery and technologies is shown in Table 2.2 (based on revenue), whereas Table 2.3 depicts the distribution between the various core technologies.

Table 2.1. Biotechnology companies 1997–1999 (third quarter) partnering revenue breakdown (revenue = US$ 4.5 billion). Source: Recombinant Capital (Walnut Creek, CA, USA, www.recap.com) [Refs in 462].

Technology platform	54%
Disease	39%
Drug delivery	7%

Table 2.2. Technology platform companies 1997–1999 (third quarter) partnering revenue breakdown (revenue = US$ 2.4 billion). Source: Recombinant Capital (Walnut Creek, **CA, USA,** www.recap.com) [Refs in 462].

First-generation genomics	40%
Screening	21%
Chemistry	19%
Second-generation genomics	12%
Gene Therapy	8%

Table 2.3. Disease specialized companies 1997–1999 (third quarter) partnering revenue breakdown (revenue = US$ 1.7 billion) Source: Recombinant Capital (Walnut Creek, CA, USA, www.recap.com) [Refs in 462].

Cancer	27%
Infection	19%
Autoimmune	18%
CNS	14%
Cardiovascular	11%
Metabolic	11%

2.3 Outsourcing

Outsourcing is the contractual arrangement to carry out processes which are part of the total effort to bring a drug from discovery to the patient. It has several advantages for the outsourcing corporation. It enables the choice of the most advanced, fastest or cost-effective process providers for each segment, and thus ascertains (at least theoretically) the availability of the best resources for a given task. Since the specialized companies often also make best use of available human resources, physical and financial capital, technologies, and organizational means, the total productivity of the pharmaceutical and healthcare industry is increased.

These advantages are partly counter-balanced by the need to manage a complex task in real time, to control the timely provision of services and the achievement of milestones, the partly simultaneous execution of subprocesses, and quality control of the services, products, devices and results.

Various segments of the total effort to discover, develop, test, produce and market active principles, medical devices and therapies are outsourced to specialized companies or other organizations, which cover a certain part of the discovery-to-patient chain. So-called contract research organizations (CROs) are used for toxicity testing, clinical studies and galenics. Even drug discovery is outsourced, as seen by the huge number of biotechnology companies, which have drug discovery as their core competence. Marketing may be outsourced to specialized marketing organizations or 'big pharma' companies with huge numbers of sales and marketing staff.

The concept of 'virtual pharma' carries the notion to the (present) extreme, where all activities are outsourced except strategy and financing.[152]

The cost of calculating the discovery of a new drug is yielding astonishing figures, since each successful drug has to cover the cost burden of the failures. Approximately 50 novel drugs enter the market each year at a current combined cost of approximately US\$ 60 billion in R & D, resulting in a cost of approximately US\$ 1.2 billion per drug. The time requirement for development to bring the drug to market at present is on average in excess of 12 years.

These unsustainable figures require increases at every stage of drug discovery, drug development, clinical trials and marketing. Thus, strategic outsourcing is potentially a cost-effective solution for accelerating drug discovery.

Pharmaceutical research organizations can benefit from outsourcing discovery activities that are not core competencies of the organization. The core competencies for a discovery operation are the expertise and systems that give the organization an advantage over its competitors. Successful outsourcing ventures result in cost reduction, increased operation efficiency and optimization of resource allocation. While there are pitfalls to outsourcing, including poor partner selection and inadequate implementation, outsourcing can be a powerful tool for enhancing drug discovery operations.[153]

2.4   Registrations/Permissions

The well-developed ways of validating R & D, preclinical studies and clinical studies of 'active pharmaceutical ingredients' (APIs) are applied with the appropriate modifications to the regulatory processes for small molecules, therapeutic proteins, proteins analogs, vectors for delivery and gene therapy, synthetic lipids, and other compounds derived from molecular biology.

Special consideration is given to the effects on the increasingly better understood biological, biochemical and genetic networks as well as the chemical and biotechnological systems of manufacture, recovery and formulation.

The different glycosylation patterns (depending on the eukaryotic cell chosen for producing therapeutic or analytical proteins or protein analogs) are an example of the influence of the manufacturing system on the structure of the product.

3   Markets and Factors

The demand for pharmaceutical intervention to achieve freedom from disease and the aim of delaying or preventing the onset of degenerative diseases are shaping the targets of drug discovery. Additionally, the patterns of diseases and the demands for therapies are influenced by the changing demographic composition of the population and the increasing affluence of the economically developing areas of the world.

The increasing prevalence (with age) of different cancers is prompting several pharmaceutical companies to step up their R & D efforts in pertinent fields, such as the profiling of different tumor cells and the development of specific antibodies.

The competitive outcome of drug discovery is influenced by the size of the R & D budget (which is a major factor driving mergers in the pharmaceutical field) as well as by the availability of human resources, the network of collaborative institutions and biotech companies, and the supportive regulatory environment (such as speedy registration processes and orphan drug regulation).

3.1   Products and Services

The pharmaceutical industry has long relied on the outsourcing of parts of the development and manufacturing process, e.g., clinical trials and contract manufacture. The increasing diversity and multitude of competencies and technologies required have already led to the increased outsourcing of functions like marketing, development, registration, manufacture and, most remarkably, the drug discovery process itself.

A vast industry, including CROs, fine chemicals manufacturers, biotechnology companies, equipment suppliers, human resource providers, software developers and contract manufacturers is providing an increasing variety and increasing share of the pharmaceutical industry's activities.

3.2   Economies

The increasing body of knowledge and the resulting requirements for generating, handling, managing and evaluating data plus the increasing need to take into account the effects of APIs and therapies on different parts of a biological system have increased the cost of drug discovery and drug development.

3.3 Manpower

The rapidly expanding requirements of the pharmaceutical industry, and the broadening areas of expertise contributing to the processes of R & D, development, manufacturing, marketing, validation, surveillance, and in particular its innovative potential are increasing the demand for skilled manpower.

The educational systems of different countries are providing a crucial competitive advantage (or disadvantage) for essentially R & D-based industries like the pharmaceutical and the biotechnological industries.

3.4 Resources

The range of resources required for the pharmaceutical industry has expanded from its traditional areas (chemists, physicians, pharmacists, epidemiologists, etc.) to include biotechnologists, bioinformatics specialists, software developers, chip designers, and micro- and nanotechnology experts.

The classical resources such as industrial enterprises to organize the R & D, trials, and development, manufacturing, and marketing are still necessary, but the providers of these resources are becoming increasingly fragmented.

Another crucial resource is the financial base upon which to develop the API or therapy. In this field, too, the changing structures of markets, financial markets, and the industrial landscape have opened additional sources for drug discovery and development, such as direct access to capital markets and alliances with larger companies.

4 Biotechnology and Medicine

Genomic technologies and computational advances are leading to an information revolution in biology and medicine. Simulations of molecular processes in cells and predictions of drug effects in humans will advance pharmaceutical research and speed-up clinical trials. Computational prognostics and diagnostics that combine clinical data with genotyping and molecular profiling will soon cause fundamental changes in the practice of health care.

Computational elucidation of protein structure and function, genotyping of individual susceptibilities, and simulation of cellular and organismal processes will lead to a 'personalized medicine'.[154]

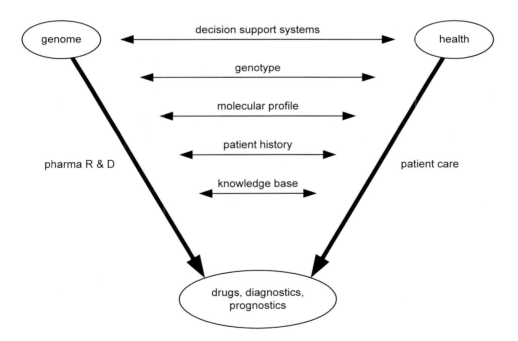

Figure 4.1. Genomic and computational technologies have an increased impact on pharmaceutical R & D (arrow at left). The same technologies will be adapted to directly serve patients (arrow right), leading to personalized, information-driven medical care [Refs in 154].

4.1 Diagnostics

The expanding availability of information on genetic sequences, polymorphisms, SNPs, protein variations and interactions, metabolic and phenotypical variants as well as the sensor technologies plus micro- and nanotechnologies in combination with the communication network spanning the globe will allow increasingly precise measurements of the:

- Susceptibilities for disease
- Strengths of genetic and metabolic networks
- Status of vital organs and tissues
- Regimen of pharmaceuticals and therapies
- Developments in surgery and implantation

Diagnostics is an essential tool for providing basic data for medical therapies, and biological and industrial processes, e.g., process control, and food and environmental monitoring. One of the first biomedical applications is the measurement of blood glucose in diabetics. Due to the expanding numbers of diabetics (projected to increase from 175 million persons in 2000 to 239 million afflicted in 2010 according to World Health Organization data), this is a large and still growing diagnostic market segment.

The sensitivity and versatility of diagnostic devices is increasing due to refinements in established technologies such as enzymatic assays and enzyme-linked immunoassays, the ongoing expansion of the analytical repertoire to include DNA and RNA diagnostics such as PCR, and physico-chemical analytics such as MALDI-TOF, affinity sensors and techniques such as surface plasmon resonance (SPR). Neural network software is being applied to assess complex mixtures such as in olfactory analyses. The technique of molecular imprinting is enabling the design of molecule-specific assays, separations and sensors.[501]

Table 4.1 lists typical examples of molecular imprinting in design and applications from amino acids, pharmaceuticals, herbicides, and chemicals to proteins, steroids and alkaloids.

4.2 Therapeutics

Immunotherapy is the focus of significant efforts in the development of cancer therapies. Several approaches have tried to create tumor cell vaccines for cancer treatment. An effective cancer cell vaccine is created by expressing MHC class II molecules without the invariant chain protein (Ii) that normally blocks the antigenic peptide-binding site of MHC class II molecules at their synthesis in the endoplasmic reticulum. Such cell–cell constructs are created either by the transfer of genes for MHC class II α and β chains or by the induction of MHC class II molecules and Ii protein with a transacting factor, followed by Ii suppression using antisense methods. Preclinical validation of this approach has been reviewed with the goal of using this immunotherapy for metastatic human cancers.[155]

Table 4.1. Typical examples of MIPS design and application [Refs in 501].

Template	Application	Refs
Amino acids and amino acid derivatives	Separation and binding	6
	Synthesis	7
	Assay and sensors	8
Aniline, phenol and their derivatives	Assay and sensors	9
Drugs	Separation and binding	10
	Assay and sensors	11, 12
Gases and vapors	Sensors	13
Herbicides	Separation and binding	14
	Assay and sensors	15, 16
Heterocycles	Separation and binding	17, 18
Metal ions	Separation and binding	19
	Assays and sensors	20
Micro-organisms	Separation and binding	21
Nucleic acids and nucleic acid derivatives	Separation and binding	22
	Assays and sensors	23
Pokynuclear aromatic hydrocarbons	Separation and binding	24
	Assays and sensors	25
Proteins	Separation and binding	26
Steroids	Separation and binding	27, 28
	Detection	29
Sugars and sugar derivatives	Separation and binding	1
	Assays and sensors	30
Alkaloids, toxins and narcotics	Separation and binding	4
	Assays and sensors	31

A novel way of approaching cancer therapy is the use of viruses to attack tumor cells. Rapid advances are being made in the engineering of replication-competent viruses to treat cancer. Adenovirus is a mildly pathogenic human virus that propagates prolifically in epithelial cells – the origin of most human cancers. While virologists have revealed many details about its molecular interactions with the cell, applied scientists have developed powerful technologies to genetically modify or regulate every viral protein. In tandem, the limited success of non-replicative adenoviral vectors in cancer gene therapy has brought the old concept of adenovirus oncolysis back into the spotlight. Major efforts have been directed toward achieving selective replication by the deletion of viral functions dispensable in tumor cells or by the regulation of viral genes with tumor-specific promoters. The predicted replication selectivity has not been realized because of incomplete knowledge of the complex virus–cell interactions and the leakiness of cellular promoters in the viral genome. Capsid modifications are being developed to achieve tumor targeting and enhance infectivity. Cellular and viral functions that confer greater oncolytic potency are also being elucidated. Ultimately, the interplay of the virus with the immune system will likely dictate the success of this approach as a cancer therapy.[156]

Biomaterials play an important role in a variety of clinical applications in wound healing, regeneration, and tissue engineering. Important features of such materials include the potential to be remodeled and replaced by the proteolytic activity associated with cells during migration and invasion, as well as the ability to display a variety of adhesive ligands that directly bind to cell-surface receptors to provide adhesive and morphogenetic clues.

Fibrin plays an important role in wound healing and regeneration, and enjoys widespread use in surgery and tissue engineering. The enzymatic activity of Factor XIIIa was employed to covalently incorporate exogenous bioactive peptides within fibrin during coagulation. Fibrin gels were formed with incorporated peptides from laminin and *N*-cadherin alone and in combination at concentrations up to 8.2 mol peptide/mol fibrinogen. Neurite extension *in vitro* was enhanced when gels were augmented with exogenous peptide, with the maximal improvement reaching 75%. When this particular fibrin derivative was evaluated in rats in the repair of the severed dorsal root within polymeric tubes, the number of regenerated axons was enhanced by 85% relative to animals treated with tubes filled with unmodified fibrin. These results demonstrate that it is possible to enhance the biological activity of fibrin by enzymatically incorporating exogenous oligopeptide domains of morphoregulatory proteins.[157]

4.3   Gene Therapy

Gene therapy promises to revolutionize medicine by treating the (genetic) causes of diseases rather than the symptoms. This radical improvement is possible because the gene-based approach can provide superior targeting and prolonged duration of action. Moreover, gene therapy is a platform technology applicable to a wide range of diseases. Ongoing clinical studies are addressing a wide range of diseases and target cells, such as cardiovascular diseases, inherited monogenic disorders, rheumatoid arthritis (RA), cancer and cubal tunnel syndrome. The vectors utilized include adeno-associated virus (AAV), pox viruses, Herpes simplex virus (HSV), naked DNA, naked RNA, adenoviruses and retroviruses.

The most promising gene-therapy concepts concern the direct killing of tumor cells with genes delivered by adenovirus vectors, delivery of naked DNA for preventive vaccination against infectious diseases, naked DNA delivery of genes promoting angiogenesis for cardiovascular disorders, and AAV delivery for chronic disorders such as hemophilia and anemia. The efficiency in clinical studies has been clearly demonstrated.[158]

The genomic technologies will provide an increasing range of genes and information about their causal relationship with diseases and the genetic susceptibility for diseases, rendering many more amenable to gene therapy.

Gene therapy may be segmented into somatic gene therapy and germline therapy. Somatic gene therapy promises to be more successful when targeting selected cell populations rather than total changes.

Examples of cell populations are osteoblasts, which undergo apoptosis and are depleted with increasing age. A directed change in that program will prevent osteoporosis by reducing bone loss and increasing bone mass. The classical approach, i.e. application of parathyroid hormone (PTH), will be superseded by gene therapy.

Pancreas failure, Alzheimer's disease and Parkinson's disease are urgent targets for pharmaceutical intervention, where tissue engineering and gene therapy offer the best solutions.

An intriguing example of gene therapy in cardiovascular diseases is the combined application of angiopoietin-1 (Ang-1) and vascular endothelial growth factor (VEGF), both of which are endothelial cell-specific growth factors. Direct comparison of transgenic mice overexpressing these factors in the skin revealed that VEGF-induced blood vessels were leaky, whereas those induced by Ang-1 were non-leaky. Vessels in Ang-1-overexpressing mice were resistant to leaks caused by inflammatory agents. Co-expression of Ang-1 and VEGF had an additive effect on angiogenesis but resulted in leakage-resistant vessels typical of Ang-1. Thus, Ang-1 may be useful in reducing microvascular leakage in diseases in which the leakage results from chronic inflammation or elevated VEGF and in combination with VEGF for promoting the growth of non-leaky vessels. Since plasma leakage is a key pathophysiological feature of various diseases, such as chronic inflammatory, and degenerative and neoplastic diseases, inhibition of the leakage could have important therapeutic benefits. Ang-1 has the potential to reduce plasma leakage in such conditions, whether the leak is due to inflammatory mediators or VEGF.[159]

Gene therapy requires the introduced genes to act in concert with the existing genomic environment without upsetting the genetic, transcriptional and metabolic balances and networks. Regulated hormone expression has been demonstrated in pigs. Ectopic expression of a new serum protease-resistant porcine growth hormone-releasing hormone, directed by an injectable muscle-specific synthetic promoter plasmid vector (pSP-HV-HGRH), elicits growth in pigs. A single muscular injection of 10 mg followed by electroporation in 3-week-old piglets elevated serum GHRH levels by 2- to 4-fold, enhanced growth hormone secretion and increased serum insulin-like growth factor I (IGF-I) by 3- to 6-fold over control pigs. Evaluation of body composition indicated a uniform increase in mass, with no organomegaly or associated pathology.[160]

In utero fetal gene transfer could provide an alternative to transgenic approaches for functional analysis of stage- and tissue-specific, *cis*-acting DNA regulatory elements. Stage-specific *in utero* gene transfer and expression might also prove useful for conditional rescue of lethal targeted genes, either *in utero* or in the immediate postnatal period. Moreover, studies of fetal gene transfer may help in developing strategies to ameliorate or correct human genetic deficiencies *in utero*.

In utero injection of cationic liposome–DNA complexes (CLDCs) containing chloramphenicol acetyltransferase, β-galactosidase (β-gal) or human granulocyte colony-stimulating factor (hG-CSF) expression plasmids produced high-level gene expression in fetal rats. Tissues adjacent to the injection site exhibited the highest levels of gene expression. Chloramphenicol acetyltransferase expression persisted for at least 14 days and was re-expressed following postnatal re-injection of CLDCs. Intraperitoneal administration of the hG-CSF gene produced high serum hG-CSF levels. X-gal staining demonstrated widespread β-gal expression in multiple fetal tissues and cell types. No toxic or inflammatory responses were observed nor was there evidence of fetal–maternal or maternal–fetal gene transfer, suggesting that CLDCs may provide a useful alternative to viral vectors for *in utero* gene transfer.[161]

Gene therapy of the brain is hindered by the presence of the blood–brain barrier (BBB), which prevents the brain uptake of blood-borne gene formulations. Exogenous genes have been expressed in the brain after invasive routes of administration, such as craniotomy or intra-carotid arterial infusion of noxious agents causing BBB disruption. Studies were presented describing the expression of an exogenous gene in brain after non-invasive i.v. administration of a 6–7 kb expression plasmid encoding either luciferase or β-gal packaged in the interior of neutral pegylated immunoliposomes. The latter are conjugated with the OX26 mAb to the rat transferrin receptor, which enables targeting of the plasmid DNA to the brain via the endogenous BBB transferrin receptor. Unlike cationic liposomes, this neutral liposome formulation is stable in blood and does not result in selective entrapment in the lung. Luciferase gene expression in the brain peaks at 48 h after a single i.v. administration of 10 μg of plasmid DNA per adult rat, a dose that is 30- to 100-fold lower than that used for gene expression in rodents with cationic liposomes. β-Gal histochemistry demonstrated gene expression throughout the CNS, including neurons, choroid plexus epithelium and the brain microvasculature. Widespread gene expression in the brain can be achieved by using a formulation that does not employ viruses or cationic liposomes, but instead uses endogenous receptor-mediated transport pathways at the BBB.[162]

Lysosomal storage disorders (LSDs) constitute an important group of conditions in which the potential of gene manipulation as therapy can be assessed. They are monogenic defects, often with severe manifestations for which there are limited treatment options. Over-expression of the lysosomal hydrolase by gene-corrected cells results in secretion of some of the enzyme and its uptake by uncorrected bystander cells (metabolic cooperativity). Gene therapy strategies, enzyme therapy and bone marrow transplantation have all been described.

Fabry disease is a compelling target for gene therapy as a treatment strategy. A deficiency in the lysosomal hydrolase α-galactosidase A (α-gal A; EC 3.2.1.22) leads to impaired catabolism of α-galactosyl-terminal lipids such as globotriaosylceramide (b3). Patients develop vascular occlusions that cause cardiovascular, cerebrovascular and renal disease. Unlike for some lysosomal storage disorders, there is limited primary nervous system involvement in Fabry disease. The enzyme defect can be corrected by gene transfer. Overexpression of α-gal A by transduced cells results in secretion of this enzyme. Secreted enzyme is available for uptake by non-transduced cells, presumably by receptor-mediated endocytosis. Correction of bystander cells may occur locally or systemically after circulation of the enzyme in the blood. A long-term genetic correction in an α-gal A-deficient mouse model of Fabry disease was reported. α-Gal A-deficient bone marrow mononuclear cells (BMMCs) were transduced with a retrovirus encoding α-gal A, and transplanted into sublethally and lethally irradiated α-gal A-deficient mice. α-gal A activity and Gb3 levels were analyzed in plasma, peripheral blood mononuclear cells, BMMCs, liver, spleen, heart, lung, kidney and brain. Primary recipient animals were followed for up to 26 weeks. BMMCs were then transplanted into secondary recipients. Increased α-gal A activity and decreased Gb3 storage were observed in all recipient groups in all organs and tissues except the brain. These effects occurred even with a low percentage of transduced cells. Genetic correction of bone marrow cells derived from patients with Fabry disease may have utility for phenotypic correction of patients with this disorder.[163]

In vivo gene therapy may be a method of choice for compounds that are difficult to produce or to deliver, such as endostatin. Tumors require ongoing angiogenesis to support their growth. Inhibition of angiogenesis by production of angiostatic factors should be a viable approach for cancer gene therapy. Endostatin, a potent angiostatic factor, was expressed in mouse muscle and secreted into the bloodstream for up to 2 weeks after a single intramuscular administration of the endostatin gene. The biological activity of the expressed endostatin was demonstrated by its ability to inhibit systemic angiogenesis. Moreover, the sustained production of endostatin by intramuscular gene therapy inhibited both the growth of primary tumors and the development of metastatic lesions. These results demonstrate the potential utility of intramuscular delivery of an antiangiogenic gene for the treatment of disseminated cancers.[164]

Gene therapy requires appropriate vectors with the necessary regulatory and coding regions for integration, regulation and synthesis of required genes (respectively, proteins). Artificial chromosomes are DNA molecules of defined structure, which are assembled *in vitro* from defined constituents that behave with the properties of natural chromosomes. Artificial chromosomes were first assembled in budding yeasts and have proved useful in many aspects of yeast genetics. Several attempts have been made to build artificial chromosomes in mammals. Particularly, mini-chromosomes of defined structure have been developed to address questions regarding mammalian chromosome function and for biotechnological applications.[165]

Major targets for gene therapy are the diseases involving mutated globins, such as β-thalassemia and sickle-cell disease. The stable introduction of a functional β-globin gene in HSC could be a powerful approach to treat β-thalassemia and sickle-cell disease. Genetic approaches aiming to increase normal β-globin expression in the progeny of autologous HSC might circumvent the limitations and risks of allergenic cell transplants. Low-level expression, position effects and transcriptional silencing hampered the effectiveness of viral transduction of the human β-globin gene when it was linked to minimal regulatory sequences. It was shown that the use of a recombinant lentivirus enables efficient transfer and faithful integration of the human β-globin gene together with large segments of its locus control region. In long-term recipients of unselected transduced bone marrow cells, tetramers of two murine α-globin and two human β(A)-globin molecules account for up to 13% of total hemoglobin in mature red cells of normal mice. In β-thalassemic heterozygous mice higher percentages are obtained (17–24%), which are sufficient to ameliorate anemia and red cell morphology. Such levels should be of therapeutic benefit in patients with severe defects in hemoglobin production.[166]

Gene silencing is defined as a state of complete transcriptional repression that is epigenetically inherited through subsequent cell divisions. This phenomenon has great impact on gene therapy, as transferred genes may become silenced to various degrees, depending on the presence of *cis*-acting elements opposing the forces of silencing (e.g., enhancers, locus control regions, matrix attachment sites and insulators), the sites of chromosomal integration, and the state of differentiation the cells initially transduced and of their subsequent progeny. During development, a widespread wave of global silencing takes place at the time of implantation. Subsequently, progressive tissue-specific gene activation occurs. In pre-implantation embryos, ES cells and HSC, transferred genes are exposed to a high frequency of gene silencing. Irreversible HSC gene silencing remains one of the main obstacles to the gene therapy of hematological disorders.

Transcriptional silencing of genes transferred into HSC thus poses one of the most significant challenges to the success of gene therapy. If the transferred gene is not completely silenced, a progressive decline in gene expression as the mice age often is encountered. These phenomena were observed to various degrees in mouse transplant experiments using retroviral vectors containing a human β-globin gene, even when *cis*-linked to locus control region derivatives. *Ex vivo* preselection of retrovirally transduced stem cells was investigated on the basis of expression of the GFP driven by the CpG island phosphoglycerate kinase promoter being able to ensure subsequent long-term expression of a *cis*-linked β-globin gene in the erythroid lineage of transplanted mice. All mice ($n = 7$) engrafted with pre-selected cells concurrently expressed human β-globin and the GFP in 20–95% of their red blood cells (RBCs) for up to 9.5 months post-transplantation, the longest time point assessed. This expression pattern was successfully transferred to secondary transplant recipients. In the presence of β-locus control region hypersensitive site 2 alone, human β-globin mRNA expression levels ranged from 0.15 to 20% with human β-globin chains detected by HPLC. Neither the proportion of positive blood cells nor the average expression levels declined with time in transplanted recipients. Although suboptimal expression levels and heterocellular position effects persisted, *in vivo* stem cell gene silencing and age-dependent extinction of expression were avoided, thus demonstrating the potential of this vector for the gene therapy of human hemoglobinopathies.[167]

Most gene therapy strategies involving HSC require both a high level of gene transfer and persistent transgene expression in specific target lineages. Gene transfer rates of approximately 10% in reconstituting HSC can now be routinely achieved with virus vectors based on murine leukemia virus and related oncoretroviruses. Achieving persistent, uniform gene expression from murine leukemia virus-based vectors remains a challenge. Focus has been on the definition of elements of the virus long terminal repeat (LTR) that are responsible for provirus silencing *in vivo*, and the identification of appropriate promoters and enhancers. Expression of integrated provirus is also affected by chromatin structure. Since the bulk of the mammalian genome is packaged into transcriptionally silent heterochromatin and murine leukemia virus-based vectors insert at random sites in the genome, a large portion of murine leukemia virus insertions result in gene silencing. The progeny of a single clone containing a unique integration event can also be affected by the surrounding chromatin to varying degrees, a phenomenon known as position effect variegation.

The mammalian chromosome is organized into discrete chromosomal domains, in part through the use of sequences termed chromatin insulators. These elements, first described in *Drosophila* and also in several vertebrate species, help define the boundary between differentially regulated loci and serve to shield promoters from the influence of neighboring regulatory elements. Insulators function in a polar manner (e.g., they must be located between the *cis* effectors and promoter) and do not have stimulatory or inhibitory transcriptional effects on their own, distinguishing them from classical enhancers and silencers. The best characterized vertebrate chromatin insulator is located within the chicken β-globin locus control region, and contains a DNase I hypersensitive site (cHS4) and appears to constitute the 5′ boundary of the chicken β-globin locus. A 1.2-kb fragment containing the cHS4 element displays classic insulator activities, including the ability to block the interaction of globin gene promoters and enhancers in cell lines, and the ability to protect expression cassettes in *Drosophila*, transformed cell lines and transgenic mammals from position effects. Much of this activity is contained in a 250-bp fragment, which comprises a 49-bp cHS4 core that interacts with the zinc finger DNA-binding protein CTCF implicated in enhancer-blocking assays.

Recombinant murine retroviruses are widely used as delivery vectors for gene therapy. Once integrated into a chromosome, these vectors often suffer from profound position effects, with vector silencing observed *in vitro* and *in vivo*. To overcome this problem, the HS4 chromatin insulator from the chicken β-globin locus control region was studied with respect to protection of retrovirus vector from position effects. When used to flank a reporter vector, this element significantly increased the fraction of transduced cells that expressed the provirus in cultures and in mice transplanted with transduced marrow. A chromatin insulator can improve the expression performance of a widely used class of gene therapy vectors by protecting these vectors from chromosomal position effects.[168]

A combination of gene therapy and ribozyme drug technology was reported which involved the generation of a novel ribozyme ('maxizyme') to specifically cleave the Bcr–Abl mRNA thus inducing apoptosis in chronic myelogenous leukemia (CML) cells. CML is a hematopoietic malignant disease associated with the expression of a chimeric Bcr–Abl gene. A retroviral system was used to transduce a line of CML cells (BV173) which were injected into NOD-SCID mice. These mice are produced by crossing SCID and NOD mice, and are a valuable tool for growing human hematopoietic cells *in vivo*. As could be observed by examining the spleens of injected and control animals, the maxizyme functions successfully in animals by cleaving Bcr–Abl mRNA with high efficiency. Maxizymes could be a new class of gene-inactivating agents that can cleave any type of chimeric mRNA.[169]

Mitochondrial DNA (mtDNA) mutations underlie many rare diseases and may contribute to human ageing. Gene therapy is a tempting future possibility for intervening in mitochondriopathies. Expression of the 13 mtDNA-encoded proteins from nuclear transgenes (allotopic expression) might be the most effective gene-therapy strategy. Its only confirmed difficulty is the extreme hydrophobicity of these proteins, which prevents their import into mitochondria from the cytosol. Inteins (self-splicing 'protein introns') might offer a solution to this problem: their insertion into such transgenes could greatly reduce the encoded proteins' hydrophobicity, enabling import, with post-import excision restoring the natural amino acid sequence.[170]

Precise targeting of genetic modifications is a prerequisite for safe and successful gene therapy by modification of DNA sequences.

Mobile group II intron RNAs insert directly into DNA target sites and are then reverse transcribed into genomic DNA by the associated intron-encoded protein. Target site recognition involves modifiable base-pairing interactions between the intron RNA and a longer than 14-nucleotide region of the DNA target site, as well as fixed interactions between the protein and flanking regions. A highly efficient *E. coli* genetic assay was developed to determine detailed target site recognition rules for *Lactococcus lactis* group II intron Ll.LtrB and to select introns that insert into desired target sites. Using HIV-1 proviral DNA and the human CCR5 gene as examples, it was shown that group II introns can be retargeted to insert efficiently into virtually any target DNA and that the retargeted introns retain activity in human cells. The work provides the practical basis for potential applications of targeted group II introns in genetic engineering, functional genomics and gene therapy.[171]

Chimeric RNA–DNA oligonucleotides have formed the basis of genetic strategies for correcting mutations, while maintaining genomic organization important for the appropriate expression and regulation of genes. Visualization of pigmentation in a live animal and easy accessibility makes skin an attractive system for *in vivo* testing of chimeric oligonucleotides as novel skin gene therapeutics.

An RNA–DNA oligonucleotide corrected a point mutation in the mouse tyrosinase gene, resulting in permanent and inheritable restoration of tyrosinase enzymatic activity, melanin synthesis and pigmentation changes in cultured melanocytes. Gene correction was extended from tissue culture to live animals, using a chimeric oligonucleotide designed to correct a point mutation in the tyrosinase gene. Both topical applications and intradermal injection of this oligonucleotide to albino BALB/c mouse skin resulted in dark pigmentation of several hairs in a localized area. The restored tyrosinase enzymatic activity was detected by dihydroxyphenylacetic acid (DOPA) staining of hair follicles in the treated skin. Tyrosinase gene correction was also confirmed by restriction fragment length polymorphism (RFLP) analysis, and DNA sequencing from skin that was positive for DOPA staining and melanin synthesis. Localized gene correction was maintained 3 months after the last application of the chimeric oligonucleotides. These results demonstrated correcting of the tyrosinase gene point mutation by chimeric oligonucleotides *in vivo*.[172]

The suitability of using the tRNA import pathway for the correction of respiratory deficiencies in the mitochondrial DNA and the applicability of this system for human therapeutic application were investigated. Mitochondrial import of nuclear encoded tRNAs has been described in yeasts, plants and protozoans. The complexity of the imported tRNA pool varies among organisms, from a complete set required for reading all codons of the mitochondrial genetic code in trypanosomatids to a single tRNA in the yeast *S. cerevisiae*.

Mitochondrial import of a cytoplasmic tRNA in yeast requires the preprotein import machinery and cytosolic factors. Cytoplasmic tRNAs with altered aminoacylation identity can be specifically targeted to the mitochondria and participate in mitochondrial translation. Human mitochondria, which do not normally import tRNAs, are able to internalize yeast tRNA derivatives *in vitro* and this import requires an essential yeast import factor.[173]

Adoptive immunotherapy is developing into a tool for combating diseases like infections and cancers. The infusion of antigen-specific T lymphocytes is a potential therapy against certain cancers and infectious diseases. One limitation to its broad usage is the generation of autologous T cells directed against well-defined epitopes. The induction and expansion of antigen-specific cells require optimal antigen presentation and T cell co-stimulation. These requirements are met by antigen-presenting cells (APC) such as Epstein–Barr virus-transformed B cells and dendritic cells (DC), which constitutively express high levels of co-stimulatory, adhesion and MHC molecules. Despite a cumbersome process, the use of autologous cells to present cell-defined epitopes is mandated to obviate strong allergenic responses.

The adoptive transfer of antigen-specific cytotoxic T lymphocytes (CTLs) is a promising therapeutic approach for a number of diseases. To overcome the difficulty in generating specific CTLs, stable artificial APCs (AAPCs) were established that can be used to stimulate T cells of any patient of a given HLA type. Mouse fibroblasts were retrovirally transduced with a single HLA–peptide complex along with the human accessory molecules B7.1, ICAM-1 and LFA-3. These AAPCs consistently elicit strong stimulation and expansion of HLA-restricted CTLs. Due to the high efficiency of retrovirus-mediated gene transfer, stable AAPCs can be readily engineered for any HLA molecule and any specific peptide.[174]

Activation of the immune system in response to challenge by a foreign organism is a complex process. In the case of T-dependent antigens, naive T cells are activated in response to novel peptides presented by APCs. To activate T cells the APCs must acquire antigen, and then present the antigen in the context of self-MHC molecules and the appropriate co-stimulatory signals. Tissue damage and/or cell death may facilitate antigen presentation and may be necessary for activation of the immune system. Following cell death, dying cells transfer their antigens (by ingestion or otherwise) to 'professional' APCs, which then prime T cells. Antigen from necrotic cells was shown to prime MHC II-restricted T cell responses *in vitro*. Antigen transfer and priming have been demonstrated by showing that bone marrow-derived APCs can acquire viral antigens from apoptotic cells and prime antigen-specific CTLs. Thereby, DCs were observed to engulf intact macrophages as well as their cellular fragments, suggesting that apoptosis can prime CTLs through DCs.

Immunity to tumors as well as to viral and bacterial pathogens is often mediated by CTLs. Thus, the ability to induce a strong cell-mediated immune response is an important requirement of novel immunotherapies. APCs, including DCs, are specialized in initiating T cell immunity. Harnessing this innate ability of these cells to acquire and present antigens, antigen presentation was improved by targeting antigens directly to DCs *in vivo* through apoptosis. Fas-mediated apoptotic death of antigen-bearing cells was engineered *in vivo* by co-expressing the immunogen and Fas in the same cell. The death of antigen-bearing cells results in an increased antigen acquisition by APCs including DCs. This *in vivo* strategy led to enhanced antigen-specific CTLs, and the elaboration of T helper 1 (T_h1)-type cytokines and chemokines. This adjuvant approach has important implications for viral and non-viral delivery strategies for vaccines or gene therapies.[175]

The topical delivery of transgenes to hair follicles has potential for treating disorders of the skin and hair. The topical administration of liposome–DNA mixtures (lipoplex) to mouse skin and to human skin xenografts resulted in efficient *in vivo* transfection of hair follicle cells. Transfection depended on liposome composition and occurred only at the onset of a new growing stage of the hair cycle. Manipulating the hair follicle cycle with depilation and retinoic acid treatment resulted in nearly 50% transfection efficiency – defined as the proportion of transfected, newly growing follicles within the xenograft. Transgenes administered in this fashion are selectively expressed in hair progenitor cells and therefore have the potential to affect the characteristics of the follicle. These findings form a foundation for the future use of topical lipoplex applications to alter hair follicle phenotype, and treat diseases of the hair and skin.[176]

Gene therapy for cystic fibrosis (CF) has focused on correcting electrolyte transport in airway epithelia. Success has been limited by the failure of vectors to attach to and enter into airway epithelia, and may require redirecting vectors to targets on the apical membrane of airway cells that mediate these functions. The G-protein-coupled P2Y2 receptor (P2Y2-R) is abundantly expressed on the airway lumenal surface and internalizes into coated pits upon agonist activation. A small-molecule agonist (UTP) was tested whether it could direct vectors to P2Y2-R and mediate attachment, internalization, and gene transfer. Fluorescein–UTP studies demonstrated that P2Y2-R agonists internalized with their receptor and biotinylated UTP (BUTP) mediated P2Y2-R-specific internalization of fluorescently labeled streptavidin (SAF) or SAF conjugated to biotinylated Cy3 adenoviral vector (BCAV). BUTP conjugated to BACV-mediated P2Y2-R-specific gene transfer in (i) adenoviral-resistant A9 and polarized MDCK cells by means of heterologous P2Y2-R, and (ii) well-differentiated human airway epithelial cells by means of endogenous P2Y2-R. Targeting vectors with small-molecule ligands to apical membrane GPCRs may be a feasible approach for successful CF gene therapy.[177]

In the canine model of Duchenne muscular dystrophy in golden retrievers (GRMD), a point mutation within the splice acceptor site of intron 6 leads to deletion of exon 7 from the dystrophin mRNA and the consequent frameshift causes early termination of translation. A DNA and RNA chimeric oligonucleotide was designed to induce host cell mismatch repair mechanisms and correct the chromosomal mutation to wild-type. Direct skeletal muscle injection of the chimeric oligonucleotide into the cranial tibialis compartment of a 6-week-old affected male dog, and subsequent analysis of biopsy and necropsy samples, demonstrated *in vivo* repair of the GRMD mutation that was sustained for 48 weeks. RT-PCR analysis of exons 5–10 demonstrated increasing levels of exon 7 inclusion with time. An isolated exon 7-specific dystrophin antibody confirmed synthesis of normal-sized dystrophin product and positive localization to the sarcolemma. Chromosomal repair in muscle tissue was confirmed by RFLP-PCR and sequencing the PCR product. This work provides evidence for the long-term repair of a specific dystrophin point mutation in muscle of a live animal using a chimeric oligonucleotide.[178]

The use of neurotrophic viruses as vectors for targeted gene delivery to the CNS has many applications for the development of new therapies for neurological diseases and spinal cord trauma. Poliovirus is attractive for the development of such a gene delivery vector because it has been established in humans that once poliovirus invades the CNS, infection is restricted to the motor neurons of the hindbrain and the spinal cord. To exploit the unique features of poliovirus tropism, poliovirus genomes (referred to as replicons) were constructed to encode foreign proteins in place of the capsid proteins. Because replicons do not encode capsid proteins, they undergo only a single round of infection, without spreading to neighboring cells. Since no infectious poliovirus is generated during production of replicons, the use of replicons for gene delivery purposes following worldwide poliovirus eradication will not be a concern. Replicons maintain the tropism of poliovirus in the CNS and exclusively infect spinal cord and brainstem motor neurons. Replicons can mediate gene delivery in animals that have previously been immunized with poliovirus, indicating that pre-existing immunity in humans from vaccination will not be a limitation for the use of replicons.

Poliovirus replicon vectors transiently express foreign proteins selectively in motor neurons of the anterior horn of the spinal cord. Mice transgenic for the poliovirus receptor (PVR) were intraspinally inoculated with replicons encoding murine tumor necrosis factor (mTNF)-α. High-level expression of mTNF-α was detected in spinal cords of these animals at 8–12 h post-inoculation and this returned to background by 72 h. The mice exhibited ataxia and tail atony, whereas animals given a replicon encoding GFP exhibited no neurological symptoms. Histology of spinal cords from mice given the replicon encoding mTNF-α revealed neuronal chromatolysis, reactive astrogliosis, decreased expression of myelin basic protein and demyelination. These animals recovered with only slight residual damage. This showed replicon vectors to have potential for targeted delivery of therapeutic proteins to the CNS and provide a new approach for treatment of spinal cord trauma and neurological disease.[179]

Clinical trials of gene therapy for CF have demonstrated encouraging steady progress, indicating that transfer of the CF transmembrane conductance regulator (CFTR) gene can partially correct the chloride transport defect in human subjects. Current levels of gene transfer are apparently too low for clinical effectiveness – in large part as a result of the barriers faced by gene transfer vectors within the airways. Clinical studies of gene therapy for CF suggest that the key problem is the efficiency of gene transfer to the airway epithelium. The availability of relevant vector receptors, the transient contact time between vector and epithelium, and the barrier function of airway mucus contribute significantly to this problem. Recombinant Sendai virus (SeV) was developed as a new gene transfer agent. SeV produces efficient transfection throughout the respiratory tract of both mice and ferrets *in vivo*, as well as in freshly obtained human nasal epithelial cells *in vitro*. Gene transfer efficiency was several log orders greater than with cationic liposomes or adenovirus. Even very brief contact time was sufficient to produce this effect and levels of expression were not significantly reduced by airway mucus. These investigations suggest that SeV may provide a useful new vector for airway gene transfer.[180]

Retroviral vectors can insert genes stably into the chromosomes of mammalian cells and have been used in clinical human gene therapy trials to introduce genes into a variety of cells and tissues. All retroviral vectors currently used in such clinical trials have been derived from murine leukemia virus (MLV), a C-type retrovirus, which is considered non-pathogenic for humans. In contrast to lentiviruses, C-type retroviruses have a rather simple genomic organization and contain only two gene units, which code for the inner core structural proteins and the envelope protein. As safe gene delivery systems, they require only the retroviral vector, a genetically modified viral genome containing the gene of interest in place of all retroviral protein coding sequences, and a helper cell that provides the retroviral proteins for the encapsidation of the vector genome into retroviral particles. Gene therapy applications of retroviral vectors derived from C-type retroviruses have been limited to introducing genes into dividing target cells. Genetically engineered C-type retroviral vectors were constructed by derivation from spleen necrosis virus (SNV), which are capable of infecting non-dividing cells. This was achieved by introducing a nuclear localization signal (NLS) sequence into the matrix protein (MA) of SNV by site-directed mutagenesis. This increased the efficiency of infecting non-dividing cells, and was sufficient to endow the virus with the capability to efficiently infect growth-arrested human T lymphocytes and quiescent primary monocyte-derived macrophages. This vector was demonstrated to actively penetrate the nucleus of a target cell and has the potential for a gene therapy vector to transfer genes into non-dividing cells.[181]

AAV vectors have demonstrated considerable promise for gene therapy of inherited diseases. With a packaging size of less than 5 kb, applications have been limited to relatively small disease genes. Based on the finding that AAV genomes undergo inter-molecular circular concatemerization after transduction in muscle, a paradigm was developed to increase the size of delivered transgenes with this vector through *trans*-splicing between two independent vectors co-administered to the same tissue. When two vectors encoding either the 5' or 3' portions of the erythropoietin genomic locus were used, functional erythropoietin protein was expressed in muscle subsequent to the formation of intermolecular circular concatemers in a head-to-tail orientation through *trans*-splicing between these two independent vector genomes. This allows the AAV technologies to be applied to a wider variety of diseases for which therapeutic transgenes exceed the packaging limitation of prior AAV vectors.[182]

An intriguing use of gene therapy by using genetically engineered cells in cancer therapy was demonstrated by applying encapsulated endostatin-secreting cells for the effective treatment of human glioblastoma xenografts[703] in mice and of BT4C glioma cells upon intracerebral implantation in rats.[704] The continuous release of endostatin, a sensitive anti-angiogenic protein, by implanted bioreactors, demonstrates an effective approach to tumor therapy with cellular and gene therapy.[705] The other extreme of application of gene therapy, oral delivery, is subject to intensive studies due to its advantages with respect to compliance, cost, and patient convenience.[706]

4.4 Implantates

Small implants acting as artificial neurons serve to stimulate muscles and thus prevent them from wasting away,[183] and offer the prospect of helping patients with stroke or spinal cord injury.

Biodegradable rods to stabilize bone fractures are absorbed once the bone has healed, whereby clearly the timing of resorption is crucial.[184]

An efficient porous chamber system to convert Factor VII to VIIa was developed and implantation of this chamber, *in vivo*, was shown to be effective. The chamber containing immobilized Factor XIIa was able to generate Factor VII-dependent bypass activity for at least 3 days as a peritoneal implant in guinea pigs and for up to 1 month when tested in rhesus monkeys.

Hemophilia A and B coagulation defects, which are caused by deficiencies of Factor VIII and Factor IX, respectively, can be bypassed by administration of recombinant Factor VIIa. However, the short half-life of recombinant Factor VIIa *in vivo* negates its routine clinical use. An *in vivo* method for the continuous generation of Factor VIIa is reported which depends on the implantation of a porous chamber that contains Factor Xa or XIIa and continuously generates Factor VIIa bypass activity from the subject's own Factor VII, which enters the chamber by diffusion. Once inside, the Factor VII is cleaved to Factor VIIa by the immobilized Factor Xa or XIIa. The newly created Factor VIIa diffuses out of the chamber and back into circulation, where it can bypass the deficient Factors VIII or IX and enable coagulation to occur. *In vitro*, this method generates sufficient Factor VIIa to substantially correct Factor VIII-deficient plasma when assessed by the classical aPTT coagulation assay. *In vivo*, a Factor XIIa peritoneal implant generates bypass activity for up to 1 month when tested in rhesus monkeys. Implantation of such a chamber in a patient with hemophilia A or B could eventually provide a viable alternative to replacement therapies using exogenous coagulation factors.[185]

4.5 Medical Devices and Technology

Medical devices include drug-delivery systems, neurological stimulators, cardiovascular devices such as implantable cardiac defibrillators as well as monitoring devices for blood glucose and heart function. Intelligent devices like the implantable cardiac defibrillators are endowed with sufficient intelligence to analyze the signals from the heart, and apply the right power and type of impulse for correction. Likewise, implantable devices for blood constituents are capable of monitoring, for example, glucose levels and blood factors, and deliver the right amounts of insulin or therapeutic proteins. The market for medical devices amounts to approximately US$ 160 billion per year and is thus about half the size of the market for pharmaceuticals at around US$ 350 billion (in the year 2000).

The diagnosis of biological systems with lasers, such as blood disorders, cancer cells, and cellular constituents like proteins, RNA, DNA is becoming a reality with the availability of biocavity lasers.[477]

The development of MEMS enables the expansion of biomedical applications in chemical and biochemical analysis, and molecular detection and health monitoring. The biological microcavity laser or biocavity laser is such a system on the nanometer scale. Nanolasers have numerous applications in fiber optic communications, optical computers, and nano- or microscale system sensors, actuators and systems-on-a-chip. The nanolaser can probe the cell and nuclear dimensions of lymphocytes and thus shed light on the status of the human immune system or it can diagnose genetic disorders by measuring, for example, red blood cell shapes and hemoglobin content. The semiconductor laser is based on silicon, which is a material suitable for microfabrication, microfluidics and sensors, thus allowing the integrated design of test arrays on chips and implantable devices. Compound semiconductors based on III–V materials (from columns III and V of the periodic table), such as GaAs, are especially suited to the generation, modulation and transmission of light, and thus can combine illumination, data transmission, optical computing and optical analysis plus biomedical applications such as photodynamic therapy, optical tomography, cell micro-manipulation and laser cytometry.

Table 4.2 correlates various materials and (bio-)medical products.

Microsystems in healthcare can generally be divided into
- Sensor systems for diagnostics and patient monitoring.
- Endoscopes and instrumentation for minimally invasive therapy.
- Active implantable devices.
- Systems for bioanalytics and pharmaceutical screening.

Table 4.2. Materials and (bio-)medical products [Refs in 478].

Application technologies new materials	Medical equipment	Other (bio-)medical applications
Structural ceramics and mechanically resistant materials	Long-term implants Prosthetic devices	
Functional ceramics	Neurostimulation in case of paralysis Heat treatment of tumors through implanted, magnetically excitable ceramics Ultrasound converter arrays Microsensors and -actuators for ceramic radio-logical luminous substances	Microsensors and -actuators Health monitoring
Polymers	Controlled drug delivery Skin transplant Biological membranes	
Diamond (-layers)	Eye lenses with a high index of refraction Scalpels	
a-SiC:H	Biocompatible coatings for stents	
Titanium based alloys	Endoprotheses (artificial joints)	
Multifunctional, Adaptive materials	Release from implanted drug depots as required	"Intelligent" instruments
Biomimetic materials	Artificial skin, Dialysis membranes, Artificial parts of organs	Biosensors

Biomaterials are structural or functional materials in the construction of medical devices for implantation or application to living tissues or organs. They need to be biocompatible, i.e. to perform their function without negative side effects such as interference with the circulation or coagulation, impairment of the immune system, facilitation of infections or scar-formation. Bioactive materials such as enzymes, antibodies or living cells can be integrated as functional parts of sensors, actuators or total systems.[478]

The increasing need for novel delivery devices for the application of minute amounts of potent, macromolecular biochemicals at defined intervals without inflicting pain (thus improving compliance) has led to new drug delivery systems employing microstructures. An array of microstructures connected to a microfluidic device can be used to deliver fluids transdermally without pain.[479]

Microstructures permit precise fluid handling and dosing in the microliter to picoliter range. Microstructural design enables the mixing, dosing, resuspension of chemicals, and actuation of valves and micropumps. Examples of microstructured systems are a nebulizer with a highly precise micronozzle for oral drug delivery, a miniaturized microtiter plate (laboratory-on-a-chip) for the determination of microbial sensitivities and a microspectrometer for analysis and color measurement, e.g., in bilirubin analysis.[480]

Living cells acting as parts of bioelectronic hybrid systems offer intriguing avenues for the development of biosensors, bioinformatics and implantable devices for the restoration of function. A crucial issue is the functional coupling of the output signal from the cell system to a micro-electronic or opto-electronic transducer unit. The coupling of excitable cells with an array of field-effect transistors (FETs) integrated into the bottom of a cell culture dish allows the measurement of action potentials.[481]

Controllable prosthetics are possible with the creation of interfaces between neural tissue and machines. Electrical signals from five regions in the cerebral cortex of monkeys have already been used to drive the movement of robotic arms.[482]

The direct electrical interfacing of a recombinant ion channel to a field-effect transistor on a silicon chip has been reported. In the bio-electronic hybrid, an ion current through activated maxi-K_{Ca} channels in human embryonic kidney (HEK293) cells gives rise to an extracellular voltage between cell and silicon chip which controls the electronic source-drain current. The channels at the cell/chip interface are fully functional as is shown by patch-clamp recording. Moreover, the channels are accumulated at the cell/chip interface. The direct coupling of potassium channels to a semiconductor on the level of an individual cell is a prototype for an iono-electronic interface of ligand-gated or G protein-coupled ion channels. It allows the development of bio-electronic hybrids for implantates and of biosensors for biochemical or biopharmaceutical screening arrays with numerous cells on a silicon chip combined with an array of transistors.[506]

These developments will provide amputees and patients with a variety of motor disorders such as paralysis, amyotropic lateral sclerosis with the means to act and communicate by replacing the control of muscles with the control of artificial devices by brain activity.

The study of neural processes for the development of neural–robotic hybrid systems is facilitated by experimental research on visual processing using functional MRI (fMRI). The fMRI method was used to test key predictions of the proposed object-based theory wherein pre-attentive mechanisms segment the visual array into discrete objects, groups or surfaces serving as targets for visual attention. The magnetic resonance signal was recorded from subjects viewing stationary versus moving objects. The signals were recorded from each subject's fusiform face area, parahippocampal place area and MT/MST area, providing a measure of the processing of faces, houses and visual motion. Attending to one attribute of an object enhanced the neural representation of that attribute as well as of the other attribute of the same object. The experiments provide physiological evidence of the selection of whole objects with one relevant visual attribute and may lead to vision-directed steering of prosthetics.[483]

The involvement of the visual cortex in tactile discrimination of orientation was demonstrated experimentally,[484] thus pointing to the necessity to take the wider concept of perceptions into account when developing advanced hybrid interfaces for prosthetics.

Transcranial magnetic stimulation (TMS), whereby a pulsed magnetic field creates current flow in the brain and can temporarily excite or inhibit specific areas, is being developing as an analytically and therapeutically useful non-invasive tool for studying the human brain. The application of TMS to the motor cortex can produce a muscle twitch or may block the movement of a muscle. TMS of the occipital cortex can produce visual phosphenes or scotomas. Moreover, TMS can be used to alter the functioning of the brain beyond the time of stimulation, and thus offers potential for the analysis of neural processes, and research in and therapy of neural disorders.[485]

TMS has been applied to studies of the cortical physiology, to elucidate the processes underlying the mature brain's plasticity such as those involved in repair, learning, memory, and in research on neurologic disorders such as epileptic seizures, Parkinson's disease, Huntington's disease and dystonia.

Figure 4.19 illustrates the physics and mechanism of action of TMS, and specifically emphasizes adherence to safety guidelines in the application of this technique.

A microversion of a positron emission tomography (PET) scanner called microPET was developed and applied to studies of the metabolic activity in different regions of the conscious rodent brain using [^{18}F]fluorodeoxyglucose (FDG) as a tracer to monitor changes in neuronal activity. Limbic seizures result in significantly elevated metabolic activity in the hippocampus and vibrissal stimulation causes modest increases in FDG uptake in the contralateral neocortex. MicroPET is also useful for studying lesion-induced plasticity of the brain, as shown by cerebral hemidecortication resulting in diminished relative glucose metabolism in the neostriatum and thalamus ipsilateral to the lesion, with subsequent recovery of metabolic function. Thus, microPET is useful for serial assessment of metabolic function of individual awake rats with minimal invasiveness, and is applicable to studies of neural disorders and brain repair.[486]

The experimental tagging of expressed genes and the resulting observation of gene activation and gene down-regulation enable the generation of a three-dimensional atlases for gene expression.[186]

In *Xenopus laevis* and *Caenorhabditis elegans*, organism-wide gene expression and down-regulation has been measured with a mutant (E5) of the red fluorescent protein drFP583 which changes its fluorescence from green to red over time. The rate of color conversion is independent of protein concentration and can be applied to trace time-dependent gene expression. *In vivo* labeling with E5 was used to measure expression from the heat shock-dependent promoter in *C. elegans* and from the *Otx-2* promoter in developing *Xenopus* embryos.[487]

4.6 Complex Traits

Increasingly, multifactorial diseases are the targets of medical and biotechnological intervention. A particularly important area is the study of the molecular, biochemical and genetic basis of senescence.

Senescence appears from certain studies to be a multigenetic trait. The mutated form of a single protein, p66^{shc}, which controls the oxidative stress response, also significantly shapes the mammalian life span.[513] Ageing was studied in a mouse mutant *klotho* showing premature ageing. The syndrome resembling human ageing includes reduced liefspan, decreased activity, infertility, osteoporosis, arteriosclerosis is caused by the disruption of a single gene, *klotho*. The klotho gene codes for a protein apparently functioning outside of the cells, exhibits homology to β-glucosidase enzymes, and may function through a signaling pathway involving circulating humoral factor(s).[516]

The evolutionary theory of ageing points to the complex cellular and molecular processes shaping senescence, in particular stress resistance phenomena.[519] A link is observed between advanced age and increased inciidence of cancer. Accumulated experimental evidence indicates that the cancer-prone phenotype is a result of the combined pathogenetic effects of mutation load, epigenetic regulation, telomere dysfunction, and altered stromal milieu.[521] Lessons may be drawn from the study of human progeroid syndromes, wherein a number of genes have been identified in whichg mutations can lead to the accelerated emergence of senescence.[523] The intervention in the ageing process will have profound societal, biomedical, philosophical, psychological, and economic consequences.[524]

Profiling of gene expression is providing transcriptional patterns and pattern changes as well as targets for identification of genes (respectively, gene sequences) for further studies.

Most multicellular organisms exhibit a progressive and irreversible physiological decline that characterizes senescence, the molecular basis of which remains unknown. Postulated mechanisms include cumulative damage to DNA leading to genomic instability, epigenetic alterations that lead to altered gene expression patterns, telomere shortening in replicative cells, oxidative damage to critical macromolecules by reactive oxygen species (ROS) and non-enzymatic glycation of long-lived proteins.

Genetic manipulation of the ageing process in multicellular organisms has been achieved in *Drosophila* through the overexpression of catalase and Cu/Zn superoxide dismutase, in the nematode *C. elegans* through alterations in the insulin receptor (IR) signaling pathway and through the selection of stress-resistant mutants in either organism. In mammals, mutations in the Werner syndrome (WS) locus (WRN) accelerate the onset of a subset of ageing-related pathology in humans, but the only intervention that appears to slow the intrinsic rate of ageing is caloric restriction.

The gene expression profile of the aging process was analyzed in skeletal muscle of mice. Use of high-density oligonucleotide arrays representing 6347 genes revealed that ageing resulted in a differential gene expression pattern indicative of a marked stress response and lower expression of metabolic and biosynthetic genes. Most alterations were either completely or partially prevented by caloric restriction – the only intervention known to retard ageing in mammals. Transcriptional patterns of calorie-restricted animals suggest that caloric restriction retards the ageing process by causing a metabolic shift towards increased protein turnover and decreased macromolecular damage.[188]

DNA repair plays a crucial role in the maintenance of genetic integrity and thus the basis for organismal homeostasis. Cellular DNA is under continuous attack by reactive species inside cells and by environmental agents. Toxic and mutagenic consequences are minimized by distinct repair pathways and the presently 130 known human DNA repair genes were described. These include four enzymes that remove uracil from DNA, seven recomibantion genes related to RAD51, and several DNA polymerases that bypass damage. Only one system is described which removes the main DNA lesions induced by ultravilet light. The number of DNA repair genes is likely to be extended by comparative studies of different organisms and the structural, functional protein and genomics studies. Modulation of DNA repair has potential clinical applications in radiotherapy, anticancer treatment, and understanding and possibly modulation of the aging process.[637]

WS is an inherited disease characterized by premature onset of ageing, increased cancer incidence and genomic instability. The WS gene encodes a 1432 amino acid polypeptide (WRN) with a central domain homologous to the RecQ family of DNA helicases. Purified WRN unwinds DNA with 3′–5′ polarity and also possesses 3′–5′ exonuclease activity. Elucidation of the physiologic function(s) of WRN may be aided by the identification of WRN-interacting proteins. WRN functionally interacts with DNA polymerase δ (pol δ), a eucharistic polymerase required for DNA replication and DNA repair. WRN increases the rate of nucleotide incorporation by pol δ in the absence of proliferating cell nuclear antigen (PCNA) but does not stimulate the activity of eukaryotic DNA polymerases α or ε, or a variety of other DNA polymerases. Functional interaction with WRN is mediated through the third subunit of pol δ, i.e. Pol32p of *S. cerevisiae*, corresponding to the recently identified p66 subunit of human pol δ. Absence of the third subunit abrogates stimulation by WRN and stimulation is restored by reconstituting the three-subunit enzyme. WRN may facilitate pol δ-mediated DNA replication and/or DNA repair, and disruption of WRN-pol δ interaction in WS may contribute to the observed S-phase defects and/or the unusual sensitivity to a limited number of DNA damaging agents.[189]

An intriguing observation in yeast sheds some light on the effects of caloric restriction on life span. Yeast silent information regulator 2 (Sir2) is a heterochromatin component that silences transcription at silent mating loci, telomeres and the ribosomal DNA, and also suppresses recombination in the rDNA and extends replicative life span. Mutational studies indicate that lysine 16 in the N-terminal tail of histone H4 and lysines 9, 14 and 18 in H3 are critically important in silencing, whereas lysines 5, 8 and 12 of H4 have more redundant functions. Lysines 9 and 14 of histone H3 and lysines 5, 8 and 16 of H4 are acetylated in active chromatin and hypoacetylated in silenced chromatin, and overexpression of Sir2 promotes global deacetylation of histones, indicating that Sir2 may be a histone deacetylase. Deacetylation of lysine 16 of H4 is necessary for binding the silencing protein, Sir3. It was shown that yeast and mouse Sir2 proteins are nicotinamide adenine dinucleotide (NAD)-dependent histone deacetylases, which deacetylate lysines 9 and 14 of H3 and specifically lysine 16 of H4. The analysis of two Sir2 mutations supports the idea that this deacetylase activity accounts for silencing, recombination suppression and extension of life span *in vivo*. These findings provide a molecular framework of NAD-dependent histone deacetylation that connects metabolism, genomic silencing and ageing in yeast and, perhaps, in higher eukaryotes.[425]

Yeast Sir2 protein functions in transcriptional silencing of the silent mating loci, telomeres and rDNA. The Sir2 family of enzymes catalyze a NAD–nicotinamide exchange reaction that requires the presence of acetylated lysines such as are found in the N-termini of histones and there appears to be no evidence of ADP-ribosylation activity.[190]

Yeast Sir2 is a heterochromatin component that silences transcription at silent mating loci, telomeres and the ribosomal DNA, and that also suppresses recombination in the rDNA and extends replicative life span. Mutational studies indicate that lysine 16 in the amino-terminal tail of histone H4 and lysines 9, 14 and 18 in H3 are critically important in silencing, whereas lysines 5, 8 and 12 of H4 have more redundant functions. Lysines 9 and

14 of histone H3 and lysines 5,8 and 16 of H4 are acetylated in active chromatin and hypoacetylated in silenced chromatin, and overexpression of Sir2 promotes global deacetylation of histones, indicating that Sir2 may be a histone deacetylase. Deacetylation of lysine 16 of H4 is necessary for binding the silencing protein, Sir3. It was shown that yeast and mouse Sir2 proteins are nicotinamide adenine dinucleotide (NAD)-dependent histone deacetylases, which deacetylate lysines 9 and 14 of H3 and specifically lysine 16 of H4. The analysis of two Sir2 mutations supports the idea that this deacetylase activity accounts for silencing, recombination suppression and extension of life span *in vivo*. The findings provide a molecular framework of NAD-dependent histone deacetylation that connects metabolism, genomic silencing and ageing in yeast and possibly in higher eukaryotes.[204]

Severe dietary restriction, catabolic states and even short-term caloric deprivation impair fertility in mammals. Likewise, obesity is associated with infertile conditions such as polycystic ovary syndrome. The reproductive status of lower organisms such as *C. elegans* is also modulated by availability of nutrients. Thus, fertility requires the integration of reproductive and metabolic signals. Deletion of IR substrate-2 (IRS-2), a component of three insulin/IGF-1 signaling cascade, causes female infertility. Mice lacking IRS-2 have small, anovulatory ovaries with reduced numbers of follicles. Plasma concentrations of luteinizing hormone, prolactin and sex steroids are low in these animals. Pituitaries are decreased in size and contain reduced numbers of gonadotrophs. Females lacking IRS-2 have increased food intake and obesity, despite elevated levels of leptin. Insulin, together with leptin and other neuropeptides, may modulate hypothalamic control of appetite and reproductive endocrinology. Coupled with findings on the role of insulin-signaling pathways in the regulation of fertility, metabolism and longevity *in C. elegans* and *Drosophila*, an evolutionarily conserved mechanism in mammals that regulates both reproduction and energy homeostasis has been identified.[191]

Ageing in budding yeast is measured by the number of mother cell divisions before senescence. Genetic studies have linked ageing in this organism to the Sir genes which mediate genomic silencing at telomeres, mating type loci and the repeated ribosomal RNA. Sir2p determines life span in a dose-dependent manner by creating silenced rDNA chromatin, thereby repressing recombination and the generation of toxic rDNA circles. Silencing is triggered by the deacetylation of certain lysines in the N-termini of histones H3 and H4. Sir2p has a NAD-dependent histone deacetylase activity that is conserved in Sir2p homologs.

Caloric restriction extends life span in a wide variety of organisms. Although it has been suggested that calorie restriction may work by reducing the level of ROS produced during respiration, the mechanism by which this regimen slows ageing is uncertain. Calorie restriction was mimicked in yeast by physiological or genetic means and showed a substantial extension in life span. This extension was not observed in strains mutant for Sir2 (which encodes the silencing protein Sir2p) or NPT1 (a gene in a pathway in the synthesis of NAD, the oxidized form of nicotinamide adenine dinucleotide). The increased longevity induced by calorie restriction requires the activation of Sir2p by NAD.[192]

In view of the involvement of IRs in energy metabolism, ageing and homeostasis, the studies of these IRs bears on several connected phenomena. IRs are expressed in most tissues of the body, including classic insulin-sensitive tissues (liver, muscle and fat), as well as 'insulin-insensitive' tissue such as RBCs and the neuronal tissue of the CNS. In fact, IR and insulin signaling proteins are widely distributed throughout the CNS. To study the physiological role of insulin signaling in the brain, mice were created with a neuron-specific disruption of the IR gene (NIRKO mice). Inactivation of the IR had no impact on brain development or neuronal survival. Female NIRKO mice showed increased food intake, and both male and female mice developed diet-sensitive obesity with increases in body fat and plasma leptin levels, mild insulin resistance, elevated plasma insulin levels, and hyper-triglyceridemia. NIRKO mice also exhibited impaired spermatogenesis and ovarian follicle maturation because of hypothalamic dysregulation of luteinizing hormone. IR signaling in the CNS plays an important role in regulation of energy disposal, fuel metabolism, and reproduction.[193]

The study of quantitative trait loci (QTLs) requires several methodologies to study the connection between genotype and phenotype associated with certain normal or aberrant conditions.[194]

Recent studies have intriguingly shown that complex traits may be controlled by few genes or even a single gene.

In natural populations, most phenotypic variation is continuous and is effected by alleles at multiple loci. Although this quantitative variation fuels evolutionary change and has been exploited in the domestication and genetic improvement of plants and animals, the identification and isolation of the genes underlying this variation have been difficult.

Domestication of many plants has correlated with dramatic increases in fruit size. In tomato, one QTL, *fw2.2*, was responsible for a large step in this process. When transformed into large-fruited cultivars, a cosmid derived from the *fw2.2* region of a small-fruited wild species reduced fruit size by the predicted amount and had the gene action expected for *fw2.2*. The cause of the QTL effect is a single gene, *ORFX*, that is expressed early in floral development, controls carpel cell number and has a sequence suggesting structural similarity to the human oncogene c-H-*ras* p21. Alterations in fruit size, imparted by *fw2.2* alleles, are most likely due to changes in regulation rather than in the sequence and structure of the encoded protein.[195]

Key proteins and their genes may influence whole organisms and thus determine overall physiological traits. Uncoupling protein-3 (UCP-3) is a member of the mitochondrial transporter superfamily that is expressed predominantly in skeletal muscle. Its close relative UCP-1 is expressed exclusively in brown adipose tissue, a tissue whose main function is fat combustion and thermogenesis. Studies on the expression of UCP-3 in animals and humans in different physiological situations support a role for UCP-3 in energy balance and lipid metabolism. Transgenic mice were created that overexpress human UCP-3 in skeletal muscle. These mice are hyperphagic but weigh less than their wild-type littermates. MRI analysis shows a striking reduction in adipose tissue mass. The mice also exhibit lower fasting plasma glucose and insulin levels, and an increased glucose clearance rate. This provides evidence that skeletal muscle UCP-3 has the potential to influence metabolic rate and glucose homeostasis in the whole animal.[196]

This observation may have relevance for a systemic approach to deal with a multitude of age-related pathological developments.

The reaction of glucose with the amino groups of proteins such a collagen and elastin to form advanced glycosylation end-products (AGEs) leads to gradual loss of elasticity in the cardiovascular system and plays a role in the ageing process as exemplified by atherosclerosis, stroke and heart failure. Cardiac stiffening, which is accelerated in diabetics, can be prevented by inhibitors of AGE formation. A new class of therapeutic agents has been developed which can reverse the cross-linking process and restore the cardiovascular system to a more youthful state. A lead compound is ALT711 (4,5-dimethyl-3-(2-oxo-2-phenylethyl)-thiazolium chloride) which interacts with the cross-linked proteins, separating them by cleaving the cross-link.

AGE cross-link breakers could also be beneficial for many other conditions such as nephropathy, retinopathy, neuropathy, and urinary elastic dysfunction.[197]

Another example where one gene influences the whole organism in an apparently concerted fashion is the size of the whole organism which appears to be controlled by few key genes and proteins. Suppressor of cytokine signaling-2 (SOCS-2) is a member of the suppressor of cytokine signaling family, a group of related proteins implicated in the negative regulation of cytokine action through inhibition of the Janus kinase (JAK) signal transducers and activators of transcription (STAT) signal-transduction pathway. Mice unable to express SOCS-2 were used to examine its function *in vivo*. SOCS-2$^{-/-}$ mice grew significantly larger than their wild-type littermates. Increased body weight became evident after weaning, and was associated with significantly increased long bone lengths and the proportionate enlargement of most organs. Characteristics of deregulated growth hormone and IGF-I signaling, including decreased production of major urinary protein, increased local IGF-I production and collagen accumulation in the dermis, were observed in SOCS-2-deficient mice, indicating that SOCS-2 may have an essential negative regulatory role in the growth hormone/IGF-I pathway.[198]

With the escalation of obesity-related disease, there is a great interest in defining the mechanisms that control appetite and body weight. A link between anabolic energy metabolism and appetite control was identified. Both systemic and intracerebroventricular treatment of mice with fatty acid synthase (FAS) inhibitors (cerulenin and a synthetic compound C75) led to inhibition of feeding and dramatic weight loss. C75 inhibited expression of the prophagic signal neuropeptide Y in the hypothalamus and acted in a leptin-independent manner that appears to be mediated by malonyl-CoA. Fas may represent an important link in feeding regulation and may be a potential therapeutic target.[199]

The study of complex phenomena involving hundreds or thousands of genes is facilitated by genome profiling (gene expression analysis). An example of a temporally and spatially complex process is metamorphosis. Metamorphosis is an integrated set of developmental processes controlled by a transcriptional hierarchy that coordinates the action of hundreds of genes. In order to identify and analyze the expression of these genes, high-density DNA microarrays containing several thousand *D. melanogaster* gene sequences were constructed. Many differentially expressed genes were assigned to developmental pathways known to be active during metamorphosis, whereas others can be assigned to pathways not previously associated with metamorphosis. Additionally, many genes of unknown function were identified that may be involved in control and execution of metamorphosis. The utility of this genome-based approach is demonstrated for studying a set of complex biological processes in a multicellular organism, such as the coordination between exdysone-regulated pathways and developmental pathways controlling differentiation of particular cell types and tissues.[200]

The genetic basis of senescence is the target of intense studies at the molecular, differential gene expression level. Recent studies point towards the machinery of chromosome separation and repair as major contributors to senescence. Fibroblasts derived from donors of different ages (normal young, normal middle and normal old) and from pathological source (Hutchinson–Gilford progeria) were studied by examining their different mRNA transcription profiles for significant expression changes. The comparative analysis of gene expression in natural and accelerated human ageing at the fibroblast level revealed a limited set of genes that are differentially expressed.

mRNA levels were measured in actively dividing fibroblasts isolated from human donors of various ages as well as from an individual suffering from progeria, a rare genetic disorder of accelerated ageing. Genes whose expression is associated with age-related phenotypic changes and with disease were identified. The results indicate that an underlying mechanism of the ageing process involves increasing errors in the mitotic machinery of dividing cells in the post-reproductive stage of life. This dysfunction may lead to chromosomal pathologies resulting in misregulation of genes involved in the ageing process.[201]

Evidence that somatic mutations are causally related to the degenerative aspects of the ageing process has been derived from human syndromes of accelerated ageing, such as WS. This disease is caused by a heritable mutation in the WRN gene, encoding both a helicase and an exonuclease, and is considered to play a role in suppressing genomic instability. Cultured somatic cells from patients with WS display an increased rate of somatic mutations and a variety of cytogenetic abnormalities, such as deletions and translocations. Other so-called progeroid syndromes, such as ataxia telangiectasia and Bloom syndrome, also show genomic instability.

Mouse models with inactivated genes involved in double-strand break repair show enhanced chromosomal instability and prematurely exhibit symptoms of age-related degeneration in various tissues. Telomere erosion in telomerase-deficient mice results in an early initiation of genetic instability, accelerating the age-related loss of cell viability and increased tumor formation. Genetic defects promoting gross genomic instability are associated with symptoms of accelerated ageing. Ageing affects both proliferative and post-mitotic organs. Mutation accumulation in post-mitotic cells is difficult to study because most methods require cycling cells. With the development of transgenic mouse models harboring bacterial transporter genes which can be retrieved from chromosomal DNA, mutation accumulation in post-mitotic tissues can be quantitated and characterized.

Somatic mutation accumulation has been implicated as a major cause of cancer and ageing. By using a transgenic mouse model with a chromosomally integrated *lacZ* reporter gene, mutational spectra were characterized at young and old age in two organs greatly differing in proliferative activity, i.e. the heart and small intestine. At young age the spectra were nearly identical, mainly consisting of GC → AT transitions and 1-bp deletions. At old age, however, distinct patterns of mutations had developed. In the small intestine, only point mutations were found to accumulate, including GC → TA, GC → CG and AT → CG transversions and GC → AT transitions. In contrast, in heart about half of the accumulated mutations appeared to be large genome arrangements, involving up to 34 cM of chromosomal DNA. Virtually all other mutations accumulating in the heart appeared to be GC $\rightarrow$ AT transitions at CpG sites. Distinct mechanisms lead to organ-specific genome deterioration and dysfunction at old age.[202]

Induction of the cyclin-dependent kinase (CDK) inhibitor $p21^{Waf1/Cip1/Sdi1}$ is a common mechanism of growth arrest in different physiological situations. p21 is transiently induced in the course of replicative senescence, reversible and irreversible forms of damage-induced growth arrest, and terminal differentiation of post-mitotic cells. Its induction is regulated through p53-dependent and -independent mechanisms. Ectopic overexpression of p21 leads to cell growth arrest in G_1 and G_2; this arrest is accompanied by phenotypic markers of senescence in some or all cells.

Induction of CDK inhibitor $p21^{Waf1/Cip1/Sdi1}$ triggers cell growth arrest associated with senescence and damage response. Overexpression of p21 from an inducible promoter in a human cell line induces growth arrest and phenotypic features of senescence. cDNA array hybridization showed that p21 expression selectively inhibits a set of genes involved in mitosis, DNA replication, segregation and repair. The kinetics of inhibition of these genes on p21 induction parallels the onset of growth arrest and their re-expression on release from p21 precedes the re-entry of cells into the cell cycle, indicating that inhibition of cell cycle progression genes is a mechanism of p21-induced growth arrest. P21 also up-regulates multiple genes that have been associated with senescence or implicated in age-related diseases, including atherosclerosis, Alzheimer's disease, amyloidosis and arthritis. Most of the tested p21-induced genes were not activated in cells that had been growth arrested by serum starvation, but some genes were induced in both forms of growth arrest. Several p21-induced genes encode secreted proteins with paracrine effects on cell growth and apoptosis. In agreement with the overexpression of such proteins, conditioned media from p21-induced cells were found to have anti-apoptotic and mitogenic activity. The effects of p21 induction on gene expression in senescent cells may contribute to the pathogenesis of cancer and age-related diseases.[203]

Studies of ageing involve numerous model systems, such as cell cultures, yeast, nematodes (*C. elegans*), *Drosophila* and mice. Some revealing genetic studies in yeast have provided clues to the fundamental mechanisms.

Studies of the replicative potential of rat cells in vivo have shown that rat Schwann cells and most oligodendrocyte precursor cell purified from postnatal rat optic nerve are capable of indefinite proliferation in culture.[508,509] The rat Schwann cells maintain checkpoints otherwise lost in the immortalization process. The rat oligodendrocyte precursor cells can proliferate indefinitely in serum-free culture provided they are prevented from differentiating. The cells maintain high telomerase activity and p53- and Rb-dependent cell cycle checkpoint responses. Serum or genotoxic drugs induce the cells to develop a senescence-like phenotype. The experiments suggest that for rat cells senescence is not inevitable during cultural proliferation and that some normal rodent precursor cells have an unlimited proliferative capacity in conditions avoiding differentiation and the activation of cell-cycle arresting checkpoint responses.

Rodent cells thus appear not to have the cell counting mechanism of human cells, where the shortening of telomeres with each cell division counts the number of cell divisions they

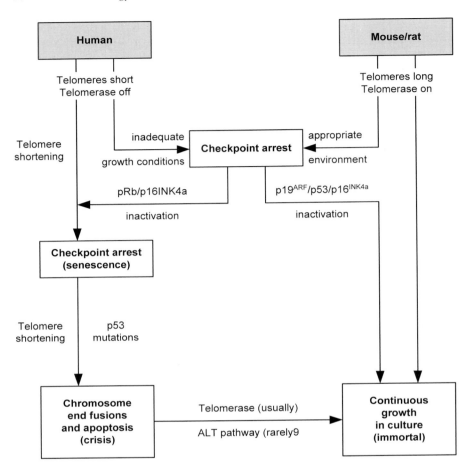

Figure 4.2. The long and short of aging. Most rodent cells contain the enzyme telomerase and have long telomeres. Under the appropriate tissue culture conditions, rodent cells are capable of continuous growth *(1,2,8)*. However, an inadequate culture environment (for example, that in which appropriate survival factors are missing or where different cell types cannot interact) may result in DNA damage or other stresses that induce arrest of cell division at cell cycle checkpoints. The spontaneous inactivation of $p19^{ARF}$, p53, and perhaps $p16^{INK4a}$ under standard culture conditions frequently enables normal rodent cell to grow continuously. Human fibroblasts have short telomeres and are satiny telomerase silent. As is the case with rodent cells, inadequate culture conditions may induce human cells to activate checkpoint pathways that lead to their early (and telomere-independent) growth arrest (9, 10). In contrast to rodent cells, bypass of the p53 and/or $pRb/p16^{INK4a}$ checkpoints is insufficient to immortalize human fibroblasts but does prolong their lifespan. Continuous telomere shortening of human fibroblasts leads to chromosome fusions, crisis, and apoptosis. Only a rare human cell (1 in 10 million) can bypass crisis either through telomerase reactivation or through ALT (the rare alternative pathway for telomere lengthening). Although methods for counting the number of cell divisions do not limit the growth of cultured rodent cells, in human cells telomere shortening does provide a record of the number of doublings and does control replicative senescence [Refs in 510].

have undergone.[510] The difference between human and rodent cellular aging is depicted in Figure 4.2.

Telomeres play a crucial role in the viability of cells. Experimental disruption of telomerase action causes telomere shortening and cellular senescence[525] and genetic defects of telomerase have significant impact. The X-linked form of the human disease dyskeratosis congenita (DKC) is due to mutations in the gene coding for dyskerin and causes defects in regenerative tissues such as skin and bone marrow, chromosome instability, and a predisposition towards certain malignancies. Dyskerin is associated with human telomerase RNA which contains an H/ACA RNA motif and mutations lead to defective telomere maintenance.[526]

In *Caenorhabditis elegans*, an insulinlike signaling pathway controls aging, metabolism, and development. Mutations in the *daf-2* insulin receptor-like gene or the downstream *age-1* phosphoinositide 3-kinase extend adult life-span by two- to three-fold. In order to identify the regulation of aging and metabolism by this pathway, the *daf-2* pathway signaling was restored to only neurons, muscle, or intestine. Insulin like signaling in neurons alone was sufficient to specify wild-type life span, whereas restoration of muscle or intestinal signaling was insufficient to do so. The restoration of the *daf-2* pathway signaling in muscle rescued metabolic defects, indicating a decoupling of life-span and metabolism. The nervous system appears to be a central regulator of longevity in *C. elegans*.[511] The study of aging and developmental genetic regulation in *C. elegans* is supported by the availability of genome-wide data on gene expression profiles.[588]

Research for genes involved in aging carried out in genetically tractable organisms such as yeast, the nematode *Caenorhabditis elegans, Drosophila melanogaster* fruitflies, and mice has established that aging is regulated by specific genes. The pertinent pathways are linking physiology, signaltransduction, and gene regulation. The mutations identified and the correlations of phenotypes indicate that these genes are in control of aging in higher organisms.[522]

Mutations in *Caenorhabditis elegans* affecting sensory cilia or their support cells, or in sensory signal transduction, extend lifespan. *C. elegans* senses environmental signals through ciliated sensory neurons located primarily in specific sensory organs in head and tail. The cilia are membrane-bound microtubule-based structures found at the dendritic endings of sensory neurons in *C. elegans*. The findings of increased lifespan in mutants with defects in sensory cilia implies that lifespan may be regulated by environmental signals.[512]

Biochemical and genetic studies in the long-lived *Caenorhabditis clk-1* mutants demonstrated that these mutants are Q auxotrophs. CLK-1 is a mitochondrial polypeptide with sequence and functional conservation from human to yeast. The *Saccharomyces cerevisiae* homolog Coq7p is essential for ubiquinone synthesis and thus respiration. Development of the *clk-1* mutants requires a dietary source of Coenzyme Q.[527]

In the filamentous fungus *Podospora anserina* the impairment of the nuclear *COX5* gene encoding subunit V of the cytochrome c oxidase complex leads to the use of the alternative respiratory pathway and a decrease in production of reactive ion species with a concomitant increase in lifespan associated with stabilization of the mitochondrial genome. This provides evidence for a causal link between mitochondrial metabolism and longevity in *Podospora anserina*.[518]

Table 4.3. Mutations revealing links between longevity and stress resistance in multicellular organisms [Refs in 520].

Species/ mutations	Gene description	Phenotypic influence*	
		Life span[†]	Stress resistance
C. elegans			
age-1	Human Pl (3) K homologue	65% increase	Enhanced (UV, paraquat, heat)
daf-2	Human insulin-receptor homologue	100% increase	Enhanced (UV, paraquat, heat)
daf-16	Forkhead transcription factor	Suppresses longevity conferred by *age-1* and *daf-2* mutations	Suppresses stress resistance of *age-1* and *daf-2* mutants
clk-1	Homologue of yeast gene associated with coenzyme Q biosynthesis	40% increase	Enhanced increased (UVC)
spe-10	Unknown (sperm defective	40% increase	Enhanced (UV, paraquat, but not heat)
spe-26	Unknown (sperm defective)	65% increase	Enhanced (UV, paraquat, heat)
old-1	Putative receptor tyrosine kinase	65% increase	Enhanced (UV, heat)
ctl-1	Cytostatic catalase	25% decrease; suppresses longevity conferred by *daf-2, age-1* and *clk-1*	Not determined
mev-1	Cytochrome *b* subunit of succinate dehydrogenase	37% decrease	Hypersensitive to oxygen
Drosophila			
mth	Putative G-protein-coupled receptor	35% increase	Enhanced (UV, paraquat, heat)
Mouse			
shc[66]	Cytoplasmic signal-transduction adaptor protein	30% increase	Enhanced (UV, H_2O_2)

- See [ref. 83] for original references describing the phenotypes of the *daf-2, daf-16, clk-1* and *spe-26* mutants. References for other mutants are as follows: *age-1* [refs 49,50,83]; *spe-10* [ref. 84]; *old-1*
- [ref. 85]; *ctl-1* [ref. 56]; *mev-1* [ref. 55]; *mth* [ref. 60]; and *shc*[66] [ref. 6].
- [†]   Numbers reflect changes in mean life span of the mutants relative to wild-type animals.

A genome-wide study of aging and oxidative stress response in *Drosophila melanogaster* involving measurements of genome-wide changes in transcript levels as a function of age in comparison with the effects of a free-radical generator molecule revealed significant responses in a number of genes. Free radicals appear to play an important role in regulating transcript levels in addition to other factors. The studies identify several candidate genes as molecular markers for aging and potential targets for biopharmaceutical intervention in the aging process.[514]

Another genetic study in *Drosophila melanogaster* found that five independent P-element insertional mutations in a single gene resulted in a near doubling of the average adult lifespan without a decline in fertility or physical activity. The product of the pertinent gene *Indy* is closely rrelated to a mammalian sodium dicarboxylate cotransporter, a membrane protein transporting Krebs cycle intermediates. Excision of the P element caused a reversion to normal lifespan. The mutations in this gene may mimic the metabolic state of caloric restriction, which is known to extend lifespan.[515]

The ability to respond to oxidative stress is a critical determinant of life-span. The production of oxidants and the scavenging of reactive oxidant species involve cellular strategies for detection and detoxification of reactive oxygen species and are linked to longevity.[520] Table 4.3 lists mutations with links to stress resistance in multicellular organisms and Table 4.4 shows disease systems related to oxidant stress.

Table 4.4. Oxidants, antioxidants and diseases of ageing [Refs in 520].

Disease System	Laboratory/animal studies	Clinical data
Cardiovascular	Pre-atherosclerotic blood vessels have increased levels of ROS. Vitamin E protects against development of atherosclerosis Disruption of SOD leads to heart failure and overexpression protects against injury	PHSI: no overall benefit of beta-carotene on CVD? Benefit in high-risk subgroup. CHAOS trail: vitamin E reduces rate of non-fatal myocardial infarct. ATBC study: no overall benefit on CVD rate with Vitamin E or beta-carotene? Increase in CVD deaths with beta-carotene.
Opththalmological	Offspring of pregnant mice depleted of glutathione develop cataracts. Retinal pigments produce ROS after light exposure. Retinal degeneration in primates with Vitamin A or E deficiencies.	PHSI: non-significant reduction in cataracts and macular degeneration with Vitamin E and multivitamins NHS: carotenoids intake may decrease risk of cataract.
Neurological	Mutations in SOD1 result in human ALS and transgenic animal models rescued by antioxidants. NMDA-receptor stimulation produces superoxide. Defects in the function of complex 1 seen in Parkinson's disease.	Vitamin E not protective in early Parkinson's disease. Vitamin E beneficial in Alzheimer's disease. *N*-acetylcysteine does not effect survival in ALS.

Altered gene expression in young and senescent cells was catalogued by using enhanced differential display[517] and high-density oligonucleotide arrays were used to examine differences in gene expression in the hypothalamus and cortex of young and aged mice.[187] The hypthalamus plays a key role in regulating metabolism and hypothalamic modulation may influence the aging process. Table 4.5 shows the various gene expression changes in the hyopthalamus from young and old mice, whereas Table 4.6 lists the gene expression changes in the cortex.

Eukaryotic genomes are packaged into nucleosomes, which are thought to repress gene expression generally. Repression is particularly evident at yeast telomeres, where genes within the telomeric heterochromatin appear to be silenced by the histone-binding Sir complex (Sir2, Sir3 and Sir4) and Rap1. To investigate how nucleosomes and silencing factors influence global gene expression, high-density arrays were used to study the effects of depleting nucleosomal histones and silencing factors in yeast. Reducing nucleosome content by depleting histone H4 caused increased expression of 15% of genes and reduced expression of 10% of genes, but it had little effect on expression of the majority (75%) of yeast genes. Telomere proximal genes were found to be de- repressed over regions extending 20 kb from the telomeres, well beyond the extent of the Sir protein binding and the effects of loss of Sir function. These results indicate that histones make Sir-independent contributions to telomeric silencing, and that the role of histones elsewhere in chromosomes is gene specific rather than generally repressive.[205]

Genomic imprinting is characterized by allele-specific expression of multiple genes within large chromosomal domains that undergo DNA replication asynchronously during S phase. Using both FISH analysis and S-phase fractionation techniques, differential replication time was shown to be associated with imprinted genes in a variety of cell types and that it is already present in the pre-implantation embryo soon after fertilization. This pattern is erased.

Table 4.5. Gene expression changes in the hypothalamus from young (2 months) and old (22 months) BALB/c mice [Refs in 187].

Accession number	Name	Fold change
Metabolic enzymes		
W33716	NADH-Ubiquinone oxidoreductase KFYI subunit	2.8
Y07708	NAOH oxidoreductase subunit MWFE	3
AA109866	NADH-ubiquinone oxidoreductase chain 4L	2.1
AA219829	NADH-ubiquinone oxidoreductase SGDH subunit	2.5
AA521794	Cytochrome c oxidase subunit VIIb	2.2
AA672840	ATP synthase 0 subunit	2.2
C76507	ATP synthase μ-subunit	2.7
AA003458	**Sarco/endoplasmic reticulum Ca2+-ATPase 2**	**−2.9**
X56007	**Na/K-ATPase â 2**	**−11**
AA106307	**H+ ATPase subunit E**	**−6.1**
AA087605	H(+)-ATPase (mvp)	−6.9

Table 4.5. Gene expression changes in the hypothalamus from young (2 months) and old (22 months) BALB/c mice (Cont'd).

Accession number	Name	Fold change
Protein processing		
U05333	Cochaperonin 'cofactor' A'	6.8
U09659	Mitochondrial chaperonin 10	2.6
Z31557	Chaperonir containing TCP-1	3.2
AA027544	Ubiquitin-activating enzyme E1	2.8
AA146437	Cathepsin S precursor, a cysteine proteinase	3.4
Z31297	Sorting nexin 2-like protein, protein degradation	4.8
AA013993	**Prolyl olig-peptidase,**	**11**
AA020512	Caspase 6	3
Neuronal growth/structure		
U95116	**Lissencephaly-1 protein (LIS-1)**	**3**
V00835	Metallothionein-I	3.3
X61452	Cell division control-related protein 2b	2.8
AA709861	**A-X actin-like protein**	**3.5**
L31397	**Dynamin**	**−3.5**
U27106	**Clathrin-associated AP-2**	**−5.7**
U86090	Kinesin heavy chain	−2.4
Neuronal signaling		
AA271109	Protein phosphatase 1, regulatory subunit	3.8
X15373	lnositoi-1,4,5-triphosphate receptor	3.5
X51468	Preprosomatostatin gene	3
D37792	**Synaptotagmin 1**	**−17**
D50621	PSD-95/SAP90A	−16
AA048604	**Apolipoprotein E**	**−6.7**
AA068956	**Protein phosphatase PP2A**	**−6.1**
W12204	Ca2+/calmodulin-dependent protein kinase II	−3.6
X61434	**cAMP-dependent protein kinase C â**	**−3.1**
X57497	Glutarnate receptor 1	−2.7
AA710375	*N*-ethylmaleimide-sensitive factor-attachment receptor	−3.4
A8006361	Prostaglandin D synthetase	−5.6
Stress response		
M60798	Cu(2+)-Zn2+ superoxide dismutase	2.9
AA107471	DNAJ homolog 2	−6.4
AA166139	DNA repair protein	−10.4
AA033408	Damage-specific DNA-binding protein, DNA repair	−4.6
D89787	Hif like protein	−6.5
U27830	Stress-inducible protein, STI1	−4.9

Each RNA sample was hybridized twice to two different arrays, and fold change values are averages of the duplicate measurements. Positive values indicate an increase, and negative values indicate a decrease in gene expression. Genes in bold are differentially expressed in both aged cortex and hypothalamus.

Table 4.6. Gene expression changes in the cortex from young (2 months) and old (22 months) BALB/c mice [Refs in 187].

Accession number	Gene name	Fold change
Metabolic enzymes		
M21285	Stearoyl-CoA desaturase	2.4
U27315	Adenine nucleotide translocase-1	2.2
M84145	Fumarylacetoacetate hydrolase	3.1
U13841	ATPase subunit E	−4.4
AA105755	Na+, K+-ATPase α	−4.6
AA003458	**Sarco/endoplasmic reticulum Ca2+-ATPase 2**	**−5.9**
X56007	**Na/K-ATPase β 2 subunit**	**−2.3**
AA 106307	**H(+)-ATPase E-like protein**	**3.8**
AA389346	Citrate synthase	−3.4
Protein processing		
M13500	Kallikrein gene	5.2
X61232	Carboxypeptidase H	3.6
X92665	Ubiquitin-conjugating enzyrne UbcM3	3
AA013993	**Prolyl oligopeptidase**	**2.7**
AA020512	**Caspase 6**	**2.2**
Z30970	Metalloproteinase-3 tissue inhibitor	2.1
Neuronal growth/structure		
U95116	**Lissencephaly-1 protein**	**2.6**
L20899	Cell division cycle (CDC25)	2.6
C76314	Cdc5-like protein	2.4
AA590859	**A-X actin-like protein**	**7.3**
U27106	**Clathrin-associated AP-2**	**−5.5**
AA1 18546	Actin-like protein 3	−6.2
AA050703	Defender against death 1 (DAD1)	−2.4
w 18503	Dynein heavy chain, retrograde transport	−7.7
AA1 11631	Dynactin 1, retrograde axonal transport	−3.4
L31397	Dynamin	−2.7
Neuronal signaling		
AA168959	25 kDa FK506-binding protein FKBP25	2.7
L32372	AMPA receptor subunit (GluR-B)	2.8
X79082	MDK1, a receptor tyrosine kinase	2
D37792	**Synaptotagmin 1**	**−13.2**
M73490	**Apolipoprotein E**	**−6.9**
AA124955	Casein kinase 1α	−2.9
Z67745	**Phosphatase 2A catalytic subunit**	**−6.2**
M27073	Protein phosphatase 1 β	−2.7
J02626	**cAMP-dependent protein kinase C β**	**−4.6**
W13835	*N*-ethylmaleimide-sensitive membrane protein homolog	−3.2
U10120	*N*-ethytmaleimide sensitive factor	−10.3
A8006361	**Prostaglandin D synthetase**	**−3.2**
M27844	Calmodulin	−2.7
M63436	GABA-A receptor α-1 subunit	−7.1

Table 4.6. Gene expression changes in the cortex from young (2 months) and old (22 months) BALB/c mice (Cont'd).

Accession number	Gene name	Fold change
Stress response		
AA105022	Heat-shock protein hsp84-like protein	−2.7
AA204094	HSP4O/DNAJ homolog	−2.9

Each RNA sample was hybridized twice to two different arrays, and fold change values are averages of the duplicate measurements. Genes in bold are differentially expressed in both aged cortex and hypothalamus.

The references cited above should be viewed as only representative examples derived from a much larger, relevant body literature, which is not fully presented.

Acronyms and abbreviations: PHSI, Physician's Health Study I; CHAOS, Cambridge Heart Antioxidant Study; ATBC, Alpha-Tocopherol, Beta-Carotene Cancer Prevention Study; NHS, Nurses Health Study; CVD, cardiovascular disease; ALS, amyotrophic lateral sclerosis; NMDA, *N*-methyl-D-aspartate glutamate receptors.

Before meiosis in the germline, and parent-specific replication timing is then reset in late gametogenesis in both the male and the female. Asynchronous replication timing is established in the gametes and maintained throughout development, indicating that it may function as a primary epigenetic marker for distinguishing between the parental lines.[206]

Further important mechanisms of senescence are related to tumor suppressors and the regulations between different suppressors and regulatory signals. The tumor suppressor p53 induces cellular senescence in response to oncogenic signals. p53 activity is modulated by protein stability and post-translational modification, including phosphorylation and acetylation.

The mechanism of p53 activation by oncogenes remains largely unknown. It was reported that the tumor suppressor PML regulates the p53 response to oncogenic signals. Oncogenic Ras up-regulates PML expression and overexpression of PML induces senescence in a p53-dependent manner. p53 is acetylated at lysine 382 upon Ras expression, an event that is essential for its biological function. Ras induces re-localization of p53 and the CBP acetyl-transferase within the PML nuclear bodies and induces the formation of a trimeric p53–PML–CBP complex. Ras-induced p53 acetylation, p5–CBP complex stabilization and senescence are lost in PML$^{-/-}$ fibroblasts. The data establish a link between PML and p53, and indicate that integrity of the PML bodies is required for p53 acetylation and senescence upon oncogene expression.[207]

The application of telomerase needs to be studied with the potential deleterious effects of introducing hTERT in mind. Telomerase activation extends the life span of human mammary epithelial cell (HMEC) cultures while at the same time inducing lasting overexpression of c-*myc* oncogene.[208] The relevance of telomerase in senescence is intensely debated.[209]

A thoroughly studied model for senescence in the nematode *C. elegans*. The studies of evolution of life span in *C. elegans* is consistent with few early-acting genes.[210]

A specific aspect, i.e. germline immortality and telomere integrity, was studied. The germline is an immortal cell lineage that is passed indefinitely from one generation to the next. To identify the genes that are required for germline immortality, *C. elegans* mutants with mortal germlines were isolated – worms that can reproduce for several healthy generations but eventually become sterile. One of these *mortal germline (mrt)* mutants, *mrt-2*, exhibits progressive telomere shortening and accumulates end-to-end chromosome fusions in later generations, indicating that the MRT-2 protein is required for telomere replication. In addition, the germline of *mrt-2* is hypersensitive to X-rays and to transposon activity. Therefore, *mrt-2* has defects in responding both to damaged DNA and to normal double-strand breaks present at telomeres. *Mrt-2* encodes a homolog of a checkpoint gene that is required to sense DNA damage in yeast. The results indicate that telomeres may be identified as a type of DNA damage and then repaired by the telomere-replication enzyme telomerase.[211]

The potential of cloning depends in part on whether the procedure can reverse cellular ageing and restore somatic cells to a phenotypically youthful state. The birth of six healthy cloned calves derived from populations of senescent donor somatic cells was reported. Nuclear transfer extended the replicative life span of senescent cells (zero to four population doublings remaining) to greater than 90 populations doublings. Early population doubling level complementary DNA-1 (EPC-1, an age-dependent gene) expression in cells from the cloned animals was 3.5- to 5-fold higher than that in cells from age-matched (5–10 months old) controls. Southern blot and flow cytometric analyses indicated that the telomeres were also extended beyond those of newborn (less than 2 weeks old) and age-matched control animals. The ability to regenerate animals and cells may have important implications for medicine and the study of mammalian ageing.[212]

Mice have been cloned by nuclear transfer into enucleated oocytes. In particular, the reiterative cloning of mice to four and six generations in two independent lines was reported. Successive generations show no signs of premature ageing, as judged by gross behavioral parameters and there was no evidence of shortening of telomeres at the ends of chromosomes, normally an indicator of cellular senescence. The telomeres actually increased slightly in length. This increase is astonishing, since the number of mitotic divisions greatly exceeds that of sexually produced animals and any deleterious effects of cloning might be amplified in sequentially cloned mice. These observations bear on studies of organismal ageing.[213]

Telomerase is amenable to therapeutic intervention by small-molecule effectors and thus suitable therapeutics may be developed. Telomerase, a ribonucleoprotein up-regulated in many types of cancers, possesses an RNA template necessary to bind and extend telomere ends. The intrinsic accessibility of Telomerase to incoming nucleic acids makes the RNA template an ideal target for inhibition by oligonucleotides. 2′-*O*-methyl RNA (2′-*O*-meRNA), an oligonucleotide known to exert sequence-specific effects in cell culture and animals, inhibits telomerase with potencies superior to those possessed by analogous peptide nucleic acids (PNAs). Potent inhibition relative to PNAs is surprising, because the binding affinity of 2′-*O*-meRNAs for complementary RNA is low relative to analogous PNAs. A 2′-*O*-meRNA oligomer with terminal phosphorothioate substitutions inhibits telomerase sequence-selectively within human-tumor-derived DU145 cells when delivered with cationic lipids. In contrast to the ability of 2′-*O*-meRNA oligomers to inhibit telomerase, the binding of a 2′-*O*-meRNA to an inverted repeat within plasmid DNA was not detectable, whereas binding of PNA was efficient, suggesting that the relative accessibility of the telomerase RNA template is essential for inhibition by 2′-*O*-meRNA. Inhibition of telomerase by 2′-*O*-meRNA will facilitate probing the link between telomerase activity and sustained cell proliferation, and may provide a basis for the development of chemopreventive and chemotherapeutic agents.[214]

The importance of telomeres is underscored by studies of chromosomal rearrangements and telomeres. Chromosomal rearrangement mechanisms are intimately linked to cancer development, and are thought to generate the numerous gains and losses of segments of chromosomes needed for epithelial carcinogenesis. Aged humans sustain a high rate of epithelial cancers such as carcinomas of the breast and colon, whereas mice carrying common tumor suppressor gene mutations typically develop soft tissue carcinomas and lymphomas. Among the many factors that may contribute to this species variance are differences in telomere length and regulation. Telomeres comprise the nucleoprotein complexes that cap the ends of eukaryotic chromosomes and are maintained by the reverse transcriptase, telomerase. In human cells, insufficient levels of telomerase lead to telomere attrition with cell division in culture, and possibly with ageing and tumorigenesis *in vivo*. In contrast, critical reduction in telomere length is not observed in the mouse owing to promiscuous telomerase expression and long telomeres. Telomere attrition in ageing telomerase-deficient p53 mutant mice promotes the development of epithelial cancers by a process of fusion-bridge breakage that leads to the formation of complex non-reciprocal translocations – a classical cytogenetic feature of human carcinomas. Telomere dysfunction brought about by continual epithelial renewal during life may generate the massive ploidy changes associated with the development of epithelial cancers.[215]

Life span is also influenced by external factors, such as ROS which primarily arise as by-products of normal metabolic activities and are thought to influence the etiology of age-related diseases. In order to test the theory that ROS cause ageing, the natural antioxidant systems of *C. elegans* with small synthetic superoxide dismutase/catalase mimetics. Treatment of wild-type worms increased their mean life span by a mean of 44%, and treatment of prematurely aging worms resulted in normalization of their life span (a 67% increase). It appears that oxidative stress is a major determinant of life span and that it can be counteracted by pharmacological intervention.[216]

Expression profiling allows the simultaneous identification of differentially expressed genes, and by comparing different cells, tissues, disease states and metabolic states enables the pin-pointing of condition-associated and/or causative gene expression. In tumor therapy, the identification of tumor type and genetic make-up of the malignant cells allows the early and often decisive choice of therapy.

Diffuse large B cell lymphoma (DLBCL), the most common subtype of non-Hodgkin's lymphoma, is clinically heterogeneous: 40% of patients respond well to current therapy and have prolonged survival, whereas the remainder succumb to the disease. This variability in natural history may reflect unrecognized molecular heterogeneity in the tumors. Using DNA microarrays, a systematic characterization of gene expression in B cell malignancies was conducted. It was shown that there is diversity in gene expression among the tumors of DLBCL patients, apparently reflecting the variation in tumor proliferation rate, host response and differentiation state of the tumor. Two molecularly distinct forms of

DLBCL were identified which had gene expression patterns indicative of different stages of B cell differentiation. One type expressed genes characteristic of germinal center B cells ('germinal center B-like DLBCL'); the second type expressed genes normally induced during *in vitro* activation of peripheral blood B cells ('activated B-like DLBCL'). Patients with germinal center B-like DLBCL had a significantly better overall survival than those with activated B-like DLBCL. The molecular classification of tumors on the basis of gene expression can thus identify previously undetected and clinically significant subtypes of cancer.[217]

Similarly, the technique of gene expression profiling was applied to another difficult to categorize type of tumor, malignant melanoma. The most common human cancers are malignant neoplasms of the skin. Incidence of cutaneous melanoma is rising especially steeply, with minimal progress in non-surgical treatment of advanced disease. Despite significant effort to identify independent predictors of melanoma outcome, no accepted histopathological, molecular or immunohistochemical marker defines subsets of this neoplasm. Accordingly, although melanoma is thought to present different 'taxonomic' forms, these are considered part of a spectrum rather than discrete entities. The discovery of a subset of melanomas identified by mathematical analysis of gene expression in a series of samples was reported. Many genes underlying the classification of this subset are differentially regulated in invasive melanomas that form primitive tubular networks *in vivo*, a feature of some highly aggressive metastatic melanomas. Global transcript analysis can identify unrecognized subtypes of coetaneous melanoma and predict experimentally verifiable phenotypic characteristics that may be of importance to disease progression and choice of treatment.[218]

To gain an understanding of the molecular basis of tumor angiogenesis, gene expression patterns were compared of endothelial cells derived from blood vessels of normal and malignant colorectal tissues. Of over 170 transcripts predominantly expressed in the endothelium, 79 were differentially expressed, including 46 that were specifically elevated in tumor-associated endothelium. Several of these genes encode extracellular matrix proteins, but most are of unknown function. Most of these tumor endothelial markers were expressed in a wide range of tumor types, as well as in normal vessels associated with wound healing and corpus luteum formation. These studies demonstrated that tumor and normal endothelium are distinct at the molecular level – a finding that may have significant implications for the development of antiangiogenic therapies.[219]

SAGE analysis reveals previously characterized and novel pan endothelial markers. The most abundant characterized or novel tags derived by summing the tags from normal EC (N-ECs) and tumor EC (T-ECs). SAGE libraries are listed in descending order. For comparison, the corresponding number of SAGE tags found in HUVEC and HMVEC endothelial cell cultures, and several nonendothelial cell lines (14), are shown. Tag numbers for each group were normalized to 100.000 transcripts. A description of the gene product corresponding to each tag is given, followed by alternative names in parentheses. Some uncharacterized genes have predicted full-length coding sequence. The sequence CATG precedes all tags, and the 15th base (11th shown) was determined as described (38) [References in 219].

Human breast tumors are diverse in their natural history and in their responsiveness to treatment. Variation in transcriptional programs accounts for much of the biological diversity of human cells and tumors. In each cell, signal transduction and regulatory systems transduce information from the cell's identity to its environmental status, thereby controlling the level of expression of every gene in the genome. Variation was characterized in gene expression patterns in a set of 65 surgical specimens of human breast tumors from 42 different individuals, using complementary DNA microarrays representing 8102 human genes. These patterns provided a distinctive molecular portrait of each tumor. Twenty of the tumors were sampled twice, before and after a 16-week course of doxorubicin chemotherapy, and two tumors were paired with a lymph node metastasis from the same patient. Gene expression patterns in two tumor samples from the same individual were almost always more similar to each other than either was to any other sample. Sets of co-expressed genes were identified for which variation in mRNA levels could be related to specific features of physiological variation. The tumors could be classified into subtypes distinguished by pervasive differences in their gene expression patterns.[220]

The elucidation of genetic contributions to senescence in *Drosophila melanogaster* and *Caenorhabditis elegans* and their relationship points towards fundamental evolutionary conserved genetic and biochemical mechanisms.[655]

The *Drosophila melanogaster* gene *insulin-like receptor (InR)* was found homologous to mammalian insulin receptors as well as to *Caenorhabditis elegans daf-2*, a signal transducer regulating worm dauer formation and adult longevity. A heteroallelic, hypomorphic genotype of mutant *InR*, which results in dwarf females with up to an 85% extension of adult longevity and dwarf males with reduced late age-specific mortality. Treatment of the long-lived *InR* dwarfs with a juvenile hormone analoge restores the life expectancy toward that of wild-type controls thus demonstrating that juvenile hormone deficiency resulting from the *InR* signal pathway mutation is sufficient to extend life-span in flies. The results also show that insulin-like ligands nonautonomously mediate aging in flies through retardation of growth or activation of specific endocrine tissue.[654]

The *Drosophila melanogaster* gene *chico* encoding an insulin receptor substrate which functions in an insulin/insulin-like growth factor (IGF) signaling pathway was found to extend upon mutation the fruit fly median life-span by up to 48% in homozygotes and 36% in heterozygotes. The extension of life-span was neither a result of impaired oogenesis in *chico* females nor correlated with increased stress resistance nor was the dwarf phenotype of *chico* homozygotes necessary for extension of life-span.[653]

Genetic linkage studies are required to identify the multiple genes involved in complex traits and population genetic studies involving single-nucleotide polymorphisms and other genetic variations such as different alleles are the basis for pharmaco-genomics and the identification of genetic disease associations.[691] [692]

Table 4.7 shows a listing of the pan-endothelial markers identified by SAGE analysis.

Table 4.7. Gene determined by SAGE analysis to be potential pan-endothelial markers [Refs in 219].

Tag sequence	N-ECs	T-ECs	HUVEC	HMVEC	Cell lines	Description	
Known genes							
CATATCATTAA	247	501	130	87	2	Angiomodulin (ANG, IGFBP-7, IGFBP-rP1, Mac25, TAF)*	
TGCACTTCAAG	328	141	0	0	0	Hevin*	
TTTGCACCTTT	165	84	191	115	4	Connective tissue growth factor (CTGF, IGFBP-rP2)*	
TTGCTGACTTT	73	131	2	14	1	Collagen, type VI, alpha 1*	
ACCATTGGATT	102	67	0	0	2	Interferon induced transmembrane protein 1 (9-27, Leu 13)*	
ACACTTCTTTC	104	44	60	62	2	Guanine nucleotide binding protein 11	
TTCTGCTCTTG	71	67	118	72	0	Von Willebrand factor*	
TCCCTGGCAGA	66	68	3	13	3	Cysteine-rich protein 2 (CRP-2, ESP-1, SmLIM)	
TAATCCTCAAG	26	106	34	16	1	Collagen, type XVIII, alpha 1*	
ATGTCTTTTCT	58	65	17	17	3	Insulin-like growth factor-binding protein 4*	
GGGATTAAAGC	40	67	30	14	2	CD146 (S-Endo 1, P1H12, Muc18, MCAM. Mel-CAM)*	
TTAGTGTCGTA	38	69	9	13	0	SPARC (osteonectin,BM-40)*	
TTCTCCCAAAT	20	86	16	64	2	Collagen, type IV, alpha 2*	
GTGCTAAGCGG	24	74	0	10	2	Collagen, type VI, alpha 2*	
GTT'TATGGATA	35	56	11	11	1	Matrix Gla protein (MGP)	
Novel genes							
CCCTTGTCCGA	131	104	1	1	0	PEM1	EST
CCCTTTCACAC	52	33	0	0	0	PEM2	ESTs
CAACAATAATA	42	25	13	6	0	PEM3	ESTs
GGCCCTACAGT	26	13	2	3	0	PEM4	ESTs/KIAA0821 protein
GCTAACCCCTG	7	31	0	1	0	PEMS	ESTs
GGCACTCCTGT	22	13	19	12	0	PEM6	ESTs
TCACAGCCCCC	20	15	8	5	0	PEM7	ESTs
CAACAATAATA	42	25	13	6	0	PEM3	ESTs
GGCCCTACAGT	26	13	2	3	0	PEM4	ESTs/KIAA0821 protein
GCTAACCCCTG	7	31	0	1	0	PEMS	ESTs
GGCACTCCTGT	22	13	19	12	0	PEM6	ESTs
TCACAGCCCCC	20	15	8	5	0	PEM7	ESTs
TTTCATCCACT	20	13	0	2	0	PEM8	ESTs. KIAA0362 protein
ATACTATAATT	25	6	2	0	0	PFM9	ESTs
AATAGGGGAAA	13	19	4	1	0	PEM10	KIAA1075 protein

*   Characterized genes that have previously been shown to be expressed predominantly in endothelium (10, 15–26).

5   Drug Discovery

The drug discovery process is the fundamental source of active principles, i.e. compounds which interact as agonists or antagonists, functionally or structurally, with constituents of living organisms.

Drugs arise form either natural sources, with an almost infinite variety of molecular structures, or are designed with the target effect or molecule in mind [using quaternary structure–activity relationships (QSAR)]. The combination and creative utilization of several areas of scientific studies, i.e. genomics, proteomics, bioinformatics, structural analysis and functional experiments plus advanced chemistries, will provide significant results in the next decades.[221]

Modern drug discovery often involves screening of small molecules for their ability to bind to a pre-selected protein target. Target-oriented syntheses of these small molecules, individually or as collections (focused libraries), can be planned effectively with retrosynthetic analysis. Drug discovery can also involve screening small molecules for their ability to modulate a biological pathway in cells or organisms, without regard for any particular protein target. This process is likely to benefit in the future from an evolving forward analysis of synthetic pathways, used in diversity-oriented synthesis that leads to structurally complex and diverse small molecules. One goal of diversity-oriented syntheses is to synthesize efficiently a collection of small molecules capable of perturbing any disease-related biological pathway, leading eventually to the identification of therapeutic protein targets capable of being modulated by small molecules. Several synthetic planning principles for diversity-oriented synthesis and their role in the drug discovery process have been reviewed.[222]

Recent drug discovery has moved from biology-directed chemical synthesis towards an area where large-scale zero-knowledge screening dominates. As the cost of carrying out large-scale diversity high-throughput screening (HTS) continues to fall, driven by extreme miniaturization, a paradigm shift towards less centralized screening activities distributed more closely to end-user groups or individuals is a possible consequence. The power of the methodology therefore becomes more responsive to a more diverse range of individual requirements. This model is a logical extrapolation of current practice, taking into account the history of other technologies such as computing, DNA sequencing and other physico-chemical analytical techniques. The era of personal or desktop screening is not far off, but it will require a considerable degree of miniaturization to be applied to the whole screening system, not just to the sample carrier.

In parallel with these approaches, the power of computational predictive methodologies has now reached the stage where a third strategy, 'chemical space definition', can be applied to the discovery portfolio to aid in finding pharmaceutical interventions. Information-based screening, be it through the application of a focussed, systems-based method or the use of iterative screening, will become more efficient, selecting fewer false-positives and, more importantly, fewer false-negatives. This will be achieved not only through the development of better algorithms, but through established algorithms made accessible by an increase in computing power. The other major contribution to the development of computational methods is the data generated over the years, particularly for target families such as kinases. In the next few years, predictive models for these targets might negate the need for random screening against them, as we will have learnt the essential features required for recognition.

Diversity and focussed approaches are likely to continue to provide complementary benefits. The portfolio of techniques will together provide a still more effective and comprehensive approach to the effective identification of novel medicines for the multitude of novel targets that are being revealed through genomics and proteomics.

The complementarity of diversity and focussed approaches can be seen in clinical practice. HIV protease inhibitors are all from the rational, structure-based design school of drug discovery, whereas the non-nucleoside HIV reverse transcriptase inhibitors have almost all originated from diversity screening. As combination therapy is the best way to treat HIV infection, this is evidence of the benefits of complementary technologies to fight disease.[223]

Knowledge of cellular signaling pathways and knowledge of the human genetics of a specific disease provide powerful validation and rational targets for disease identification and development of drugs and therapies.

Cancer is going to become the most prevalent disease-related cause of death because it is a disease of multiple accumulating mutations that are manifesting in human populations with an increasingly prolonged life span. Chemotherapies and gene therapies make use of increasing molecular knowledge in cancer, and are supported by tumor genotyping and patient gene profiling.[224]

Table 5.1 gives examples of mechanism-based cancer therapies in development.

A specific feature of CML is a reciprocal chromosomal translocation involving the long arms of chromosomes 9 and 22. This somatic mutation fuses a segment of the *bcr* gene, from chromosome 9, to a region upstream of the second exon of the c-*abl* (Abelson tyrosine kinase) gene from chromosome 22. c-*abl* encodes a non-receptor tyrosine kinase that has tightly controlled activity in normal cells. In contrast, Bcr–Abl fusion proteins have constitutive catalytic activity, despite the fact that the amino acid sequence of the Abl segment of Bcr–Abl is identical to that of c-Abl. The reason for the elevated catalytic activity of the Bcr–Abl fusion protein is poorly understood, but it is clear that this activity of the kinase domain is necessary for the ability of the Bcr–Abl protein to transform cells and cause malignancy.

A small-molecule inhibitor of Abl (STI-571) is effective in the treatment of chronic myelogenous leukemia CML, which is caused by the inadvertent activation of Abl. The crystal structure of the catalytic domain of Abl, complexed to a variant of STI-571, was reported. Critical to the binding of STI-571 is the adoption by the kinase of an inactive conformation, in which a centrally located 'activation loop' is not phosphorylated. The conformation of this loop is distinct from that in active protein kinases, as well as in the inactive form of the closely related Src kinases. These results suggest that compounds that

Table 5.1. Examples of mechanism-based cancer therapies in development [Refs in 224].

Drug	Target	Drug	Target
Receptor antagonists		**Other enzyme inhibitors**	
Raloxifene	Estrogen receptor	R115777	FTase
LY353381	Estrogen receptor	SCH 66336	FTase
GW5638	Estrogen receptor	L-778,123	FTase
CP336156	Estrogen receptor	BMS-214662	FTase
EM-800	Estrogen receptor	CP-609,754	FTase
EMD-121974	Integrin	Marimastat	MMPs
		AG-3340 MMPs	
Protein kinase inhibitors		BMS-275291	MMPs
CGP 57148/STI 571	Bcr-abl	CGS-27023A	MMPs
ZD-1839	EGF receptor	BAY 12-9566	MMPs
CP-358,774	EGF receptor		
SU-5416	KDR	**Therapeutic antibodies**	
SU-6668	KDR; FGFR; and	C225	EGF receptor
	PDGFR	Anti-VEGF	VEGF
CGP 60474	Cyclin-dependent	IMC-1C11	KDR
	kinases	Vitaxin	Integrin
Flavopiridol	Cyclin-dependent		
	kinases	**Antisense oligonucleotides**	
CGP 41251	Protein kinase C	ISIS-2503	Ras
UCN-01	Protein kinase C	ISIS-5132	Raf
		ISIS-3521	Protein kinase C
Viruses		G3139	Bcl2
SCH 58500 (AD-*p53*)	p53		
Onyx-015	p53		

exploit the distinctive inactivation mechanism of individual protein kinases can achieve both high affinity and high specificity.[225]

The studies were the result of the identification of a series of inhibitors, based on the 2-phenylaminopyrimidine class of pharmacophores that have exceptionally high affinity and specificity for Abl. The most potent of these is STI-571. Figure 5.1 shows the structural formula of STI-571 (formerly Novartis compound CGP 57148), which has been success- fully tested in clinical trials as a therapeutic agent for CML. The compound has led to complete hematological response in 96% of patients treated for more than 4 weeks at a dose level of 300 mg and is well tolerated. The possibility of taking advantage of distinct inactive conformations of protein kinase points towards further development of specific protein kinase inhibitors.

Protein kinases have proved to be largely resistant to the design of highly specific inhibi- tors, even with the aid of combinatorial chemistry. The lack of these reagents has complicated efforts to assign specific signaling roles to individual kinases. A chemical genetic strategy was described for sensitizing protein kinases to cell-permeable molecules that do not inhibit wild-type kinases. From two inhibitor scaffolds, potent and selective inhibitors for sensitized

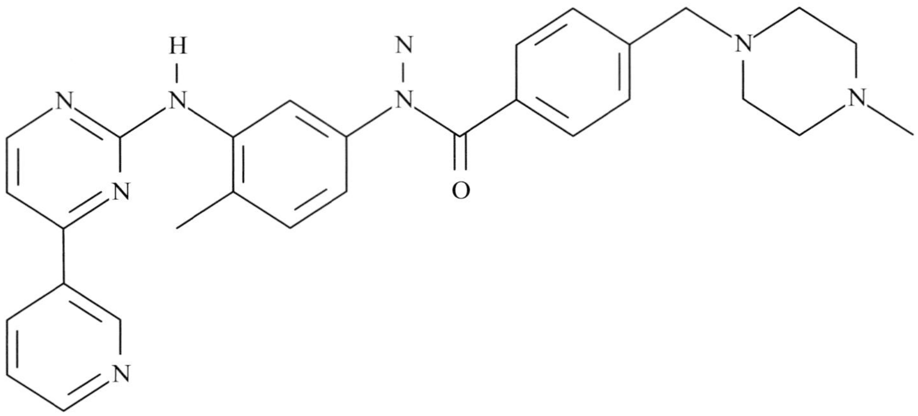

(Panel 1)

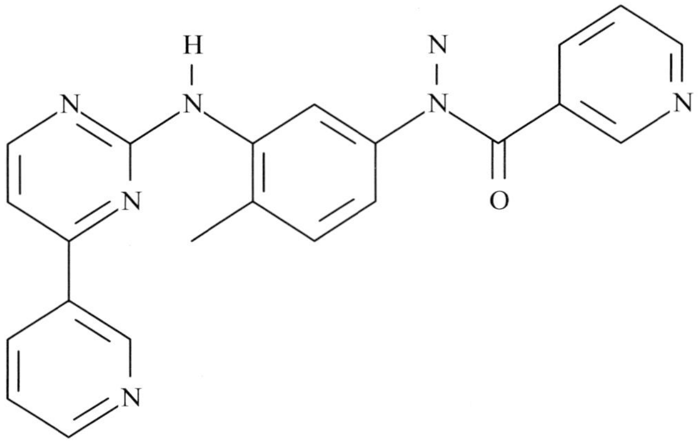

(Panel 2)

Figure 5.1. Structural formula of the Abl inhibitor STI-571 (panel 1) and the variant (panel 2) [Refs in 225].

kinases from five distinct subfamilies were identified. Tyrosine and serine/threonine kinases are equally amenable to this approach. A budding yeast strain carrying an inhibitor-sensitive form of the CDK cdc28 (CDK1) in place of the wild-type protein was analyzed. Specific inhibition of cdc28 *in vivo* caused a pre-mitotic cell-cycle arrest that is distinct from the G_1 arrest typically observed in temperature-sensitive cdc28 mutants. The mutation that confers inhibitor sensitivity is easily identifiable from primary sequence alignments. This approach can be used to systematically generate conditional alleles of protein kinases, allowing for rapid functional characterization of members of this important gene family.[226]

Protein tyrosine kinases and phosphatases are involved in a range of intracellular activities, including cell proliferation, migration and differentiation. Many such proteins can selectively bind their cognate targets and initiate a cascade of signaling events through key modular domains that control protein–protein interactions. One such domain, the Src homology 2 (SH2) domain, has been determined to play a pivotal role in many signaling pathways by recognizing phosphotyrosine (pTyr) sequences of cognate proteins. The SH2 domain of the non-receptor protein tyrosine kinase Src has been shown to interact with focal adhesion kinase, p130^{cas}, p85, phosphatidylinositol 3-kinase and p68^{sam}. Small molecules designed to inhibit SH2-mediated protein–protein interactions have promise as pharmaceutical agents to block specific intracellular pathways critically involved in the pathogenesis of certain diseases.

Targeted disruption of the pp60^{src} (Src) gene has implicated tyrosine kinase in osteoclast-mediated bone resorption and as a therapeutic target for the treatment of osteoporosis and other bone-related diseases. The discovery of a non-peptide inhibitor (AP22408) of Src has been described that demonstrates *in vivo* antiresorptive activity. Based on a co-crystal structure of the non-catalytic SH2 domain of Src complexed with citrate (in the pTyr binding pocket), 3′,4′-diphosphonophenylalanine (Dpp) was designed as a pTyr mimic. In addition to its design to bind Src SH2, the Dpp moiety exhibits bone-targeting properties that confer osteoclast selectivity, hence minimizing possible undesired effects on other cells that have Src-dependent activities. The chemical structure AP22408 also illustrates a bicyclic template to replace the post-pTyr sequence of cognate Src SH2 phophonopeptides such as Ac-pTyr–Glu–Glu–Ile (1). The X-ray structure of AP22408 complexed with Lck (S164C) SH2 confirmed molecular interactions of both the Dpp and bicyclic template of AP22408 as predicted from molecular modeling. Relative to the cognate phosphopeptide, AP22408 exhibits significantly increased Src SH2 binding affinity (IC$_{50}$ = 0.30 µM for AP22408 and IC$_{50}$ = 5.5 µM for 1). AP22408 inhibits rabbit osteoclast-mediated resorption of dentine in a cellular assay, exhibits bone-targeting properties based on a hydroxyapatite adsorption assay and demonstrates *in vivo* antiresorptive activity in a parathyroid hormone-induced rat model.[227]

The fibroblast growth factors (FGFs) are a large family of pleiotropic growth factors that promote survival, proliferation and differentiation of a variety of cell lines and tissues. Treatment with FGF induces formation of new blood vessels, and improves blood flow and organ function in animal models of peripheral vascular disease and myocardial ischemia. In addition, FGF treatment can reduce infarct size following experimental coronary and cerebral artery occlusion.

Clinical studies with FGF and other angiogenic molecules imply the need to develop therapeutic molecules with improved properties. For example, the plasma half-life of FGFs is short, implying that repeated administration of recombinant protein, or gene therapy, may be necessary for optimal therapeutic effects. The existence of multiple FGF receptor subtypes in various tissues suggests that molecules with greater receptor specificity may have fewer systemic toxicities. FGFs are being investigated in human clinical trials as treatments for angina, claudication and stroke. A molecule was designed which is structurally unrelated to all FGFs, which potently mimicked basic FGF activity, by combining domains that (i) bind FGF receptors, (ii) bind heparin and (iii) mediate dimerization. A 26-residue peptide identified by phage display specifically bound FGF receptor (FGFR) 1c extracellular domain but had no homology with FGFs. When fused with the c-*jun* leucine zipper domain, which binds heparin and forms heterodimers, the polypeptide specifically reproduced the mitogenic and morphogenic activities of basic FGF with similar potency (EC_{50} = 240 pM). The polypeptide required interaction with heparin for activity, demonstrating the importance of heparin for FGFR activation, even with designed ligands structurally unrelated to FGF. It is thus possible to engineer potent artificial ligands for the receptor tyrosine kinases, which has important implications for the design of non-peptidic ligands for FGF receptors. Artificial FGFR agonists may be useful alternatives to FGF in the treatment of ischemic vascular disease.[228]

The heparin-binding FGF family consists of more than 20 members that play essential roles in regulating cellular differentiation, growth, and development. Because FGF signal transduction requires formation of three-member complexes made up of FGF, heparan sulfate/heparin and FGFRs, heparin sulfate/heparin-bearing proteoglycans (PGs) located on cell surfaces or in the extracellular matrix are important regulators of FGF biological activity. Several FGF isoforms, including FGF-1, are known to require heparin/heparin sulfate in order to manifest their full mitogenic activity *in vitro*, heparin facilitates formation of biologically active FGF-1 dimers and high-affinity binding of FGF-1 to FGFR and also stabilizes the structure of FGF-1 and protects it from proteolytic inactivation.

In the absence of heparan sulfate on the surface of target cells or free heparin in the vicinity of their receptors, FGF family members cannot exert their biological activity and are easily damaged by proteolysis. This limits the utility of FGFs in a variety of applications including treatment of surgical, burn and periodontal tissue wounds, gastric ulcers, segmental bony defects, and ligament and spinal cord injury. An FGF analog was engineered to overcome this limitation by fusing FGF-1 with heparin sulfate PG core protein. The fusion protein (PG–FGF-1), which was expressed in Chinese hamster ovary cells and collected from the conditioned medium, possessed both heparin sulfate and chondroitin sulfate sugar chains. After fractionation, intact PG–FGF-1 proteins with little affinity to immobilized heparin and high-level heparin sulfate modification, but not their heparitinase or heparinase digests, exerted mitogenic activity independent of exogenous heparin toward heparin sulfate-free Ba/F3 transfectants expressing FGFR. Although PG–FGF-1 was resistant to tryptic digestion, its physiological degradation with a combination of heparitinase and trypsin augmented its mitogenic activity toward human endothelial cells. The same treatment abolished the activity of simple FGF-1 protein. By constructing a biologically active PG–FGF-1 fusion protein, an approach was demonstrated that may be effective for engineering other FGF family members and other heparin-binding proteins as well.[229]

Protein–protein interactions are crucial to many biological processes. Such interactions can be identified by phage display and the yeast two-hybrid system, as well as by biochemical methods. The identification of protein–protein interaction is only the first step in elucidating process and function. Modulation of interaction is necessary to produce true insight into the biological purpose.

The processes of protein–protein interaction include signal transduction, cell cycle regulation, gene regulation, and viral assembly and replication. Many proteins and enzymes manifest their function as oligomers. An efficient means has been developed to sift through large combinatorial libraries and identify molecules that block the interaction of nine-residue peptides from a combinatorial library that inhibit the intracellular dimerization of HIV-

1 protease. Fewer than 1 in 10^6 peptides do so. *In vitro* biochemical analyses of one such peptide demonstrate that it acts by dissociating HIV-1 protease into monomers, which are inactive catalysts. Inhibition is further enhanced by dimerizing the peptide. This approach enables the facile identification of new molecules that control cellular processes.[230]

Studies of protein–protein interactions and their specific interference are the basis of developing small molecules as therapeutic agents in the area of protein networks and their regulation. Mutations introduced into human growth hormone (hGH) and the extracellular domain of the hGH receptor created a cavity at the protein–protein interface that resulted in binding affinity to be reduced by a factor of 10^6. A small library of indole analogs was screened for small molecules that bind the cavity created by the mutations and restore binding affinity. The ligand 5-chloro-2-trichloromethylimidazole was found to increase the affinity of the mutant hormone for its receptor more than 1000-fold. Cell proliferation and JAK2 phosphorylation assays showed that the mutant hGH activates growth hormone signaling in the presence of added ligand. This approach may allow other protein–protein and protein-nucleic acid interactions to be switched on or off by addition or depletion of exogenous small molecules.[231]

Eukaryotic transcriptional activators are minimally comprised of a DNA binding domain and a separable activation domain; most activator proteins also bear a dimerization module. These proteins modules were replaced with synthetic counterparts to create artificial transcription factors. One of these, at 4.2 kDa, mediates high levels of DNA site-specific transcriptional activation *in vitro*. This molecule contains a sequence-specific DNA binding polyamide in place of the typical DNA binding region and a non-protein linker in place of the usual dimerization peptide. The protein–DNA binding module was replaced with a hairpin polyamide composed of *N*-methylpyrrole (Py) and *N*-methylimidazole (Im) amino acids that binds in the minor groove of DNA. Thus the activating region, a designed peptide, functions outside of the archetypal protein context, as long as it is tethered to DNA. Because synthetic polyamides can, in principle, be designed to recognize any specific sequence, this represents a key step toward the design of small molecules that can up-regulate any specified gene.[232]

Mutations in the p53 tumor suppressor gene are the most common specific genetic changes in human tumors. In over 90% of cervical cancers and cancer-derived cell lines, the p53 tumor suppressor pathway is disrupted by human papillomavirus (HPV). The HPV E6 protein promotes the degradation of p53, and thus inhibits the stabilization and activation of p53 that would normally occur in response to HPV E7 oncogene expression. Restoration of p53 function in these cells by blocking this pathway should promote a selective therapeutic effect. Treatment with the small-molecule nuclear export inhibitor, leptomycin B and actinomycin D leads to the accumulation of transcriptionally active p53 in the nucleus of HeLa, CaSki and SiHa cells. Northern blot analysis showed that both leptomycin B and actinomycin D reduced the amount of HPV E6–E7 mRNA, whereas combined treatment with the drugs showed almost complete disappearance of the viral mRNA. The combined treatment activated p53-dependent transcription, and increases in both p21$^{Waf1/Cip1}$ and Hdm2 mRNA were observed. The combined treatment resulted in apoptotic death in the cells, as evidenced by nuclear fragmentation and PARP-cleavage indicative of caspase 3 activity. These effects were greatly reduced by expressing a dominant negative p53 protein. Small molecules can re-activate p53 in cervical carcinoma cells and this reactivation is associated with an extensive biological response, including the induction of apoptotic death of the cells.[233]

Triple-helix-forming oligonucleotides (TFOs) bind in the major groove of double-stranded DNA at oligopyrimidine–oligopurine sequences and therefore are candidate molecules for artificial gene regulation, *in vitro* and *in vivo*. Oligonucleotide analogs containing N3′–P5′ phosphoramidate (np) linkages were described that exhibited efficient inhibition of transcription elongation *in vitro*. Conclusive evidence was provided that np-modified TFOs targeted to the HIV-1 polypurine tract (PPT) sequence can inhibit transcriptional elongation in cells, either in transient or stable expression systems. The same constructs were used in transient expression assays (target sequence on transfected plasmid) and in the generation of stable cell lines (target sequence integrated into cellular chromosomes). In both cases the only distinguishable feature between the cellular systems is the presence of an insert containing the wild-type PPT/HIV-1 sequence, a mutated version with two mismatches, or the absence of the insert altogether. The inhibitory action induced by np-TFOs was restricted to the cellular systems containing the complementary wild-type PPT/HIV-1 target and consequently can be attributed only to a triple-helix-mediated mechanism. An imaging technique was applied to quantitatively investigate the dynamics of TFO-mediated specific gene silencing in single cells.[234]

The drug discovery process in its non-rational form often begins with massive screening of compound libraries to identify modest affinity leads. The screening process targets not specific sites for drug design but only those sites where high-throughput assays are available. Many traditional screening methods rely on inhibition assays that are often subject to artifacts caused by reactive chemical species or denaturants.

A strategy (called tethering) was developed to discover low molecular weight ligands (around 250 Da) that bind weakly to targeted sites on proteins through an intermediary disulfide tether. A native or engineered cysteine in a protein is allowed to react reversibly with a small library of disulfide-containing molecules (around 1200 compounds) at concentrations typically used in drug screening (10–200 μM). The cysteine-captured ligands, which are readily identified by MS, are among the most stable complexes, even though in the absence of the covalent tether the ligands may bind very weakly. This method was applied to generate a potent inhibitor for thymidylate synthase, an essential enzyme in pyrimidine metabolism with therapeutic applications in cancer and infectious diseases. The affinity of the untethered ligand (K_i ~ 1 mM) was improved 3000-fold by synthesis of a small set of analogs with the aid of crystallographic structures of the tethered complex. Such site-directed ligand discovery allows one to nucleate drug design from a spatially targeted lead fragment.[235]

Influencing genetic regulatory networks may be achieved by small-molecule effects, by regulatory proteins and by gene therapeutic methods. The complexities to be considered are illustrated by the phenomenon of RNA interference and co-suppression. Originally discovered in plants, the phenomenon of co-suppression by transgenic DNA has since been observed in many organisms from fungi to animals: introduction of transgenic copies of a gene result in reduced expression of the transgene as well as the endogenous gene. The effect depends on sequence identity between transgene and endogenous gene. Some cases of co-suppression resemble RNA interference (the experimental silencing of genes by the introduction of double-stranded RNA), as RNA seems to be both an important initiator and a target in these processes. It was shown that co-suppression in *C. elegans* is also probably mediated by RNA molecules. Both RNA interference and co-suppression have been implicated in the silencing of transposons. Mutants of *C. elegans* that are defective in transposon silencing and RNA interference (*mut-2*, *mut-7*, *mut-8* and *mut-9*) are in addition resistant to co-suppression. Thus RNA interference and co-suppression in *C. elegans* may be mediated at least in part by the same molecular machinery, possibly through RNA-guided degradation of the mRNA molecules.[236]

RNA-mediated interference (RNAi) is a recently discovered method to determine gene function in a number of organisms, including plants, nematodes, *Drosophila*, zebra fish and mice. Injection of double-stranded RNA (dsRNA) corresponding to a single gene into an organism silences expression of the specific gene. Rapid degradation of mRNA in affected cells blocks gene expression. Despite the promise of RNAi as a tool for functional genomics, injection of dsRNA interferes with gene expression transiently and is not stably inherited. Consequently, use of RNAi to study gene function in the late stages of development has been limited. It is particularly problematic for development of disease models that rely on postnatal individuals. To circumvent this problem in *Drosophila*, a method was developed to express dsRNA as an extended hairpin-loop RNA. This method has recently been successful in generating RNAi in the nematode *C. elegans*. The hairpin RNA is expressed from a transgene exhibiting dyad symmetry in a controlled temporal and spatial pattern. It was reported that the stably inherited transgene confers specific interference of gene expression in embryos, and tissues that give rise to adult structures such as the wings, eyes and brain. Thus, RNAi can be adapted to study late-acting gene function in *Drosophila*. The success of this approach in *Drosophila* and *C. elegans* suggests that a similar approach may be useful to study gene function in higher organisms for which transgenic technology is available.[237]

RNAi of gene expression facilitates reverse genetic studies in *C. elegans* (reviews references 1–4 of Clemens[238]). RNAi of gene expression in generating knock-out phenotypes for specific proteins in several *Drosophila* cell lines was demonstrated. The technique was used for studying signaling cascades by dissecting the well-characterized insulin signal transduction pathway. Inhibiting the expression of the DSOR1 [mitogen-activated protein kinase kinase (MAPKK)] prevents the activation of the down-stream ERK-A (MAPK). In contrast, blocking ERK-A expression results in increased activation of DSOR1. *Drosophila* AKT (DAKT) activation depends on the IR substrate, CHICO (IRS1–4). Blocking the expression of *Drosophila* PTEN results in the activation of DAKT. In all cases, the interference of the biochemical cascade by RNAi is consistent with the known steps in the pathway. The technique was extended to study two proteins, DSH3PX1 and *Drosophila* ACK (DACK). DSH3PX1 is an SH3, phox homology domain-containing protein and DACK is homologous to the mammalian activated cdc42 tyrosine kinase, ACK. Using RNAi, DACK is demonstrated to be upstream of DSH3PX1 phosphorylation, making DSH3PX1 an identified downstream target/substrate of ACK-like tyrosine kinases. RNAi is useful in dissecting complex biochemical signaling cascades and a highly efficient method for determining function of the identified genes arising from *Drosophila* genome sequencing.[238]

In *A. thaliana*, reverse genetic techniques for isolating mutants corresponding to known sequences, such as antisense suppression, co-suppression by overexpression of the target gene, targeted gene disruption or the PCR approach of screening for T-DNA insertion libraries, have been developed. The widespread identification of differentially expressed genes, homologous genes and interacting proteins has created a need for potent and efficient methods for obtaining their loss-of-function or reduction-of-function mutants. RNAi with expression of specific genes has been observed in a number of organisms including *C. elegans*, plants, *Drosophila*, *Trypanosoma brucei*, and a planarian.

The potential of RNAi with gene activity in *A. thaliana* was investigated. To construct transformation vectors that produce RNAs capable of duplex formation, gene-specific sequences in the sense and antisense orientations were linked, and placed under the control of a strong viral promoter. When introduced into the genome of *A. thaliana* by *Agrobacterium*-mediated transformation, dsRNA-expressing constructs corresponding to four genes, AGAMOUS (AG), CLAVATA3, APETALA1 and PERIANTHIA, caused specific and heritable genetic interference. The severity of phenotypes varied between transgenic lines. *In situ* hybridization revealed a correlation between a declining AG mRNA accumulation and increasingly severe phenotypes in AG (RNAi) mutants, suggesting that endogenous mRNA is the target of RNAi. The ability to generate stably heritable RNAi and the resultant specific phenotypes allows to selectively reduce gene function in *A. thaliana*.[239]

RNA, unlike DNA, folds into a myriad of tertiary structures that are responsible for its diverse function in cells. In addition to its role as a mediator of genetic information from DNA to protein, RNA serves as genetic material (in some viruses), as a structural component of many ribonucleoprotein (RNP) particles and in some cases as a catalytic subunit of RNPs. RNA folds into complex structures that can interact specifically with effector proteins. These interactions are essential for various biological functions. In order to discover small molecules that can affect important RNA–protein complexes, a thorough analysis of the thermodynamics and kinetics of RNA–protein binding is required. This can facilitate the formulation of high-throughput screening strategies and the development of structure–activity relationships for compound leads. In addition to traditional methods, such as filter binding, gel mobility shift assays and various fluorescence techniques, newer methods such as SPR and MS are being used for the study of RNA–protein interactions.[240]

Intriguingly, RNA serves several functions and in addition it was shown that related protein or RNA sequences with the same folded conformation can often perform very different biochemical functions, indicating that new biochemical functions can arise from pre-existing folds. A single RNA sequence was described that can assume either of two ribozyme folds and catalyze the two respective reactions. The two ribozyme folds share no evolutionary history and are completely different, with no base pairs (and probably no hydrogen bonds) in common. Minor variants of this sequence are highly active for one or other of the reactions and can be accessed from prototype ribozymes through a series of neutral mutations. Thus, in the course of evolution, new RNA folds could arise from pre-existing folds, without the need to carry inactive intermediate sequences. This raises the possibility that biological RNAs having no structural or functional similarity might share a common ancestry. Furthermore, functional and structural divergence might, in some cases, precede rather than follow gene duplication.[241]

Regulation at the transcriptional and translational level requires a molecular understanding of the processes, and the design of specific effectors to influence transcription and translation. A tRNA-like property has been speculated for release factors in reading stop codons and experiments were designed to identify putative peptide anticodons equivalent to release factors. The two translational release factors of prokaryotes, RF1 and RF2, catalyze the termination of polypeptide synthesis at the UAG/UAA and UGA/UAA stop codons, respectively. How these polypeptide release factors read both non-identical and identical stop codons is puzzling. The basis of this recognition has now been elucidated. Swaps of each of the conserved domains between RF1 and RF2 in an RF1–RF2 hybrid led to the identification of a domain that could switch recognition specificity. Genetic selection among clones encoding random variants of this domain showed that the tripeptides Pro–Ala–Thr and Ser–Pro–Phe determine release factor specificity *in vivo* in RF1 and RF2, respectively. An *in vitro* release study of tripeptide variants indicated that the first and third amino acids independently discriminate the second and third purine bases, respectively. Analysis with stop codons containing base analogs indicated that the C2 amino group of purine may be the primary target of discrimination of G from A. These findings show that the discriminator tripeptide of bacterial release factors is functionally equivalent to that of the anticodon of tRNA, irrespective of the differences between protein and RNA.[242]

Integrins are a superfamily of heterodimeric transmembrane glyoproteins that function in cellular adhesion, migration, and signal transduction. These receptors consist of an α and β subunit, which associate non-covalently in defined combinations. The integrin αβ, also referred to as the vitronectin receptor, is expressed on a variety of cell types and is thought to play a key role in the initiation and/or progression of several human diseases, including osteoporosis, re-stenosis following percutaneous transluminal coronary angioplasty (PTCA), RA, cancer and ocular diseases. Antagonism of integrin αβ is therefore expected to provide an approach for the treatment and/or prevention of these diseases. A variety of potent, small-molecule αβ antagonists have been identified, several of which are active in disease models, thereby demonstrating the therapeutic potential of αβ antagonism. Recent studies have lead to the identification of small-molecule αβ antagonists and been validated by proof-of-concept studies *in vivo*.[243]

Protein tyrosine kinases have emerged as crucial targets for therapeutic intervention in cancer. Growth factor ligands and their respective receptor tyrosine kinases (RTKs) have been shown to be required for tumor cell growth. The latter aspect includes tumor angiogenesis where the growth of tumors leads to compensatory effects on host cells in the tumor microenvironment leading to the growth of microvessels. Synthetic chemical approaches are used to block RTKs associated with tumor angiogenesis as a means to limit the growth and spread of human tumors.[244]

Numerous medically relevant physiological events rely on glycoconjugates for their viability. The recognition of glycoprocessing enzymes as targets for therapeutic intervention has spurred the development of numerous drug candidates, including the influenza drugs that inhibit influenza virus neuraminidase. The discovery of selectins attracted research to the study of carbohydrate–protein interactions in the inflammatory response. The observation that cell-surface glycoforms are altered in certain cancers has served both as a diagnostic tool and as the foundation of glycoconjugate-based vaccine development. The recognition of complex carbohydrates and glycoconjugates as mediators of important biological processes has stimulated investigation into their therapeutic potential. New approaches for the simplification of glycoconjugate synthesis are overcoming the limitations of existing methods and are providing a diverse array of these biomolecules. As the accessibility of glycoconjugates increases, carbohydrate-based constructs are becoming available for analysis as medical agents in a wide range of therapies.[245]

The biological activity of peptides is of enormous interest to the pharmaceutical industry, but endogenous peptides themselves typically have some limitations regarding bioavailability and oral activity. Peptide mimicry by design used to be touted as a solution to these problems and was focused on impersonating secondary structural motifs, particularly β-turns, but this approach has yielded few pharmaceutical products. The process of identifying and optimizing peptide mimetics is driven mainly by screening to obtain hits, followed by optimization, which might include design based on arranging pharmacophores appropriately in three dimensions. As a consequence, the identification of non-peptide agonists as peptide receptors is being solved involving research in opiates, enzyme active sites, agonists at GPCRs, tyrosine kinase receptors, growth factor and cytokine receptors.[246]

Histocompatibility proteins are cell-surface proteins that bind peptides within the cell and present them for scrutiny by T cells as part of the immune system's role of recognizing and responding to foreign antigens. Class II MHC DR proteins are a set of heterodimers consisting of a monomorphic α and polymorphic β chains that bind peptide antigens in a peptide-binding groove for presentation to helper T cells. Susceptibility to autoimmune disease is strongly associated with specific alleles encoding these proteins. More than 90% of RA patients carry at least one of the class II MHC DRB1 alleles, DRB*0101, 0401, 0404 and 0405. The DRB1 locus affects several aspects of the disease, including time of onset, severity, speed of progression and pathological form.

A heptapeptide was identified with high affinity to RA-associated class II MHC molecules. Using a model of its interaction with the class II binding site, a variety of mimetic substitutions were introduced into the peptide. Several unnatural amino acids and dipeptide mimetics were found to be appropriate substituents and could be combined into compounds with binding affinities comparable to more than a 1000-fold more potent than the original peptide in inhibiting T cell responses to processed protein antigens presented by the target MHC molecules. Peptidomimetic compounds of this type could find therapeutic use as MHC-selective antagonists of antigen presentation in the treatment of autoimmune diseases.[247]

Several ligand discovery techniques have been developed that mimic the process of natural evolution. Phage display technology is the most established of these methods and has been applied to numerous technical problems including the discovery of novel drugs. Some new display technologies have emerged which, unlike phage display, operate entirely *in vitro* and have concomitant advantages. These new methods have the potential to improve the development of new drugs by, for example, the ability to generate molecular probes to the products of genes having a disease association.[248]

Highly diverse molecular libraries have become a vital tool in the search for molecules with novel properties. The antibody molecule in particular has been the focus of extensive development of different combinatorial strategies in order to obtain antibodies with novel specificities. These strategies include the re-assembly of naturally occurring genes encoding the heavy- and light-chain domains from either immune or non-immune B cell sources or introduction of synthetic variability into the individual complementarity-determining regions (CDRs), i.e. the hypervariable parts of the variable domains that interact with an antigen.

To introduce genetic diversity in antibody libraries, an approach consists of combining diverse natural CDRs into an antibody framework, using only CDR-encoding gene fragments originating from sequences formed and 'proofread' *in vivo*. Such proofread peptide sequences that have evolved and developed *in vivo* in a specific biological context may have superior functional qualities compared with those produced *in vitro*. A high degree of functional variation is achieved by means of simultaneous and random combination of six biologically derived CDRs, which can extend the genetic diversity beyond what is naturally created in the immune system. The single master framework into which these CDRs are introduced was selected to be compatible with the bacterial expression system employed. The modular design of the approach allows for the iterative shuffling to improve specificities and affinities, and the contribution of individual CDRs can be analyzed by systematic regional changes.

A single-chain Fv antibody library that permits human CDR gene fragments of any germline to be incorporated combinatorially into the appropriate positions of the variable region frameworks V_H-DP47 and V_L-DPL3. A library of 2×10^9 independent transformants was screened against haptens, peptides, carbohydrates and proteins, and the selected antibody fragments exhibited dissociation constants in the subnanomolar range. The antibody genes in this library were built on a single master framework into which diverse CDRs were allowed to recombine. These CDRs were sampled from *in vivo*-processed gene sequences, thus potentially optimizing the levels of correctly folded and functional molecules, and resulting in a molecule exhibiting a lower computed immunogenicity compared to naive immunoglobulins. Using the modularized assembly process to incorporate foreign sequences into an immunoglobulin scaffold, it is possible to vary as many as six CDRs at the same time, creating genetic and functional variation in antibody molecules.[249]

Seven-membrane segment GPCRs represent 1–2% of the total proteins encoded by the human genome and are important targets for pharmaceutical intervention. Generally low levels of expression and the dependence of the native conformation of GPCRs on the hydrophobic, intramembrane environment have complicated the study of these proteins. Analysis of ligand interactions with GPCRs and screening for inhibitors of such interaction are commonly conducted using live cells or intact cell membranes. Interpretation of these studies may be complicated by the presence of numerous cell surface proteins, many of which are expressed at much higher levels than the GPCR of interest.

The entry of HIV-1 into host cells typically requires the sequential interaction of the gp120 exterior envelope glycoprotein with the CD4 glycoprotein and a GPCR of the chemokine receptor family on the cell membrane. CD4 binding induces conformational changes in gp120 that allow high-affinity binding to the chemokine receptor. The β chemokine receptor CCR5 is the principal HIV-1 co-receptor used during natural infection and transmission. Individuals with homozygous defects in CCR5 are healthy but relatively resistant to HIV-1 infection. Thus, inhibiting the gp120–CCR5 interaction may be a useful therapeutic or prophylactic approach to HIV-1 infection.

Seven-transmembrane segment GPCRs play central roles in a wide range of biological processes, but their characterization has been hindered by the difficulty of obtaining homogeneous preparations of native protein. Paramagnetic proteoliposomes were created containing pure and oriented CCR5, a seven-transmembrane segment protein that serves as the principal co-receptor for HIV-1. The CCR5 proteoliposomes bind the HIV-1 gp120 envelope glycoprotein and conformation-dependent antibodies against CCR5. The binding of gp120 was enhanced by a soluble form of the other HIV-1 receptor, CD4, but did not require additional cellular proteins. Paramagnetic proteoliposomes are uniform in size, stable in a broad range of salt concentrations and pH, and can be used in FACS and competition assays typically applied to cells. Integral membrane proteins can be inserted in either orientation into the liposomal membrane. The magnetic properties of these proteoliposomes facilitate rapid buffer exchange useful in multiple applications. As an example, the CCR5 proteoliposomes were used to select CCR5-specific antibodies from a recombinant phage display library. Thus, paramagnetic proteoliposomes should be useful tools in the analysis of membrane protein interactions with extracellular and intracellular ligands, particularly in establishing screens for inhibitors.[250]

The seven transmembrane receptors (7TMRs) belong to a large superfamily of several thousand proteins and are involved in most major physiological processes. They share a common structure, seven hydrophobic putative transmembrane domains and a common function, the transduction of an extracellular signal to intracellular heterotrimeric G proteins. The constitutive activation of GPCRs is a major new approach to investigating their physiopathology and pharmacology. A large number of spontaneous and site-directed mutations resulting in constitutive activity have been identified, but systematic mapping of the amino acids involved for a given receptor would be useful for complete elucidation of the molecular mechanisms underlying its activation. Such mapping was carried out for the angiotensin II type 1A (AT1A) receptor by screening a randomly mutated cDNA library after expressing the mutated clones in eukaryotic cells. To test the AT1A mutants generated, an original, specific and highly sensitive assay was developed based on the properties of CPG42112A. This classical AT2 agonist is a weak partial agonist of the wild-type AT1A receptor and becomes a full agonist for constitutively active AT1A mutants, as shown experimentally and in allostery-based theoretical models. Activation of the mutated receptors by CGP42112A was monitored by using the bioluminescent protein aequorin, a very sensitive and specific sensor of intracellular calcium mobilization. The screening of 4800 clones, providing an exhaustive coverage of all of the mutations generated, led to the identification of 16 mutations in sequences encoding the transmembrane domains that were responsible for high sensitivity to CPG42112A. The constitutive activity was confirmed by agonist-independent production of inositol phosphates (IP), which showed that at least half of the clones had significantly increased basal activity. This approach is very efficient for the systematic identification of constitutively active mutants of GPCRs.[251]

In vivo genetic analysis is a powerful approach to define the relative contribution of signaling interactions to physiological processes. Small genomes, ease of mutagenesis and rapid generation time have made *D. melanogaster*, *C. elegans,* and *S. cerevisiae* the major models for genetic dissection of eukaryotic signaling pathways.

Modifier screens have been powerful genetic tools to define signaling pathways in lower organisms. The identification of modifier loci in mice has begun to allow a similar dissection of mammalian signaling pathways. Transgenic mice (Btk^{lo}) expressing 25% of endogenous levels of Bruton's tyrosine kinase (Btk) have B cell functional responses between those of wild-type and Btk$^{-/-}$ mice. It was studied whether reduced dosage or complete deficiency of genes previously implicated as Btk regulators would modify the Btk^{lo} phenotype. Two independent assays of Btk-dependent B cell function were used. Proliferative response to B cell antigen receptor cross-linking *in vitro* was chosen as an example of a relatively simple, well-defined signaling system. *In vivo* response to type II T-independent antigens (TI-II) measures complex interactions among multiple cell types over time and may identify additional Btk pathways. All modifiers identified differentially affected these two assays, indicating that Btk mediates these processes via distinct mechanisms. Loss of Lyn, PTEN (phosphatase and tensin homolog) or SH2-containing inositol phosphatase suppressed the Btk^{lo} phenotype *in vitro* but not *in vivo*, whereas CD19 and the p85α form of phosphoinositide 3-kinase behaved as Btk^{lo} enhancers *in vivo* but not *in vitro*. Effects of Lyn, PTEN or p85α haploinsufficiency were observed. Haploinsufficiency or complete deficiency of protein kinase C β, Fyn, CD22, Gαq, or Gα11 had no detectable effect on the function of Brk^{lo} B cells. A transgenic system creating a reduction in dosage of Btk can therefore be used to identify modifier loci that affect B cell responses and quantitatively rank their contribution to Btk-mediated processes.[252]

Table 5.2 gives a list of B cell phenotypes of candidate gene knock-outsA single-chain T cell receptor (scTCR) scaffold with high stability and soluble expression efficiency by directed evolution and yeast surface display was constructed. The scTCRs were evolved in parallel for either enhanced resistance to thermal denaturation at 46 °C or improved intracellular processing at 37 °C, with essentially equivalent results. This indicates that the efficiency of the consecutive kinetic processes of membrane translocation, protein folding, quality control, and vesicular transport can be well predicted by the single thermodynamic parameter of thermal stability. Selected mutations were recombined to create an scTCR scaffold that was stable for over an 1 h at 65 °C, had solubility of over 4 mg/ml and shake-flask expression levels of 7.5 mg/l, while retaining specific ligand binding to peptide–MHC (pMHCs) and bacterial superantigen. These properties are comparable to those for stable single-chain antibodies, but are markedly improved over existing scTCR constructs. Availability of this scaffold allows engineering of high-affinity soluble scTCRs as antigen-specific antagonists of cell-mediated immunity. Moreover, yeast displaying the scTCR formed specific conjugates with APCs, which could allow development of novel cell-to-cell selection strategies for evolving scTCRs with improved binding to various pMHC ligands *in situ*.[253]

NMR spectroscopy is a powerful tool for chemistry and structural biology, especially when NMR is applied to study protein–ligand interactions. In the drug discovery process, NMR is used to study whether a compound binds to a protein up to the determination of the full three-dimensional structure of the complex.[254] Rational drug discovery requires an early appraisal of all factors impacting on the likely success of a drug candidate in the subsequent preclinical, clinical and commercial phases of dug development. The study of absorption, distribution, metabolism and excretion/pharmacokinetics (ADME/PK) has de-

Table 5.2. B cell phenotype of candidate gene knockouts [Refs in 252].

Gene	Viabillty	Knockout phenotype	Ref.
Lyn	Viable	↓B cell numbers, ↓surface IgM levels, ↑BCR responses, ↑serum IgM levels, slightly ↓ T-dependent and T-independent responses, ↑ B-1 cells, autoantibodies with age	24, 28–32
Fyn	Viable	Normal B cell numbers, normal BCR responses, ↓ IL–5 response	33, 34
p85α	Perinatal lethality	Like xid with more severe block at proB to preB stage	35, 36
CD19	Viable	↓B cell numbers, ↓BCR responses, ↓serum Ig levels, ↓T-dependent responses, normal or ↑T-independent responses, ↑ B-1 cells	37–39
Gαq	Viable	No B cell phenotype described	40
Gα11	Viable	No B cell phenotype described	41
SHIP	Viable	↓B cell numbers, ↓surface IgM levels, ↑ BCR responses, ↓ FcãRIIb-mediated inhibition of BCR, ↑ serum Ig levels, normal T-dependent responses, ↑ T-independent responses	42–44
PTEN	Embryonic lethal	In +/- mice: ↑ number of activated B cells (polyclonal expansion), ↑ autoantibodies	45–49
CD22	Viable	↓B cell numbers, ↓surface IgM levels, ↑BCR responses, ↑ serum IgM levels, ↓T-independent responses, normal T-dependent responses, ↑ B-1 cells, autoantibodies with age	38, 50–52
PKCβ	Viable	Like xid	21

The B cell phenotypes of mice deficient in each candidate gene tested are indicated.

veloped into a rather mature discipline in drug discovery through the application of well-established *in vitro* and *in vivo* methodologies. The availability of improved analytical and automation technologies has dramatically increased the ability to dissect out the fundamentals of ADME/PK through the development of increasingly powerful *in silico* methods. This is fuelling a shift away from the traditional empirical nature of ADME/PK towards a more rational, *in cerebro* approach to drug design.[255]

Toxicology will greatly benefit from the application of chip-array technologies, thus reducing the need to consume the large numbers of animals, as is presently the case. The development of toxicogenomics is providing increases in speed and efficiency, thus matching the high-throughput techniques of the drug discovery process.[256]

The existence of human pathogens requires increased efforts to apply the new knowledge and methodologies to these 'extra-human' targets. Besides the development of new antibiotics against bacteria, viruses are a major threat to human health. For example, respiratory syncytical virus (RSV) was originally discovered in 1955 and soon thereafter had been detected in lung secretions from infants with pneumonia and bronchiolitis. The prevalence of RSV became evident in serological studies indicating that 80% of children under 4 years of age possessed neutralizing antibodies to RSV in their serum. RSV has emerged as the leading cause of lower respiratory tract infections in children and infants. The presently only available option for treatment is the teratogen ribavirin, since earlier attempts to develop a prophylactic RSV vaccine have been unsuccessful. Only a passive immunization by means of a monoclonal antibody is available. Thus, there is strong need for the development of small-molecule therapeutics and progress is being made in that direction.[257]

A large number of drugs are currently used therapeutically without knowledge of their target molecules or of the precise pharmacological mechanisms of action. This has resulted not only in complicated side effects and adverse reactions, but has also limited efficacy in certain patients and even in treatment failure. Identifying the target gene's function can allow predictions about the drug's efficacy and can indicate the likelihood of adverse reactions or individual sensitivity even before clinical trials have begun. Additionally, the development of a new drug depends on developing drug target molecules on the basis of their affinity for low-molecular-weight chemicals.

Developments in three crucial areas: chemical diversity, phenotype-based screening, and target identification, enable the identification of compounds inducing a specific cellular state and leading to the identification of the proteins which regulate that state. This integrated chemical genetic approach is applicable to numerous biological or disease processes.

Chemical genetic screens have identified several small molecules targeting a TGF-β-responsive reporter gene, arresting mammalian cells in mitosis, inhibiting the function of tumor suppressor p53, and acting as subtype-selective somatostatin receptor agonist.

The combinatorial synthesis may be carried out on beads to generate a large number of easily separable compounds.

Phenotype-based screens are essential to assay small synthetic molecules and more complex substances for biological activity, such as the enhancement or suppression of a particular phenotype. The different states may be morphologically identifiable, such as the spindly pathologic morphology of cells transformed with oncogenes v-ras and v-src. The pathologic state is reverted by depudecin into the wild-type state.

High-throughput screening requires appropriate test systems to study large numbers of cells simultaneously for the effects of different compounds. The automation of screening assays is facilitateed by visual detection and the combination of optical methods with computational processing of images.

To facilitate drug receptor identification, latex beads were developed that can be adapted to the small-scale purification of drug receptors.

Thus, a method using novel latex beads for rapid identification of drug receptors using affinity purification was developed. Composed of a glycidylmethacrylate (GMA) and a styrene copolmer core with a GMA polymer surface, the beads minimize non-specific protein binding and maximize purification efficiency. Their performance was demonstrated by efficiently purifying FK506-binding protein using FK506-conjugated beads and it was found that the amount of material needed was significantly reduced compared with previous methods. Using the latex beads, a redox-related factor was identified, Ref-1, as a target protein of an anti-NF-κB drug, E3330, demonstrating that the existence of a new class of receptors of anti-NF-κB drugs. The results suggest that the latex beads could provide a tool for the identification and analysis of drug receptors and should therefore be useful in drug development.[258]

Genetic screens in *S. cerevisiae* were designed to identify mammalian non-receptor modulators of G protein signaling pathways. Strains lacking a pheromone-responsive GPCRs and expressing a mammalian–yeast Gα hybrid protein were made conditional for growth upon either pheromone pathway activation (activator screen) or pheromone pathway inactivation (inhibitor screen). Mammalian cDNAs that conferred plasmid-dependent growth under restrictive conditions were identified. One of the cDNAs identified from the activator screen, a human Ras-related G protein that was termed AGS1 (for activator of G protein signaling), appears to function by facilitating guanosine triphosphate (GTP) exchange on the heterotrimeric Gα. A cDNA product identified from the inhibitor screen encodes a previously identified regulator of G protein signaling, human RGS5.[259]

5.1   Substances Derived from Bacteria, Plants, Insects, and Animals

Infectious diseases are making a comeback with menacing new features and increased resistance. Increased effort in anti-infectives research is necessary and on-going. Antibiotics are currently the third-largest selling class of drugs, with a worldwide market annually between US$ 7 and 22 billion. Most current antibiotics are derived from approximately 15 base compounds: β-lactams (e.g., penicillins, methicillin, cephalopsorin), aminoglycosides (e.g., strptomycin, getamycin, neomycin), quinolones (e.g., ciprofloxacin), macrolides (e.g., erythromycin), lincosamides (e.g., clindamycin), sulfonamides (e.g., sulfadiazine), tetracyclins (e.g., glycycline) and glycopeptides (e.g., vancomycin), and are targeting only 15 different bacterial targets.

Only two compounds presently in the clinic act on unconventional targets: oxazolidinones (which inhibit bacterial protein synthesis initiation by binding the 50S ribosomal subunit) and cationic peptides (which permeabilize bacterial membranes).[260]

Drug discovery has huge untapped reservoirs in nature as sources for novel active principles and lead structures. Molecular phylogeny has demonstrated that the majority of microbial genomes are currently inaccessible. Key objectives of future efforts are the developments of techniques for accessing 'unculturable' genomes, exploiting their biotechnologically valuable genes and products, and linking genome sequence data to molecular structure and function.[261]

An intriguing way of actually stimulating nature to deliver new biologically active molecules for drug development has been developed in the field of plants – more specifically by milking plant roots stimulated by bacterial and fungal elicitors.[262]

Easy transformation and cultivation make plants suitable for production of many recombinant proteins. Plants are capable of carrying out acetylation, phosphorylation and glycosylation as well as other post-translational protein modifications required for the biological activity of many eukaryotic proteins. Numerous heterologous (recombinant) proteins have been produced in plant leaves, fruits, roots, tubers and seeds, and targeted to different subcellular compartments, such as the cytoplasm, endoplasmic reticulum or apoplastic space. The extraction and purification of proteins from biochemically complex plant tissues is a laborious and expensive process, and is a major obstacle to large-scale protein manufacturing in plants. Secretion-based systems utilizing transgenic plant cell or plant organs aseptically cultivated *in vitro* have been investigated; however, though these *in vitro* systems are expensive, slow growing, unstable and relatively low yielding.

The problems were addressed by engineering tobacco plants to continuously secrete recombinant proteins from their roots into a simple hydroponic medium. Three heterologous proteins of diverse origins [GFP of jellyfish, human placental alkaline phosphatase (SEAP) and bacterial xylanase] were produced using the root secretion method (rhizosecretion). Protein secretion was dependent on the presence of the endoplasmic reticulum signal peptide fused to the recombinant protein sequence. All three secreted proteins retained their biological activity and, as shown for SEAP, accumulated in much higher amounts in the medium than in the root tissue.[263]

5.2 Sources of Active Principles

The huge reservoir of as-yet untapped molecular structures is evidenced in the microbial world by the assessment of the microbial diversity.[264]

Natural products are an indispensable source for novel structures and chemical diversity for drug discovery.[265] Aquatic sources and plant sources are an underutilized source of active principles or lead compounds.

Extremophiles are a source of evolution-hardened enzymes and principles, such as thermostable enzymes (Taq polymerase for PCR) and cold-adapted enzymes with applications in the detergent and food industries, fine chemicals production, bioremediation as well as broader applicable mechanisms for their high catalytic efficiency based on structural X-ray studies.[266]

Natural products are the most consistently successful source of drug leads. Natural products continue to provide greater structural diversity than standard combinatorial chemistry and offer major opportunities for finding novel low molecular weight lead structures that are active against a wide range of assay targets. Since less than 10% of the world's biodiversity has been tested for biological activity – many more useful natural lead compounds are awaiting discovery. The challenge is how to access this natural chemical diversity. Of the 520 new drugs approved between 1983 and 1994, 39% were natural products or derived from natural products, and 60–80% of antibacterials and anti-cancer drugs were derived from natural products.

Commercial evidence also supports the case for natural products. Of the 20 best-selling non-protein drugs in 1999, nine were either derived from or developed as the result of leads generated by natural products (e.g., simvastatin, lovastatin, ciprofloxacin, clarithromycin and cyclosprin) with combined annual sales of greater than US$ 16 billion. Newer developments based on natural products include the antimalarial drug artemisinin, and the anticancer agents taxol, docetaxel and camptothecin. In addition to the historical success in drug discovery, natural products are likely to continue to be sources of new commercially viable drug leads. The chemical novelty associated with natural products is higher than that of any other source: 40% of the chemical scaffolds in a published database of natural products are absent from synthetic chemistry. This is particularly important when searching for lead molecules against newly discovered targets for which there is no known small-molecule lead. Natural products can be a more economical source of chemical diversity than the synthesis of equivalent numbers of diverse chemicals.[267]

Toxins are a good source of active principles, since they have developed as part of long-term evolutionary strategies for survival, feeding and defense of organisms. Snake venoms provide a rich source of active molecules that address ion channels and receptors with high affinity and selectivity. Since high selectivity, the differentiation between pharmaco-logically relevant targets and targets that have similar structure but different functions, is a key to effective therapeutics, toxins may provide both novel structures for active principles as well as lessons for achieving selectivity.[268]

Studies of physiological responses at the protein level lead to targets for therapies. For example, heat-shock and stress proteins, synthesized as a response to external or immuno-logic stress, provide targets for immunotherapeutic interventions.[269]

Amphibians could help in the search for new substances to treat antibiotic-resistant bacteria such as the methicillin-resistant *Staphylococcus aureus*.

Peptides from the skin of the red-backed salamander (*Plethodon cinereus*) killed 90% of a culture of staphyolococci within 2 h. Peptides may be leads to design and synthesize compounds that act as novel antibiotics.[270]

The repertoire of active molecules has extended to include DNA itself. The therapeutic potential of DNA enzymes was demonstrated with a stretch designed to bind to and cleave the RNA made by a damage-sensing gene called *Egr-1*, which appears to keep ballooned arteries from closing up in rate models of heart disease.[271]

Prior to this, in an experiment of 'directed evolution', a DNA enzyme capable of efficiently cleaving RNA[426] had been generated, which has a higher cleavage efficiency than any known ribozyme. To target a specific RNA molecule, the RNA-cleaving domain (RNA 10–23) is fitted with stretches of DNA that bind to target sequences in the RNA. When the 10–23 catalytic domain is within reach, it cleaves the RNA by speeding up the background reaction making RNA inherently unstable.[272]

The rational design of active pharmaceutical compounds based on identification of genes involved in disease is demonstrated by the identification of compounds that stabilize the DNA binding domain of p53 in the active conformation. These small synthetic molecules not only promote the stability of wild-type p53 but also allowed mutant p53 to maintain an active conformation. A prototype compound caused the accumulation of conformationally active p53 in cells with mutant p53, enabling it to activate transcription and to slow tumor growth in mice. This class of compounds may thus be developed into anticancer drugs of broad utility.[273]

New strategies are also delivering novel compounds in protein engineering. A recombination strategy called iterative truncation for the creation of hybrid enzymes (ITCHY) is based on the generation of N- or C-terminal fragment libraries of two genes by progressive truncation of the coding sequences with exonuclease III followed by ligation of the products to make a single-crossover hybrid library.[427] The studies involved a methodology,

ITCHY, that creates combinatorial fusion libraries between genes in a manner that is independent of DNA homology. The ability of ITCHY and DNA shuffling was compared in creating interspecies fusion libraries between fragments of the *E. coli* and human glycinamide ribonucleotide transformylase genes, which have only 50% identity on the DNA level. Sequencing of several randomly selected positives from each library illustrated that ITCHY identified a more diverse set of active fusion points including those regions of non-homology and those with crossover points that diverged from the sequence alignment. Some of the hybrids found by ITCHY that were fused at non-homologous locations had activities that were greater than or equal to the activity of the hybrids found by DNA shuffling.[274]

Alpha interferons (IFN-α) are members of the diverse helical-bundle superfamily of cytokine genes. Although these proteins possess therapeutic value in the treatment of a number of diseases, they have not been optimized for use as pharmaceuticals. For example, dose-limiting toxicity, receptor cross-reactivity and short serum half-lives significantly reduce the clinical utility of many of the cytokines. DNA family shuffling is a method for permutation of natural genetic diversity. It is a powerful tool for rapidly evolving genes, operons, and whole viruses for desired properties.

DNA shuffling of a family of over 20 human IFN-α (Hu-IFN-α) genes was used to derive variants with increased antiviral and antiproliferation activities in murine cells. A clone with 135,000-fold improved specific activity over Hu-IFN-α2a was obtained in the first cycle of shuffling. After a second cycle of selective shuffling, the most active clone was improved 285,000-fold relative to Hu-IFN-α2a and 185-fold relative to Hu-IFN-α1. Remarkably, the three most active clones were more active than the native murine IFN-α. These chimeras are derived from up to five parental genes but contained no random point mutations. These results demonstrate that diverse cytokine gene families can be used as starting material to rapidly evolve cytokines that are more active, or have superior selectivity profiles, than native cytokine genes.[275]

Natural products are presently a major source of anti-fungal agents and lead structures. New inhibitors of cell wall biosynthesis are a major target for these efforts.[276]

For DNA as the target, a class of compounds called peptide nucleic acids (PNAs) has been developed. PNAs are synthetic homologs of nucleic acids in which the phosphate–sugar polynucleotide backbone is replaced by a flexible polyamide, an uncharged mimic consisting of repeating 2-aminoethyl-glycine units. The resulting decrease in electrostatic repulsion allows the formation of a PNA–DNA hydrogen-bonded double helix, which is more stable than the one formed by DNA–DNA interaction. Peptide nucleic acids are resistant to nucleases and proteases, and consequently are more stable in cells than oligonucleotides. Though potentially capable of blocking gene expression in a selective and specific manner, PNAs have never been shown to be effective anti-gene agents in intact live cells in culture because of their limited ability to reach cell nuclei. A study was carried out to explore the condition s that would make anti-gene PNAs effective in intact cells *in vitro*. PNA was covalently linked to PKKKRKV, a basic NLS peptide. A 17mer anti-*myc*–PNA, complementary to a unique sequence located at the beginning of the second exon of the oncogene, was covalently linked at its N-terminus to the NLS peptide (PNA–mycwt–NLS). When BL cells were exposed to PNA–mycwt–NLS, the anti-gene construct was localized predominantly in the cell nuclei and a rapid consequent down-regulation of c-*myc* expression occurred. Both completion of a productive cell cycle and apoptosis were inhibited.[277]

The backbone of PNA is a polymer of *N*-(2-aminoethyl)glycine (AEG) which is occasionally referred to as ethylendiamine monoacetic acid.

Rational approaches are yielding active pharmaceutical ingredients. For example, nuclear hormone receptors are potential targets for therapeutic approaches to many clinical conditions, including cancer, diabetes and neurological diseases. The crystal structure of the ligand-binding domain of agonist-bound nuclear hormone receptors enables the design of compounds with agonist activity. However, with the exception of the human estrogen receptor-α, the lack of antagonist-bound 'inactive' receptor structures hinders the rational design of receptor antagonists. A strategy was developed for designing such antagonists. A model of the inactive conformation of human retinoic acid receptor-α was constructed by using information derived from antagonist-bound estrogen-receptor-α and a computer-based virtual screening algorithm was applied to identify retinoic acid receptors. The available crystal structures of nuclear hormone receptors may thus be used for the rational design of antagonists, which could lead to the development of novel drugs for a variety of diseases.[278]

The increasing versatility of chemical and biotechnological syntheses are expanding the range of available molecular structures. Of special importance is access to complex carbohydrate and lipid structures since these are part of the cell surface and of membranes involved in transport, recognition and signal transduction. Biological membranes define the boundaries of the cellular compartments in higher eukaryotes and are active in many processes such as signal transduction and vesicular transport. Although post-translational lipid modification of numerous proteins in signal transduction is crucial for biological function, analysis of protein–protein interaction has mainly focused on recombinant proteins in solution under defined *in vitro* conditions. A new strategy was developed for the synthesis of such lipid-modified proteins. It involves the bacterial expression of a C-terminally truncated non-lipidated protein, the chemical synthesis of differently lipidated peptides representing the C terminus of the proteins, and their covalent coupling. The technique is demonstrated using Ras constructs, which exhibit properties very similar to fully processed Ras, but can be produced in high yields and are open for selective modifications. These constructs are operative in biophysical and cellular assay systems, showing specific recognition of effectors by Ras lipoproteins inserted into the membrane surface of biosensors and transforming activity of oncogenic variants after microinjection into cultured cells.[279]

Another example of the power of specifically synthesized carbohydrate structures is in the interplay with receptor-binding proteins such as toxins.

The diseases caused by Shiga and cholera toxin account for the loss of millions of lives each year. Both belong to the clinically significant subset of bacterial AB5 toxins consisting of an enzymatically active A subunit that gains entry to susceptible mammalian cells after oligosaccharide recognition by the B5 homopentamer. Therapies might target the obligatory oligosaccharide–toxin recognition event, but the low intrinsic affinity of carbohydrate–protein interactions hampers the development of low-molecular-weight inhibitors. The toxins circumvent low affinity by binding simultaneously to five or more cell-surface carbohydrates. The use of the crystal structure of the B5 subunit of *E. coli* O157:H7 Shiga-like toxin I (SLT-I) in complex with an analog of its carbohydrate receptor allowed to design an oligovalent, water-soluble carbohydrate ligand (named STARFISH), with subnanomolar inhibitory activity. The *in vitro* inhibitory activity is 1–10 million-fold higher than that of univalent ligands and is by far the highest molar activity of any inhibitor yet reported for Shiga-like toxins I and II. Crystallography of the STARFISH/Shiga-like toxin I complex explains this activity. Two trisaccharide receptors at the tips of each of five spacer arms simultaneously engage all five B subunits of two toxin molecules.[280]

The fortuitous nature of drug discovery is illustrated by the discovery of nitroimidazopyran as a potential therapeutic agent against tuberculosis. *Mycobacterium tuberculosis*, which causes tuberculosis, is the greatest single infectious cause of mortality worldwide, killing roughly two million people annually. Estimates indicate that one-third of the world population is infected with latent *M. tuberculosis* and the surge of multidrug-resistant clinical isolates of *M. tuberculosis* have reaffirmed tuberculosis as a primary public health threat. New antitubercular drugs with new mechanisms of action have not been developed in over 30 years. A series of compounds was reported containing a nitroimidazopyran nucleus that possesses antitubercular activity. After activation by a mechanism dependent on *M. tuberculosis* F420 cofactor, nitroimidazopyrans inhibited the synthesis of protein and cell wall lipid. In contrast to current antitubercular drugs, nitroimidazopyrans exhibited bactericidal activity against both replicating and static *M. tuberculosis*. Lead compound PA-824 showed potent bactericidal activity against multidrug-resistant *M. tuberculosis* and promising oral activity in animal infection models. Nitroimidazopyrans offer the practical qualities of a small molecule with the potential for the treatment of tuberculosis.[281]

Structure-based drug design has emerged as a powerful tool to augment the empirical and semi-empirical drug discovery processes. Successful application of this technique requires a disease target, a feasible therapeutic strategy including an assay system, a site selection strategy and the computational chemistry skills for engineering appropriate lead compounds.

The CD8–MHC class I crystal structure was used as a template for the *de novo* design of low-molecular-weight surface mimetics. The analogs were designed from a local surface region on the CD8 α chain directly adjacent to the bound MHC class I, to block the protein associations in the T cell activation cluster that occur upon stimulation of the CTLs. One small conformationally restrained peptide showed dose-dependent inhibition of a primary allogeneic CTL assay while having no effect on the CD4-dependent mixed lymphocyte reaction. The analog's activity could be modulated through subtle changes in its side chain composition. Administration of the analog prevented CD8-dependent clearance of a murine retrovirus in BALB/c mice. In C57BL/6 mice challenged with the same retrovirus, the analog selectively inhibited the antiviral CTL responses without affecting the ability of the CTLs to generate robust allogeneic responses.[282]

Aberrant apoptosis-mediated cell death is believed to result in a number of different human diseases. For example, excessive apoptosis in the liver can result in fulminant and autoimmune forms of hepatitis. The possibility was studied that inhibition of Fas expression in mice would reduce the severity of fulminant hepatitis. To do this, a chemically modified 2'-*O*-(2-methoxy)ethyl antisense oligonucleotide (ISIS 22023) inhibitor of mouse Fas expression was developed. In tissue culture, this oligonucleotide induced a reduction in Fas mRNA expression that was both concentration- and sequence-specific. In BALB/c mice, dosing with ISIS 22023 reduced Fas mRNA and protein expressions in liver by 90%. The ID_{50} for this response was 8–10 mg/kg daily dosing, and the reduction was highly dependent on oligonucleotide sequence, oligonucleotide concentration in liver and treatment time. Pretreatment with ISIS 22023 completely protected mice from fulminant hepatitis induced by agonistic Fas antibody, by a mechanism entirely consistent with an oligonucleotide antisense mechanism of action. In addition, oligonucleotide-mediated suppression of Fas expression reduced the severity of acetaminophen-mediated fulminant hepatitis, but was without effect on concanavalin A-mediated hepatitis. The results demonstrate that 2′-*O*-(2-methoxy)ethyl-containing antisense oligonucleotides targeting Fas can exert *in vivo* pharmacological activity in liver and suggest that oligonucleotide inhibitors of Fas may be useful in the treatment of human liver disease.[283]

5.3   Assay Systems and Models (e.g., Knock-out Mice)

Crucial to the processes of establishing the safety, efficacy, and mode of action of a drug is the availability and choice of appropriate and validated models for the compound, procedure, test, device, theory, or therapy under investigation.

Knowledge of gene function is required to utilize sequence information to develop new drugs to treat disease. *In vitro* techniques, such as antisense oligonucleotides (AS-ODNs), ribozymes and small molecules, can be used to inhibit gene expression and to provide clues to gene function. Although functional genomics methods analogous to those used *in vitro* exist in organisms such as *D. melanogaster*, *X. laevis* and *C. elegans*, the only methods that can currently be used to analyze gene function in mammals are recombinant transformation and the *cre/lox* conditional knock-out technology. Traditional knock-out technology is limited to the evaluation of a primary phenotype that reflects the first critical requirement of a gene during development. As many primary phenotypes are lethal, there may be little useful information at the end of the knock-out experiment. Secondary phenotypes can provide insight into the role of the gene after its first requirement during development, providing more data as to its function. A technique has been reported to define gene function using AS-ODN inhibition of gene expression in mice. A single intravenous injection of an AS-ODN targeting VEGF into pregnant mice resulted in a lack of primary angiogenesis. The critical window required to inhibit VEGF expression was defined and recapitulated the primary loss of function phenotype observed in VEGF$^{-/-}$ embryos. This phenotype was sequence specific, and time and dose dependent. Injection of an AS-ODN targeting another gene, E-adherin, into pregnant mice at E10 confirmed a hypothesized secondary phenotype. AS-ODN inhibition of gene expression *in utero* provides a strategy for target validation in functional genomics.[284]

Efficient approaches to uncover, confirm and evaluate the physiological functions of genes need to be developed continuously. At the level of molecular genetics, there are two main routes to determine the function of genes with known sequences: one is through loss of function, the other is through gain of function. Gene function must often be studied in the context of the entire organism or in a particular cell type at a specific stage of development or disease. An ideal strategy for functional studies should encompass both effectiveness and versatility. Gene therapy research has provided effective tools to achieve genetic manipulation through somatic gene transfer. In particular, the recombinant adenovirus is a widely used delivery vector for both gene therapy and functional studies.[285]

To accelerate the biological annotation of novel genes discovered in sequenced regions of mammalian genomes, large deletions in the mouse genome are created targeted to include clusters of such genes. The targeted deletion of a 450-kb region on mouse chromosome 11 is described, which, based on computational analysis of the deleted murine sequences and human 5q orthologous sequences, codes for nine putative genes. Mice homozygous for the deletion had a variety of abnormalities, including severe hypertriglyceridemia, hepatic and cardiac enlargement, growth retardation, and premature mortality. Analysis of triglyceride metabolism in these animals demonstrated a several-fold increase in hepatic very-low-density lipoprotein triglyceride secretion, the most prevalent mechanism responsible for hyper-triglyceridemia in humans. A series of mouse BAC and human YAC transgenes covering different intervals of the 450-kb deleted region were assessed for their ability to complement the deletion-induced abnormalities. These studies revealed that OCTN2, a gene recently shown to play a role in carnitine transport, was able to correct the triglyceride abnormalities. The discovery of this previously unappreciated relationship between OCTN2, carnitine and hepatic triglyceride production is of particular importance because of the clinical consequence of hypertriglyceridemia and the paucity of genes known to modulate triglyceride secretion.[286]

Another technique of interfering with and probing of genetic and biochemical networks is based on RNAi. This provides an elegant method of probing gene expression and generating desired genotypes (respectively, phenotypes) for studying genetic mechanisms and for the identification as well as the validation of drug targets. Gene-silencing mechanisms operate in regulating gene expression and cellular differentiation, and contribute to chromosomal dosage compensation, genetic imprinting, virus resistance, transposon silencing and, thus, to normal and pathological genetic and phenotypical phenomena. RNA signals have long been implied in these processes. dsRNA induces sequence-specific interference in several organisms; this process is called RNA interference (RNAi).

In *C. elegans*, the introduction of double-stranded RNA triggers sequence-specific genetic interference (RNAi) that is transmitted to offspring. The inheritance properties associated with this phenomenon were examined. Transmission of the interference effect occurred through a dominant extragenic agent. The wild-type activities of the RNAi pathway genes *rde-1* and *rde-4* genes were required for the formation of this interfering agent but were not needed for interference thereafter. In contrast, the *rde-2* and *mut-7* genes were required downstream for interference. These finding provide evidence for germline transmission of an extragenic sequence-specific silencing factor and implicate *rde-1* and *rde-4* in the formation of the inherited agent.[287]

Pathogenic effects may be caused by mutant mRNA. In the case of myotonic dystrophy (MD, prevalence 1 in 7400 live births), is the most common form of muscular dystrophy in adult humans and results from the expansion of a CTG repeat in the 3′ untranslated region of the DMPK gene. The mutant DMPK mRNA contains an expanded CUG repeat and is retained in the nucleus. An untranslated CUG repeat was expressed in an unrelated mRNA in transgenic mice. Mice that expressed expanded CUG repeats developed myotonia and myopathy, whereas mice expressing a non-expanded repeat did not. Transcripts with expanded CUG repeats are thus sufficient to generate a DM phenotype. This result argues for a role for RNA gain of function in disease pathogenesis.[288]

Epigenetic silencing by dsRNA is a widespread phenomenon for regulating gene expression. This process, RNAi, is thought to involve targeted degradation of homologous mRNAs. In *C. elegans*, seven genes have been shown to be important for RNAi: the *RNA-directed RNA polymerase* homolog *ego-1*, *mut-7*, *rde-2*, *rde-3*, *rde-4*, *mut-2* and *rde-1*, which encodes a member of the eIF2c/zwille family.

To gain insight into the mechanism of degradation, RNAi was examined with respect to the influence of mutations in the *smg* genes, which are required for nonsense-mediated decay. For three of six *smg* genes tested, mutations resulted in animals that were initially silenced by dsRNA but then recovered; wild-type animals remained silenced. The levels of target mRNAs were restored during recovery, and RNA editing and degradation of the dsRNA were identical to those of the wild-type. Thus, persistence of RNAi relies on a subset of *smg* genes.[289]

The B lymphocytes of the immune system use multiple genetic mechanisms – gene rearrangement, somatic hypermutation and gene conversion – to drive the generation of antibody diversity. RNA editing provides another contribution to antibody gene diversification. RNA editing appears to be crucial for the production of the secondary antibody repertoire in mature B cells. A deficiency in a single gene product, activation-induced deaminase (AID), is sufficient to obliterate generation of the secondary antibody repertoire in both human and mouse B cells.

Since both hypermutation and switching are abolished by AID deficiency, AID-directed mRNA editing could be required for these processes. Alternatively, AID-directed editing might occur while the immunoglobulin transcripts are still attached to their genomic template.

AID may therefore turn out to be an attractive drug target in the therapeutic modulation of antibody-dependent autoimmune diseases.[290]

The identification of functional regions in complex proteins is key to understanding their mechanisms of action and is commonly directed by conserved structural motifs. The effect of expressing distinct domains of death-associated protein kinase (DAPK) was analyzed and it was found that overexpression of the entire death domain could protect cells from apoptosis induced by the complete DAPK and that the linker region could protect cells as well, whereas overexpressing the kinase domain had no effect. A random unbiased approach to identify minimal regions in DAPK that are critical for its ability to participate in apoptotic processes was utilized. Functional domains of DAPK were identified by expression and selection of dominant-negative peptides that inhibited the activity of the protein and thus prevented cells from undergoing apoptosis. DAPK is a Ca^{2+}/calmodulin-regulated serine/threonine kinase with a multidomain structure that participates in apoptosis induced by a variety of signals. To identify regions in this protein that are critical for its proapoptotic activity, a genetic screen was performed on the basis of functional selection of short DAPK-derived fragments that could protect cells from apoptosis by acting in a dominant-negative manner. A library of randomly fragmented DAPK cDNA was expressed in

HeLa cells and these cells were treated with IFN-γ to induce apoptosis. Functional cDNA fragments were recovered from cells that survived the selection and those in the sense orientation were examined further in a secondary screen for their ability to protect cells from DAPK-dependent TNF-α-induced apoptosis. Four biologically active peptides were isolated that mapped to the ankyrin repeats, the linker region, the death domain and the C-terminal tail of DAPK. Molecular modeling of the complete death domain provided a structural basis for the function of the death-domain-derived fragment by suggesting that the protective fragment constitutes a distinct substructure. The last fragment, spanning the C-terminal serine-rich tail, defined a new regulatory region. Ectopic expression of the tail peptide (17 amino acids) inhibited the function of DAPK, whereas removal of this region from the complete protein caused enhancement of the killing activity, indicating that the C-terminal tail normally plays a negative regulatory role. This unbiased screen highlights functionally important regions in the protein and reveals an additional level of regulation of DAPK apoptotic function that does not affect the catalytic activity.[291]

Another example of utilizing the mouse system for elucidation of cellular and organismal phenomena is demonstrated in studies of bone formation, resorption and remodeling processes.

RANK (receptor activator of NF-κB) nullizygous mice were generated to determine the molecular genetic interactions between osteoprotegerin, osteoprotegerin ligand and RANK during bone resorption and remodeling processes. RANK$^{-/-}$ mice lack osteoclasts, and have a profound defect in bone resorption and remodeling and in the development of the cartilaginous growth plates of endochondral bone. The osteopetrosis observed in these mice can be reversed by transplantation of bone marrow from *rag1*$^{-/-}$ (recombinase activating gene 1) mice, indicating that RANK$^{-/-}$ mice have an intrinsic defect in osteoclast function. Calciotropic hormones and proresorptive cytokines that are known to induce bone resorption in mice and human were administered to RANK$^{-/-}$ mice without inducing hypercalcemia, although TNF-α treatment leads to the rare appearance of osteoclast-like cells near the site of injection. Osteoclastogenesis can be initiated in RANK$^{-/-}$ mice by transfer of the RANK cDNA back into hematopoietic precursors, suggesting a means to critically evaluate RANK structural features required for bone resorption. Together these data indicate that RANK is the intrinsic cell surface determinant that mediates osteoprotegerin ligand effects on bone resorption and remodeling as well as the physiological and pathological effects of calciotropoic hormones and proresorptive cytokines.[292]

Validation of drugs or therapies requires definite relationships between active agents and targets, be they genes, proteins, enzymes or receptors, preferably on the organismal level.

A new approach was developed for obtaining high resolution *in vivo* imaging of gene expression in opaque animals, previously not possible by existing techniques. Such images were obtained by MRI using an MRI contrast agent that can indicate reporter gene expression in living animals. MRI contrast agents were prepared in which the access of water to the first coordination sphere of a chelated paramagnetic ion is blocked with a substrate that can be removed by enzymatic cleavage. Following cleavage, the paramagnetic ion can interact directly with water protons to increase the MRI signal. An agent was synthesized where galactopyranose is the blocking group. This group renders the MRI contrast agent sensitive to expression of the commonly used marker gene, β-galactosidase. To cellular resolution, regions of higher intensity in the MRI correlate with regions expressing marker enzyme. These studies offer the promise of *in vivo* mapping of gene expression in transgenic animals and validate a general approach for constructing a family of MRI contrast agents that respond to biological activity.[293]

Signaling pathways are the target of a combination of *Drosophila* genetics to identify the genetic basis of diabetes type II and tumors, followed by validation of identified targets in a mouse model.[660]

6 Molecular Pharma-Biotechnology

The increasing availability of DNA sequences, structural data and analyses of interactions is opening a plethora of new targets for therapeutic intervention.

DNA may be the target of sequence changes (gene therapy), or interact with small molecules (RNA, PNA, pharmaceuticals, peptides) or large proteins; RNA may be target for small molecules, and proteins for small molecules or interacting large ones.

The advances in the determination of RNA structure and function have a significant impact on drug discovery and the pharmaceutical industry. RNA serves as a messenger between DNA and proteins and has a surprising complexity in its structure. Both RNA and proteins are potential drug-binding sites, thus the number of potential targets is more than doubled and drugs interacting with RNA may produce effects that cannot be achieved by interaction with proteins.[294]

The logical and necessary successors to the Human Genome Project, which delivers primary sequence data, are comparable efforts in bioinformatics (to understand the meaning of the sequences), in structural studies (to understand the structures and interactions of molecules derived from gene sequences) and the molecular design of interacting (small) molecules as well as (finally?) the overall biological systems for validation, verification and application (e.g., mouse, zebra fish, *Drosophila* and human).[295]

Structural genomics requires methods to determine the three-dimensional structures of proteins and complexes.

Sample damage by X-rays and other types radiation limits the resolution of structural studies on individual biomolecules or cells. Cooling can slow sample deterioration, but cannot eliminate damage-induced sample movement during the time needed for conventional measurements. Analyses of the dynamics of damage formation suggest that the conventional damage barrier (about 200 X-ray photons/Å with X-rays of 12 keV energy or 1 Å wavelength) may be extended at very high dose rates and very short exposure times. Computer simulations were used to investigate the structural information that can be recovered from the scattering of intense femtosecond X-ray pulses by single protein molecules and small assemblies. Estimations of radiation damage as a function of photon energy, pulse length, integrated pulse intensity, and sample size show that experiments using very high X-ray dose rates and ultrashort exposures may provide useful structural information before radiation damage destroys the sample. Such ultra short, high-intensity X-ray pulses from free-electron lasers that are currently under development, in combination with container-free sample handling methods based on spraying techniques, obviating the need for macroscopic crystals, will provide a new approach to structural determinations with X-rays.[296]

The application of genomic technologies to the study of infectious agents is yielding results for anti-infective therapies. The combination of sequencing, bacterial genomics, bioinformatics and proteomics provides targets for drug development and also demonstrates the efficiency of genotyping (susceptibility to infections correlated with specific gene sequences) for drug development and definition of patient subpopulations.[297]

Integrated pathway–genome databases enable and facilitate drug discovery. Such a description of the genes and the genome of an organism as well as its predicted pathways, reactions, enzymes and metabolites provide, in conjunction with visualization and analysis software, the framework for improved understanding of microbial physiology and for antimicrobial drug discovery. Pathway-based analyses of the genomes of a number of medically relevant microorganisms and a novel software tool that visualizes gene expression data on a diagram showing the whole metabolic network of the microorganism have been described.[298]

Signal transduction pathways provide a rich source of drug active principles. The rapid identification of proteins that interact with a novel gene product is an important element of functional genomics. A phage-display-based technique for interactive screening of complex cDNA libraries using proteins or synthetic peptides as bait has been developed. Starting with the epidermal growth factor receptor (EGFR) cytoplasmic tail, known protein interactions were identified that link EGFR to the Ras/MAPK signal transduction cascade and several novel interactions. This approach can be used as a rapid and efficient tool for elucidating protein networks and mapping intracellular signal transduction pathways.[299]

Dynamic protein–protein interactions are a key component of biological regulatory networks. Dimerization events – physical interactions between related proteins – represent an important subset of protein–protein interactions and are frequently employed in transducing signals from the cell surface to the nucleus. Dimerization between different members of a protein family can generate considerable functional diversity when different protein combinations have distinct regulatory properties. A survey of processes known to be controlled by dimerization illustrates the diverse physical and biological outcomes achieved through this regulatory mechanism. These include facilitated proximity and orientation, differential regulation by heterodimerization, generation of temporal and spatial boundaries, enhancement of specificity, and regulated monomer-to-dimer transitions. Elucidation of these mechanisms has led to the design of new approaches to study and to manipulate signal transduction pathways.[300]

6.1 Bioinformatics

Bioinformatics is an interdisciplinary science that includes conceptual and practical methods for the understanding, generation, processing, propagation and integration of biological information. The bioinformatics realm encompasses the input of DNA sequences, RNA sequences, protein sequences and structures plus knowledge of biochemical pathways, cellular and developmental processes, tissue and organismal physiology, normal and pathological processes. It yields knowledge about predicting gene function and structure, protein sequences, three-dimensional structures of proteins and RNAs, and simulations of gene expression networks, and of metabolic and signal transduction networks.[301]

Computational methods are essential in processing the vast data streams from sequencing efforts, structural information, protein analysis and differential gene expression. Computational approaches serve to integrate the research, development and discovery phases. An example of such efforts is the study of drug-likeness with respect to ADME parameters, in particular BBB penetration.[302]

The enormous task of determining the function and cooperation of proteins to create and maintain biological systems draws on various methods, e.g., patterns of co-occurrence identified from fused genes.[303]

Efficiency is achieved by a combination of experiments and computation, i.e. merging the results of expression profiling with DNA micro arrays with computation (co-occurrence, phylogenetic profiles and fused genes).

The large-scale effort to measure, detect, and analyze protein–protein interactions experimentally includes biochemistry (e.g., co-immunoprecipitation or cross-linking), molecular biology (e.g., the two-hybrid or phage display system) and genetics (e.g., unlinked non-complementing mutant detection). Using the two-hybrid system, a concerted effort has been put into the analysis of the whole yeast genome. These approaches tend to be tedious, labor intensive, and inaccurate. From a computational perspective, the question is how to predict that two proteins interact from structure or sequence alone. A method was developed that identifies gene-fusion events in complete genomes, based solely on sequence comparison.

Because there must be selective pressure for certain genes to be fused over the course of evolution, functional associations of proteins can be predicted. In total, 215 genes or proteins in the complete genomes of *E. coli*, *Haemophilus influenza* and *Methanococcus jannaschii* are involved in 64 unique fusion events. The approach is of general applicability and can be applied to genes of unknown function.[304]

Figure 6.1 illustrates the multitude of expertise and interaction in biology.

The sequencing of microbial genomes first focused on pathogenic organisms of medical importance such as *Haemophilus influenza* and *Mycoplasma genitalium* in order to allow researchers to develop drugs and vaccines for these pathogens. The completion elucidation of several non-pathogenic genomes such as *Thermotoga maritima*, whose genome reveals extensive potential lateral transfer with archaea, *Deinococcus radiodurans*, the most radiation-resistant microorganism known, and *Aeropyrum pernix*, the first Crearchaeota to be completely sequenced, enables progress in studies on evolution, environment, industrial synthesis, food technology and medicine.[471]

Table 6.1 lists selected genome sequencing projects of primary interest or application.

Table 6.2 gives selecting databases for extracting physiological and functional in-formation from gene sequences, and Table 6.3 describes the contribution of genomics to non-medical applications of biotechnology.

The availability of over 20 fully sequenced genomes has forced the development of new methods to find protein functions and interactions. Proteins were grouped by correlated evolution, correlated mRNA expression patterns and patterns of domain fusion to determine functional relationships among the 6217 proteins of the yeast *S. cerevisiae*. Using these methods, over 93,000 pair-wise links between functionally related yeast proteins were discovered. Links between characterized and uncharacterized proteins allow a general function to be assigned to more than half of the 2557 previously uncharacterized yeast proteins.

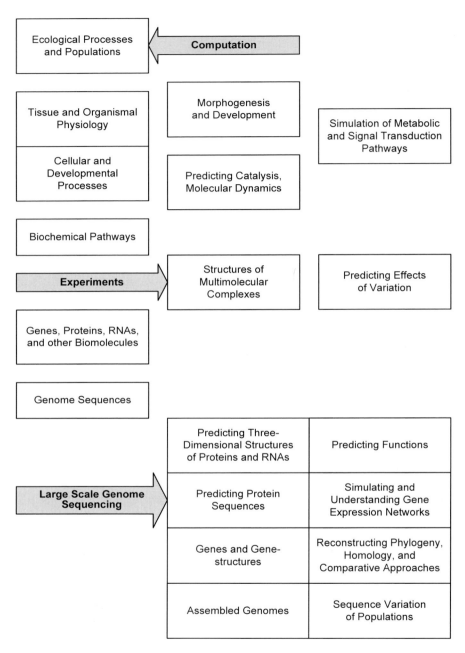

Figure 6.1. The increasing need for centralized simulation and modeling to understand biology. The biology community requires extensive, integrated computational facilities to handle the wealth of data generated by, for example, cDNA micro array analysis [Refs in 301].

Table 6.1. Selected current and past genome sequencing projects for nonpathogenic organisms [Refs in 471].

Organism[a]	Size	Completion	Interest/application[b]
Archaea			
Aeropyrum pernix	1.67	1999	Aerobic hyperthermophile/biocatalysts
Archaeoglobus fulgidus	2.18	1997	Hyperthermophile, sulfate reducer/ oil well problems
Halobacterium sp.	2.50	2000	Halophile/biocatalysts
Halobactefiurn salinarium	4.00	–	Halophile/biocatalysts
Methanococcus jannaschii	1.66	1996	Hyperthermophile, methanogen/ biogas production
Methanococcus maripaludis	?	–	Mesophile, methanogen/biogas production
M. thermoautotrophicum	1.75	1997	Hyperthermophile, methanogen/ waste digestion
Methenogenium fngidum	?	–	Psychrophile, methanogen/biogas production
Methanosarcinia mazei	2.80	1999	Hyperthermophile, methylotroph/ biogas production
Pyrobaculum aerophilum	2.22	2000	Hyperthermophile/sewage digestion/ biogas production
Pyrococcus abyssi	1.80	1999	Hyperthermophile/biocatalysts
Pyrococcus horikoshii	1.80	1998	Hyperthermophile/blocatalysts
Pyrococcus furiosus	2.10	–	Hyperthermophile/biocatalysts
Sulfolobus solfataricus	3.05	–	Hyperthermophile, sulfur-reducer/biocatalysts
Therrnoplasma acid philum	1.70	–	Thermophile, suffur- oxidizer/biomining, biocatalysts
Bacteria			
Aquifex aeolicus	1.50	1998	Hyperthermophile, chemolithoautotroph/ evolution, biocatalysts
Bacillus halodurans	4.25	1999	Alkaliphilic, deep-sea adaptations/biocatalysts
Bacillus subtilis	4.20	1997	Industrial applications
B. stearothermophilus	?	–	Extracellular xylanases
Caulobacter crescentus	3.80	2000	Cell cycle regulation
Chlorobium tepidum	2.10	2000	Photosynthetic, evolutionary implications
Clostridium acetobutylicum	4.10		Solvent production
Clostridium thermocellum	?	–	Degradation of plant polysaccharides
C. glutamicum	3.10	–	Amino acid biosynthesis
Deinocaccus radiodurans	3.20	1999	Radiation resistance/bioremediation
D. ethenogenes	1.50	2000	Tetrachloroethene degradation/bioremediation
Desulfovibrio vulgaris	1.70	2000	Transformation of sulfites/biomining, biocatalysts
Geobacter sulfurreducens	2.50	2000	Iron reduction/biomining, biocatalysts
Nitrosomas europaea	2.20	–	Nitrite removal/bloremediation, biocatalysts
Photorhabdus luminescons	5.50	2000	Biological insecticides/agbiotechnology
Pseudomonas putida	6.20	2000	Aromatic hydrocarbon metabolism/ bioremediation
Rhodobacter capsulatus	3.70	2000	Photosynthesis/bioenergy
Rhodobacter sphaeroides	4.34	–	Photosynthesis/bioenergy

Table 6.1. Selected current and past genome sequencing projects for nonpathogenic organisms (Cont'd).

Organism[a]	Size	Completion	Interest/application[b]
Shewanella oneidensis	4.50	2000	Metal reduction/biomining
Streptomyces coelicolor	8.00	2000	Polyketide synthes/natural products
Synochocystis sp.	3.57	1996	Photosynthesis/bioenergy
7hermotoga maritima	1.86	1999	Hyperthermophile/evolution
Thermus thermophilus	1.82	2000	Themophile/biocatalysts
Thiobacillus ferrooxidans	2.90	2000	Iron oxidation/biomining

a   Abbreviations: *B. stearothermophilus, Bacillus stearothermophilu; C. glutamicum, Corynebacterium glutamicum; D. ethenogenes, Dehalococcoides ethenogenes; M. thermoautotrophicum, Methano-bacterium thermoautotrophicum;*
b   Hyperthermophiles grow at/or above 90 °C. Psychrophiles grow below 18 °C.

Table 6.2. Selected databases for extracting physiological and functional information from gene sequences [Refs in 471].

Database/(URL)	Description
The EcoCyc and MetaCyc databases[30] (http://ecocyc. PangeaSystems.com/ ecocyc/)	Allow the metabolic reconstruction of microbial genomes. EcoCyc provides detailed annotation on 946 metabolic reactions comprising 139 distinct metabolic pathways. EcoCyc also includes information on the signal-transduction pathways, enzymes, and transport proteins of *E. coli*. MetaCyc describes 3,786 pathways and enzymes of many different organisms with a primary microbial focus.
University of Minnesota Biocatalysis/Biodegradation Database (http://www.labmed. umn.edu/umbbd)	Contains ~100 microbial catabolic pathways for xenobioticorganic compounds. Includes information for ~650 reactions, 600 compounds, and 400 enzymes, based on 250 microorganism entries.
The CATH database of protein structures[31] (http://www. biochem.ucl. ac.uk/bsm/dhs)	Contains ~18,000 domains organized according to their class, architecture, topology, and homologous superfamilies.
The HOBACGEN database[32] (http://pbil.univ-lyoul.fr/ databses/hobacgen.html)	Contains all available protein-encoding genes from bacteria, archaea, and yeast classified into families. Also includes multiple, alignments and phylogenetic trees built from these families. The user is allowed to select gene families based on various criteria. and homologous proteins shared among the different taxa. Multiple alignments and trees can also be generated and visualized.
Interpro (http://www.ebi.ac.uk/ interpro/)	An integrated database based on SwissProt, TrEMBL, Pfam, PRINTS and PROSitE, and thus includes data on proteins, protein families, and domains/motifs, providing useful information for predicting protein structure and function for sequenced genomes.

Table 6.2. Selected databases for extracting physiological and functional information from gene sequences (Cont'd).

Database/(URL)	Description
The E-CELL system[33] (http://www.e-cell.org/)	Resource for building models for simulating intracellular molecular processes to predict the dynamic behavior of living cells. The system allows the user to define protein functions, protein-protein interactions, protein-DNA interactions, gene regulation, and other features of cellular metabolism. Dynamic changes in concentrations of proteins and other chemical compounds can also be visualized.

Table 6.3. Contribution of genomics to nonmedical applications of biotechnology

Organism	Biotechnology application and contribution of genomics
Bioremediation	
Dehalococcoides ethenogenes	Bioremediation of tetrachloroethene-contaminated sites. Microarray studies planned to reveal requirements for growth under conditions optimal for bioremediation.
Deinococcus radiodurans	Bioremediation of contaminated sites that are radioactive. *D. radiodurans* can withstand up to 60 Gy/h of radiation. Published sequence has been used to describe compounds capable of restoring growth of *D. radiodurans* in nutritionally limiting radioactive environments.
Pseudomonas putida	Bioremediation of many toxic organic wastes, including aromatic ring-based compounds. Microarray studies underway to more closely characterize *P putida's* metabolic/degradative capacity (B. Tuemmler, personal communication). Suppressive subtractive hybridization (SSH) is also being investigated to characterize differences across many of the 200 or so strains now described, dispensing with the need to completely sequence new strains.
Thermotoga maritima	Biodegradation of simple and complex plant polymers, including xylan and cellulose, with potential as a renewable energy sources. The Institute for Genomic Research (TIGR; Rockville, MD) is using microarrays to investigate the genes involved in carbohydrate and plant polymer metabolism in *T. maritima* strain MSB8.
Enzymology *T.maritime,Aquifex aeolicus,* and *Halobacterium*	Proteins from organisms isolated in extreme *Methanogenium frigidum* environments are chemically and physically more stable and suitable for use in industrial processes. Analysis of genomic sequences will reveal new thermophilic, hyperthermophilic, psychrophilic, and salt-tolerant enzymes of biotechnological potential. *Halobacteria* and P. *putida* also have potential for the production of biodegradable plastics.
Food biotechnology *Lactococcus lactis*	Production of fermented foods, probiotics (microbial dietary supplements). Genomic analysis should yield functional information to enable the design of lactic acid bacteria (e.g., *Lactobacillus* and *Bifidobacterium)* better suited to industrial processes or tailored to provide nutritional benefit.
Natural products	
Streptomyces coelicolor A3(2)	Antibiotic production for human and veterinary medicine and agriculture. Comparative genomic analyses of the streptomycetes may assist in genome engineering of *Streptomyces* sp. to make novel and existing polyketide antibiotics more efficiently.
Photorhabdus luminescens	A source of natural insecticides to complement or substitute for the toxin genes from *Bacillus thuringiensis.* Genomic analysis should yield functional information to enable the engineering of more potent and specific toxins.

Examples of functional links are given for a protein family of previously unknown function, a protein whose human homologs is implicated in colon cancer and the yeast prion Sup35.[305]

An increasing amount of DNA sequence is becoming available with the continuing efforts in deciphering the genomes of complex organisms. Only a small fraction (about 3% of human chromosome 22) encodes proteins. Knowledge of the genetic code makes it feasible to recognize the amino acid-coding parts of genes and to perform sophisticated comparisons. The problem of recognizing gene regulatory sequences was tackled using data collected from microarray studies. The algorithm developed, *MobyDick*, reveals common motifs associated with genes whose expression changes during either sporulatin or general repression in yeast.

The availability of complete genome sequences has facilitated whole-genome expression analysis and led to rapid accumulation of gene expression data from high-density DNA microarray experiments. Correlating the information coded in the genome sequences to these expression data is crucial to understanding how transcription is regulated at the genomic scale. A great challenge is to decipher the information coded in the regulatory regions of genes that control transcription. The development of computational tools for identifying regulatory elements has lagged behind those for sequence comparison and gene discovery. One approach has been to delineate, as sharply as possible, a group of 10–100 co-regulated genes and then find a pattern common to most of the upstream regions. The analysis tools used range from general multiple-alignment algorithms yielding a weight matrix to comparison of the frequency counts of substrings with some reference set.

Although multiple copies of a single sequence motif may confer expression of a reporter gene with a minimal core promoter, real genomes are not designed as disjoint groups of genes each controlled by a single factor. To tailor a gene's expression to many different conditions, multiple control elements are required and signals are integrated. Genome-wide control is considered to be achieved by combinatorial use of multiple sequence elements. Microarray experiments for yeast have shown quantitatively that most occurrences of proven binding motifs are in non-responding genes and that many responding genes do not have the motif.

The availability of complete genome sequences and mRNA expression data for all genes creates new opportunities and challenges for identifying DNA sequence motifs that control gene expression. The algorithm *MobyDick* decomposes a set of DNA sequences into the most probable dictionary of motifs or words. This method is applicable to any set of DNA sequences, e.g., all upstream regions in a genome or all genes expressed under certain conditions. Identification of words is based on a probabilistic segmentation model in which the significance of longer words is deduced from the frequency of shorter ones of various lengths, eliminating the need for a separate set of reference data to define probabilities. A dictionary was built with 1200 words for the 6000 upstream regulatory regions in the yeast genome; the 500 most significant words (some with as few as 10 copies in all of the upstream regions) match 114 of 443 experimentally determined sites (a significance level of 18 standard deviations). When analyzing all of the genes up-regulated during sporulation as a group, many motifs are found in addition to the few previously identified by analyzing the subclusters individually to the expression subclusters. By applying *MobyDick* to the genes de-repressed when the general repressor Tup1 is deleted, both known as well as putative binding sites for its regulatory partners were found.[306]

DNA microarray technology and genome sequencing have advanced to the point that it is now possible to monitor gene expression levels on a genomic scale. These data promise to enhance the fundamental understanding of life on the molecular level, from regulation of gene expression and gene function to cellular mechanisms, and may prove useful in medical diagnosis, treatment, and drug design. Analysis of these data requires mathematical tools that are adaptable to the large quantities of data, while reducing the complexity of the data to make them comprehensible. Analysis has so far been limited to the identification of genes and arrays with similar expression patterns by using clustering methods.

The use of a singular value decomposition in transforming genome-wide expression data from genes × arrays space to reduced diagonalized eigenarrays × eigenarrays space has been described, where the eigengenes (or eigenarrays) are unique orthonormal superpositions of the genes (or arrays). Normalizing the data by filtering out the eigengenes (or eigenarrays) that are inferred to represent noise or experimental artifacts enables meaningful comparison of the expression of different genes across different arrays in different experiments. Sorting the data according to the eigengenes and eigenarrays gives a global picture of the dynamics of gene expression, in which individual genes and arrays appear to be classified into groups of similar regulation and function or similar cellular state and biological phenotype, respectively. After normalization and sorting, the significant eigengenes and eigenarrays can be associated with observed genome-wide effects of regulators or with measured samples in which these regulators are overactive or underactive, respectively.[307]

The pharmaceutical industry has embraced genomics as a source of drug targets and has recognized bioinformatics as crucial for exploiting the data produced by genome-wide sequencing and analysis. Bioinformatics is a cross-disciplinary endeavor including several aspects of molecular biology, biochemistry, cell biology, computer science, software engineering, and techno-mathematics. Bioinformatics is a key aspect of drug discovery in the genomic revolution contributing both to target discovery and target validation. In order to make sense of the mass of data from genome-wide analyses, proteomics and protein function as well as cell biology (cytomics) including expressed sequence tags, model organism sequences, microbial genome sequences, polymorphisms and gene expression data, it is essential that the data be handled in an integrated system-encompassing manner.[308]

Genome-wide transcript profiling has been used to monitor signal transduction during a yeast pheromone response and genetic manipulation allowed analysis of changes in gene expression underlying pheromone signaling, cell cycle control and polarized morphogenesis. This global transcript analysis reflects biological responses associated with the activation and perturbation of signal transduction pathways, and thus demonstrates the power of genome-wide DNA microarrays to give broad correlation of gene activity with alterations in physiological and developmental states.[309]

The assignment of function to novel genes uncovered by systematic genome-sequencing programmes is a major challenge. Part of the necessary repertoire is the analysis of patterns of gene expression via the transcriptome, proteome and metabolome. Functional genomics is partly an exercise in pattern classification and, since many genes have known functional classes, the prediction of their functional class is a learning task. The development of better-structured functional classes will facilitate the prediction of biochemically testable functions.[310]

Knowledge of gene sequences for the prediction of function does not yield descriptions of multiple functional sites. Structural descriptors for protein functional sites are essential for deciphering sequence meaning.[311]

Computational methods for recognizing coding sequences in genomic DNA are capable of detecting most genes. The identification of regulatory elements in the 95% of the genome composed of non-coding sequences is a substantial challenge.

Long-range regulatory elements are difficult to discover experimentally. They tend to be conserved among mammals, suggesting that cross-species comparisons should identify them. To search for regulatory sequences, about 1 Mb of orthologous human and mouse sequences was examined for conserved non-coding elements with greater than or equal to 70% identity over at least 100 bp. Ninety non-coding sequences that met these criteria were discovered and the analysis of 15 of these elements found that about 70% were conserved across mammals. Characterization of the largest element in yeast artificial chromosome transgenic mice revealed it to be a coordinate regulator of three genes, *IL-4*, *IL-13* and *IL-5*, spread over 120 kb.[312]

The availability of whole-genome sequence information allows the identification of protein motifs by sequence alignments, pattern recognition algorithms, protein-folding algorithms, and enables the identification of biologically relevant interacting molecules such as DNA, RNA, protein and small-molecule ligands. Computational methods can be used to identify protein-interaction motifs in open reading frames (ORFs). The coiled-coil interaction motif allows the identification of candidate ligands for other interacting molecules beyond the computational or algorithmic capabilities. The coiled-coil interaction motif consists of two or more α-helices that wrap around each other and the ligands for coiled-coil sequences are generally other coiled-coil sequences which facilitates the motif/ligand recognition determination. A two-step approach was used to identify protein–protein interactions mediated by two-stranded coiled coils that occur in *S. cerevisiae*. Coiled coils from the yeast genome were first predicted computationally by the *multicoil* program and associations were then determined experimentally by using the yeast two-hybrid assay. Between 162 putative coiled-coil sequences, 213 unique interactions were reported and the resulting interactions studied by focusing on associations between components of the spindle pole body (the yeast centromere).[470]

The increasing number of targets for therapeutic interventions requires concepts to deal with the interdependencies of numerous protein–protein, protein–DNA and protein–DNA interactions that are affected by targeting certain receptors, enzymes and control regions. The genetic and biochemical networks, which underlie such processes as homeostasis in metabolism and the developmental programs of living cells must withstand considerable variations and random perturbations of biochemical parameters. These occur as transient changes in, for example, transcription, translation, and RNA and protein degradation. The intensity and duration of these perturbations differ between cells in a population. The unique state of cells, and thus the diversity in a population is due to the different environmental stimuli the individual cells experience and the inherent stochastic nature of biochemical processes. Autoregulatory, negative feedback loops in gene circuits may provide stability, thereby limiting the range over which the concentrations of network components fluctuate. Simple gene circuits were designed and constructed consisting of a regulator and transcriptional repressor modules in *E. coli*, and a gain of stability was shown to be produced by negative feedback.[313]

The study of complex phenotypes is gaining increasing importance with the elucidation of molecular mechanisms in genomics and proteomics. Attempts to influence mesoscopic realms between the genotype and phenotype require an understanding of the networks and interactions of numerous proteins, complexes, effectors, agonists and antagonists in a systematic approach.

Chemotactic behavior and the control of metabolism are examples of complex phenotypes with complicated networks and pathways of signaling and other proteins. In the case of metabolism, the rules describing complex, context-dependent processes depend directly on thermodynamics and kinetics. Analyses of metabolic flux show the principle of distributed control governing the phenotype in mammalian and bacterial systems interacting with the environment, and explain the robust nature of these networks.

The study of network architecture allows the pinpointing of crucial interactions that are amenable to influence by genetic or pharmacological or other means to achieve desired changes in disease phenotypes. The study of complexity in embryology, development, carcinogenesis, metabolism, genetic, and protein regulatory networks is a crucial area for addressing complex traits or multifactorial disease phenomena.[314]

The complexity of intermediary metabolism and the containment of a combinatorial explosion are addressed by looking at the origins of intermediary metabolism as a network of catalytic reactions and molecules by thermodynamic and energetic considerations.

The core of intermediary metabolism in autotrophs is the citric acid cycle. In certain groups of autotrophs, the reductive citric acid cycle is an engine of synthesis, taking in carbon dioxide and synthesizing the molecules of the cycle. The chemistry of a model system of carbon, hydrogen and oxygen that starts with carbon dioxide and reductants and uses redox couples as an energy source was studied. To investigate the reaction networks that might emerge, use was made of the largest available database of organic molecules, i.e. Beilstein online, and pruned by a set of physical and chemical constraints applicable to the model system. From the 3.5 million entries in Beilstein, one comes up with 153 molecules that contain all 11 members of the reductive citric acid cycle. A small number of selection rules generates a very constrained subset, suggesting that this is the type of reaction model that will prove useful in the study of biogenesis. The model indicates that the metabolism shown in the universal chart of pathways may be central to the origin of life, is emergent from organic chemistry and may be unique.[315]

The identification of genetic networks is a crucial basis for bioinformatics and an interesting experimental technique has been developed.

Homeogenes code for homeoproteins, a large family of transcription factors characterized by their highly conserved 60-amino-acid DNA-binding homeodomain. They are highly expressed in the developing and adult nervous system, where they have been linked to many normal and pathological processes. Phenotypic analyses of gene-targeted mouse embryos have been invaluable for understanding homeogene functions during development; however, defining their precise mode of action requires that their target genes be identified.

The identification of the homeoprotein target genes is an important issue in developmental biology. A strategy was developed based on the internalization and nuclear addressing of exogenous homeodomains, using an engrailed homeodomain (EnHD) to screen an ES cell gene trap library. Eight integrated gene trap loci responded to EnHD. One is within the bullous pemphipoid antigen 1 (BPAG1) locus, in a region that interrupts two neural isoforms. By combining *in vivo* electroporation with organotypic cultures, it was shown that an already identified BPAG1 enhancer/promoter is differentially regulated by homeoproteins Hoxc-8 and Engrailed in the embryonic spinal cord and mesencephalon. This strategy can therefore be used for identifying and mutating homeoprotein targets. Because homeodomain third helices can internalize proteins, peptides, phosphopeptides and antisense oligonucleotides, this strategy should be applicable to other intracellular targets for characterizing genetic networks involved in a large number of pathophysiological states.[316]

The study of the genetic network that regulates the development of the *Drosophila* visual system has resulted in the identification of several transcription factors and other nuclear proteins that are required for the specification of early eye morphogenesis. These factors seem to act in a hierarchy in which *sine oculis* (*so*) is regulated directly by *Pax-6*, the master control function. In turn, *so* requires *eyes absent* (*eya*), encoding a nuclear protein, to induce ectopic eyes. This genetic pathway has been established in *Drosophila*, but homologous proteins also regulate eye development in vertebrates, suggesting that this regulatory network is old, is conserved in evolution and has been adapted to the control of development of different visual systems found in both clades. The identification and functional characterization of homologous genes in more primitive organisms, such as the Platyhelminthes, helps to clarify the age and extent of conservation of this genetic cascade.

A *sine oculis* gene in the planarian *Girardia tigrina* (Platyhelminthes; Turbrellaria; Tricladida) was identified. The planarian *sine oculis* gene (*Gtso*) encodes a protein with a *sine oculis* (*Six*) domain and a homeodomain that shares significant sequence similarity with So proteins assigned to the *Six-2* gene family. *Gtso* is expressed as a single transcript in both regenerating and fully developed eyes. Whole-mount *in situ* hybridization studies showed exclusive expression in photoreceptor cells. Loss of function of *Gtso* by RNA interference during planarian development inhibits eye regeneration completely. *Gtso* is also essential for maintenance of the differentiated state of photoreceptor cells. The results and the expression of *Pax-6* in planarian eyes suggest that the same basic gene regulatory circuit required for eye development in *Drosophila* and mouse is used in the prototypic eye spots of platyhelminthes and is truly conserved during evolution.[317]

Several recently developed computational approaches in comparative genomics go beyond sequence comparison. By analyzing phylogenetic profiles of protein families, domain fusions, gene adjacency in genomes and expression patterns, these methods predict many functional interactions between proteins and help deduce specific functions for numerous proteins. Although some of the resultant predictions may not be highly specific, these developments herald a new era in genomics in which the benefits of comparative analysis of the rapidly growing collection of complete genomes will become increasingly obvious.[318]

The classification of organisms was originally based on morphological and anatomical criteria. The data yielded by molecular analysis include DNA sequences, RNA sequences, protein sequences and structures. The comparison of genes and their products with respect to function in functional genomics and with respect to structure in structural genomics is a capability required for biological and pharmaceutical research and development. The application of large-scale gene expression measurement techniques yielded multi-dimensional and multi-relational data thus necessitating tools for analyzing, clustering and visualizing such data. Available packages include the Expression Profiler (www.ebi.ac.uk) and Space Explorer (www.soi.city.ac.uk).

Visualization is an efficient way to utilize the human ability to process large amounts of data. Traditional visualization methods are based on clustering and tree representation, and are complemented by projecting objects onto a Euclidean space to reflect their structural or functional differences. The data are visualized without preclustering and can be explored dynamically and interactively, e.g., in protein topology and gene expression.[463]

Developing models as a basis for simulating interactions and effects is a crucial basis for designing therapeutic intervention. Of particular importance is the modeling of key functional genes and proteins.

Programmed cell death or apoptosis plays a critical role in nature. It is a process by which the cell commits suicide when malfunctions arise from cell stress, cell damage or conflicting division signals. Cell suicide is also required, for instance, during normal embryonic development as tissues are formed. Maintaining the balance between cell death and survival is key. Many cancers are difficult to eradicate because they fail to respond to apoptotic signals. Conversely, neurodegenerative disorders such as Parkinson's, Alzheimer's and Huntingdon's disease are characterized by excessive apoptotic activity in certain classes of neurons. Autoimmune disorders and central immune system phenomena such as the elimination of virus-infected T cells and elimination of active immune cells after successful immune response are also strongly linked with caspase activity. Central to apoptotic cell death is a family of proteases termed caspases (cysteine-containing aspartate-specific proteases).

Under normal circumstances caspases are present as inactive proteins termed zymogens or pro-caspases, which must be activated. Once activated they seek out and dismantle key protein targets by making selective cuts after aspartate residues. The activation of these powerful enzymes (whose inactive forms are constitutively expressed) is regulated at a number of points. A cascade of events must occur before a cell is irreversibly committed to apoptotic death. In response to stress, damage or a signal to undergo programmed cell death, a family of caspases termed initiator caspases is activated. Active initiator caspases then activate a second group of caspases termed executioner or effector caspases. Once activated, executioner caspases seek out and cleave their respective protein targets, thereby dismantling the cell.

Due to the central role of caspases in cancer, and in neurodegenerative and autoimmune disorders, they are subject to intense studies. A mechanistic mathematical model was formulated on the basis of newly emerging information, describing key elements of receptor-mediated and stress-induced caspase activation. Mass-conservation principles were used in conjunction with kinetic rate laws to formulate ordinary differential equations that describe the temporal evolution of caspase activation. Qualitative strategies for the prevention of caspase activation are simulated and compared with experimental data. Model predictions are consistent with available information. The model could aid in better understanding caspase activation and identifying therapeutic approaches promoting or retarding apoptotic cell death.[319]

The *E. coli* MG1655 genome has been completely sequenced. The annotated sequence, biochemical information and other information were used to reconstruct the *E. coli* metabolic map. The stoichiometric coefficients for each metabolic enzyme in the *E. coli* metabolic map were assembled to construct a genome-specific stoichiometric matrix. The *E. coli* stoichiometric matrix was used to define the system's characteristics and the capabilities of *E. coli* metabolism. The effects of gene deletions in the central metabolic pathways on the ability of the *in silico* metabolic networks to support growth were assessed and the *in silico* predictions were compared with experimental observations. It was shown that, based on stoichiometric and capacity constraints, the *in silico* analysis was able to qualitatively predict the growth potential of mutant strains in 86% of the cases examined. The synthesis of *in silico* metabolic genotypes based on genomic, biochemical and strain-specific information is possible, and systems analysis methods are available to analyze and interpret the metabolic phenotype.[320]

The reconstruction of complete metabolic networks is possible from the annotated gene sequence, biochemical and physiological information. The hypothesis that *Escherichia coli* uses its metabolism to grow at a maximal rate was tested by quantitative relationship between a primary carbon source uptake rate, oxygen uptake rate, and maximal cellular growth were found consistent with the hypothesis that the *E. coli* metabolic network is optimized for maximum growth, using the *E. coli* MG1655 metabolic reconstruction. Experiments describing the quantitative relationship between a primary carbon source uptake rate, oxygen uptake rate, and maximal cellular growth were found consistent with the hypothesis that the *E. coli* metabolic network is optimized for maximum growth. The combination of *in silico* and experimental biology delivers quantitative genotype-phenotype relationship information for metabolism in bacterial cells and may be useful for studies of metabolic engineering.[596]

Digital organisms are computer programs that self-replicate, mutate and adapt by natural selection. They offer an opportunity to test generalizations about living systems that may extend beyond the organic life that is the subject of the study of biology. Two classes of digital organisms were generated: (i) simple organisms selected solely for rapid replication and (ii) complex programs selected to perform mathematical operations that accelerate replication through a set of defined metabolic rewards. To examine the differences in their genetic architecture, millions of single and multiple mutations were introduced into each organism and measured for their effects on the organism's fitness. Complex organisms are more robust than the simple ones with respect to the average effects of single mutations. Interactions among mutations are common and usually yield higher fitness than predicted from the component mutations assuming multiplicative effects. Such interactions are especially important in the complex organisms. Frequent interactions among mutations have also been seen in bacteria, fungi and fruit flies. The findings in digital organisms support the notion that interactions are a general feature of genetic systems and may offer a tool for studying such interactive networks.[321]

Analysis of published sets of DNA microarray gene expression data by singular value decomposition has uncovered underlying patterns or characteristic modes in their temporal profiles. These patterns contribute unequally to the structure of the expression profiles. The essential features of a given set of expression profiles are captured using just a small number of characteristic modes. This leads to the striking conclusion that the transcriptional response of a genome is orchestrated in a few fundamental patterns of gene expression change. These patterns are both simple and robust, dominating the alterations in expression of genes throughout the genome. The characteristic modes of gene expression that change in response to environmental perturbations are similar in such distant organisms as yeast and human cells. This analysis reveals simple regularities in the seemingly complex transcriptional transitions of diverse cells to new states and these provide insights into the operation of the underlying genetic networks.[322]

The evolution of complexity needs to be considered when studying the various networks of proteins, genes and cellular interactions. The evolution of genomic complexity was investigated in populations of digital organisms and monitored in detail for the evolutionary transitions that increase complexity. In order to study the evolution of biological complexity in biological evolution, complexity needs to be both rigorously defined and measurable. An information-theoretic definition identifies genomic complexity with the amount of information a sequence stores about its environment. Since natural selection forces genomes to behave as a natural Maxwell Demon, within a fixed environment, genomic complexity is forced to increase.[323]

Studies involving large volumes of high-dimensional data, such as gene sequence data, human or plant gene distributions, global climate patterns and stellar spectra, are confronted with the need for dimensionality reduction, i.e. the identification of meaningful low-dimensional structures hidden in their high-dimensional observations. The human brain permanently confronts this problem in its perception of the environment and extracts a manageably small number of perceptually pertinent features from high-dimensional sensory inputs from 30,000 auditory nerve fibres or 10^6 optic nerve fibers. An approach towards solving dimensionality reduction problems utilizing easily measured local metric information to learn the underlying global geometry of a data set was developed. Unlike classical techniques such as principal component analysis and multidimensional scaling, this approach is capable of discovering the non-linear degrees of freedom that underlie complex natural observations, such as irregularly shaped objects, three-dimensional structures of molecules, human handwriting or images of a face under different viewing conditions. In contrast to previous algorithms for non-linear dimensionality reduction, this algorithm efficiently computes a globally optimal solution and, for an important class of data manifolds, ascertains an asymptotical convergence to the true structure. This approach has applications in the comparison of three-dimensional biological and chemical structures, in high-volume data analysis and real time operations, and optimization of micro- and nano-devices.[464]

Another powerful algorithm was developed for the computation of low-dimensional embedding from high-dimensional inputs. The requirements for analyzing large amounts of multivariate data raises the fundamental problem of dimensionality reduction, the discovery of compact representations of high-dimensional data in exploratory data analysis and visualization. Local linear embedding (LLE) is an unsupervised learning algorithm which computes low-dimensional, neighbourhood-preserving embedding of high-dimensional inputs. Unlike clustering methods for local dimensionality reduction, LLE maps are inputs into a single global coordinate system of lower dimensionality and optimizations do not involve local minima. LLE is able to learn the global structure of non-linear manifolds, such as those generated from three-dimensional objects, representations of multi-dimensional interactions, images of faces and documents of text, based on the exploitation of local symmetries of linear reconstructions.[465]

Bioinformatics is the quest to turn huge volumes of data generated in industrial biology and chemistry research efforts into usable information in basic and applied R & D. The original bioinformatics companies and research groups evolved from biotechnology-oriented or-focussed research and business models. Increasingly, the increasingly required sophistication of information technology (IT), mathematics and software for bio- and chemoinformatics, database management and applications is leading to the movement of IT companies into the field of bioinformatics.[466]

Table 6.4 shows the technology areas where bioinformatics is used and Table 6.5 lists categories of firms issued patents covering bioinformatics technology.

Bioinformatics covers the prediction of genes from sequences, the function of genes, the structure of proteins derived from gene sequences and the function of protein motifs, and thus is faced with an increasing complexity of tasks. Genomics-based drug discovery is strongly dependent on accurate functional annotation, and functional, structural and relationship assignment. Thus, bioinformatics needs to develop and use inter-operable databases and database management, and thus enable the generation of knowledge-based inference and innovation.[467]

Table 6.4. Technology areas (in %) where bioinformatics is used [Refs in 466].

Pharmaceuticals 21	Genomics 12	Agriculture 11	Biologicals 7	Diagnostics 6
Chemicals 5	Environmental 2	Other 36		

Table 6.5. Categories of firms issued patents covering bioinformatics technology [Refs in 466].

Patent assignee	Category	Number of patents	Patent issue date	Area
IT for production *processes*				
Hitachi (Tokyo)	IT company	21	1982–1996	Bioreactors/Biochips
Promavtomatika (Baku City, Russia)	IT company	5	1981–1989	Bioreactors
Sumitomo Electrics (Tokyo)	IT company	3	1984–1988	Bioreactors/Genomics
Kiev Technology Institute for Food Industry (Kiev, Russia)	Academic	3	1979–1988	Bioreactors
Research Institute for Protein Biosynthesis (Russia)	Academic	2	1982–1987	Bioreactors
Toshiba (Tokyo)	IT company	3	1983–1998	Bioreactors

Table 6.5. Categories of firms issued patents covering bioinformatics technology [Refs in 466].

Patent assignee	Category	Number of patents	Patent issue date	Area
IT for R & D processes				
US Dept. of Health and Human Services (Bethesda, MD)	Academic	5	1986–1998	Genomics
Commissariat Atomic Energy (France)	Academic	4	1997–1998	Biochips
Johns-Hopkins University	Academic	3	1995–1998	Biochips
Univ. of Tel Aviv Ramot(Israel)	Academic	3	1995–1998	Genomics
Univ. of Southern California (Los Angeles, CA)	Academic	2	1995–1998	Genomics
Affymetrix (Santa Clara, CA)	Biotech Co.	14	1995–1998	Genomics
Affymax (Palo Alto, CA) IT for R & D processes	Biotech Co.	6	1991–1995	Genomics
BioChip Technologies (Freiburg, Germany)	Biotech Co.	4	1998–1998	Biochips
Human Genome Sciences (Rockville, MD)	Biotech Co.	4	1995–1998	Genomics
Incyte (Palo Alto, CA)	Biotech Co.	3	1994–1998	Genomics
Nexstar Pharmaceuticals (Boulder, CO)	Biotech Co.	3	1995–1997	Biochips
Shimadzu (Tokyo)	Biotech Co.	3	1987–1997	Genomics
Nanogen (San Diego)	Biotech Co.	2	1994–1995	Genomics
Biomerieux (France)	Biotech Co.	2	1998–1998	Genomics
Deutsches Krebs Forschungs-zentrum, (Heidelberg, Germany)	Biotech Co.	2	1996–1997	Genomics
Genset (Paris)	Biotech Co.	2	1998–1998	Genomics
Helix Research Institute (Chiba, Japan)	Biotech Co.	2	1997–1998	Genomics
Hyseq (Sunnyvale, CA)	Biotech Co.	2	1987–1997	Genomics
Eli Lilly (Indianapolis, IN)	Biotech Co.	2	1996–1998	Biochips
Lockheed Martin Energy Systems (Oak Ridge, TN)	Engineering	2	1995–1997	Biochips
SmithKline Beecham (Brentford, UK)	Pharma-ceutical	6	1995–1997	Genomics
Cetus (Emeryville, CA)	Biotech Co.	2	1986–1988	Genomics
DNAstar (Madison, WI)	Biotech Co.	2	1986–1990	Genomics
Genax (Piscataway, NJ)	Biotech Co.	2	1987–1987	Genomics
Tools an solutions for R & D processes				
Motorola (Chicago. IL)	IT company	4	1996–1998	Biochips
Fujitsu (Tokyo)	IT company	3	1991–1998	Genomics
Dainippon Printing (Tokyo)	Conglomerates	2	1989–1998	Genomics
Eastman-Kodak (Rochester,NY)	IT company	2	1995–1995	Biochips
Becton Dickinson (Franklin Lakes, NJ)	Supplier	2	1997–1998	Genomics
Perkin Elmer (Foster City, CA)	Supplier	5	1991–1997	Genomics

To support drug discovery efficiently, the biological information derived in bioinformatics needs to be combined with chemical data in order to facilitate the identification of lead structures and the development of small molecules as active pharmaceuticals. To this end, different databases need to be brought together for the integration of disparate results.[468]

Chemoinformatics needs to be brought to the fore and combined with bioinformatics in order to accelerate the discovery of relationships between structure and biological activity, i.e. cross-database data mining. These activities need, in view of the huge amounts of data, to include advanced visualization techniques for efficient knowledge management.[469]

Iteration between models upon introduction of experimental results aids in verification of hypotheses and the improvement of models.

The studies of interactions in the biological systems comprising genes, proteins, cellsa, tissues and respective networks leads to an integrative approach of biological and biomedical research.

6.2 Biological Systems and Models

Biological systems are the key to validating the results of sequencing, structure analysis and functional predictions.

The models employed range from *in vitro* assays to cell cultures to whole organisms. Increasingly, the use of specific differentiated cells provides new models for understanding differentiation, function, drug discovery and testing plus transplantation and therapies.[324]

Complex biological systems are defined and studied which may provide the tools for studying previously intractable problems. In the search for drugs to fight tumorigenesis, the study of tumor suppressor genes in humans and yeast with its homologs provides such a biological model. The rare familial cancer phenotype called Li–Fraumeni syndrome, which is highly penetrant, is usually associated with inherited mutations in the *TP53* gene and heterozygous germline mutations in the *hCHK2* gene are show to occur. The *hCHK2* gene encodes the human homolog of yeast Cds1 and Rad 53 G_2 checkpoint kinases, whose activation in response to DNA damage prevents cellular entry into mitosis. Thus, the *hCHK2* is a tumor suppressor gene conferring predisposition to sarcoma, breast cancer and brain tumors, and links the central role of p53 inactivation in human cancer and the G_2 checkpoint in yeast. The inactivation of this G_2 checkpoint may have therapeutic implications in rendering cancer cells sensitive to DNA-damaging agents and thus triggering a mitotic catastrophe during the attempt to segregate damaged chromosomes.[325]

Mouse models provide an important system to address studies of disease development, and for the development and testing of therapeutic compounds and strategies.

For cancer studies, several models are available, e.g., to study neurofibromatosis which affects about 1 in 3500 individuals worldwide. Neurofibromatosis type 1 (NF1) is a prevalent familial cancer syndrome resulting from germline mutations in the NF1 tumor suppressor gene. Hallmark features of the disease are the development of benign peripheral nerve sheath tumors (neurofibromas), which can progress to malignancy. Unlike humans, mice that are heterozygous for a mutation in *Nf1* do not develop neurofibromas. Chimeric mice, composed in part of Nf1$^{-/-}$ cells, do, which demonstrates that loss of the wild-type *Nf1* allele is rate limiting in tumor formation.

In addition, mice carrying linked germline mutations in *Nf1* and p53 develop malignant peripheral nerve sheath tumors (MPNSTs), which supports a cooperative and causal role for p53 mutations in MPNST development.[326]

Another resourceful biological system for studying genetic traits and elucidating functional and regulatory pathways is the zebra fish.

Defects in iron absorption and utilization lead to iron deficiency and overload disorders. Adult mammals absorb iron through the duodenum, whereas embryos obtain iron through placental transport. Iron uptake from the intestinal lumen through the apical surface of polarized duodenal enterocytes is mediated by the divalent metal transporter, DMT1. A second transporter has been postulated to export iron across the basolateral surface to the circulation. Positional cloning was used to identify the gene responsible for the hypochromic anemia of the zebra fish mutant *weissherbst*. The gene, *ferroportin1,* encodes a multiple transmembrane domain protein, expressed in the yolk sac, that is a candidate for the elusive iron exporter. Zebra fish *ferroportin1* is required for the transport of iron from maternally derived yolk stores to the circulation and functions as a iron exporter when expressed in *Xenopus* oocytes. Human Ferroportin1 is found at the basal surface of placental syncytiotrophoblasts, suggesting that it also transports iron from mother to embryo. Mammalian Ferroportin1 is expressed at the basolateral of duodenal enterocytes and could export cellular iron into the circulation. Ferroportin1 function may be perturbed in mammalian disorders of iron deficiency or overload.[327]

The *Drosophila* retina is patterned by a morphogenetic wave driven by the Hedgehog signaling protein. Hedgehog, secreted by the first neurons, induces neuronal differentiation and Hedgehog expression in nearby uncommitted cells, thereby propagating the wave. Evidence was reported that the zebra fish Hedgehog homolog, Sonic Hedgehog, is also expressed in the first retinal neurons, and that Sonic Hedgehog drives a wave of neurogenesis across the retina, similar to the wave in *Drosophila*. The conservation of this patterning mechanism supports, in view of the highly divergent structures of vertebrate and invertebrate eyes, a common evolutionary origin of the animal visual system.[328]

Genetic ablation (the elimination of genes in whole organisms by, for example, selective expression of a toxin gene or by genetic deletion) is a powerful technique to study the effects of genes in genetic and biochemical networks and pathways. This technique has yielded valuable insights in neural development and in hormone production.

During neural development in vertebrates, a spatially ordered array of neurons is generated in response to inductive signals derived from localized organizing centers. One organizing center that has been proposed to have a role in the control of neural patterning is the roof plate. To define the contribution of signals derived from the roof plate to the specification of neuronal cell types in the dorsal neural tube, a strategy was devised to ablate the roof plate selectively in mouse embryos. Embryos without a roof plate lack all the interneuron subtypes that are normally generated in the dorsal third of the neural tube. Using a genetically based lineage analysis and *in vitro* assays, it was shown that the loss of these neurons results from the elimination of non-autonomous signals provided by the roof plate. These results reveal that the roof plate is essential for specifying multiple classes of neurons in the mammalian CNS.[329]

The study of receptors and their function may be aided by genetic methods of introducing mutations or deletions in mice or other organisms and observing the phenotypical consequences. Reversing genetic defects by the introduction of transgenes validates the functional importance of the introduced gene and its product.

Targeted deletion of the metabotropic glutamate receptor subtype 1 (*mGluR1*) gene can cause defects in development and function in the cerebellum. The mGluR1 transgene was introduced into *mGluR1*-null mutant (*mGluR1$^{-/-}$*) mice with a Purkinje cell-specific promoter. *mGluR1*-rescue mice showed normal cerebellar long-term depression and regression of multiple climbing fiber innervation – events significantly impaired in *mGluR1$^{-/-}$* mice. The impaired motor coordination was rescued by this transgene, in a dose-dependent manner. MGluR1 in Purkinje cells is a key molecule for normal synapse formation, synaptic plasticity and motor control in the cerebellum.[330]

Olfactory receptors are G protein-coupled seven-transmembrane proteins encoded by a divergent multigene family in vertebrates and in *Drosophila*. Taste receptors are also thought to belong to the superfamily of GPCRs and several candidate genes have been reported, although their function as taste receptor has not been proved. In *Drosophila*, taste sensilla are present on the labelum, tarsi and wing margins. In a typical chemosensillum on the labelum, there are four taste sensory cells, each of which responds to either water, salt or sugar. There are at least three separate receptor sites for sugars in the sugar receptor cell of *Drosophila*. The *Tre* gene was identified through studies on natural variants.

Employing a differential screening strategy, a taste receptor gene, *Tre1*, was identified that controls the taste sensitivity to trehalose in *D melanogaster*. The *Tre1* gene encodes a novel protein with similarity to G protein-coupled seven-transmembrane receptors. Disruption of the *Tre1* gene lowered the taste sensitivity to trehalose, whereas sensitivities to other sugars were unaltered. Overexpression of the *Tre1* gene restored the taste sensitivity to trehalose in the *Tre1* deletion mutant. The *Tre1* gene is expressed in taste sensory cells. The results provide direct evidence that *Tre1* encodes a putative taste receptor for trehalose in *Drosophila*.[331]

The parathyroid glands are the only known source of circulating parathyroid hormone (PTH), which initiates an endocrine cascade that regulates serum calcium concentration. *Glial cells missing2 (Gcm2)*, a mouse homolog of *Drosophila Gcm*, is the only transcription factor whose expression is restricted to the parathyroid glands. It could be shown that *Gcm2*-deficient mice lack parathyroid glands and exhibit a biological hypoparathyroidism, identifying *Gcm2* as a master regulatory gene of parathyroid gland development. Unlike *PTH-receptor*-deficient mice, however, *Gcm2*-deficient mice are viable and fertile and have only a mildly abnormal bone phenotype. Despite their lack of parathyroid glands, Gcm2-deficient mice have PTH serum levels identical to those of wild-type mice, as do parathyroidectomized wild-type animals. Expression and ablation studies identified the thymus, where *Gcm1*, another Gcm homolog, is expressed, as the additional, down-regulatable source of PTH. Thus, *Gcm2* deletion uncovers an auxiliary mechanism for the regulation of calcium homeostasis in the absence of parathyroid glands. This backup mechanism may be a general feature of endocrine regulation.[332]

Based on the similarity of biochemical, phenotypic and genetic regulatory pathways, models may be derived from different spheres of biology and can contribute to research in other fields.

Parkinson's disease is a common neurodegenerative syndrome characterized by loss of dopaminergic neurons in the substantia nigra, formation of filamentous intraneuronal inclusions (Lewy bodies) and an extrapyramidal movement disorder. Mutations in the α-*synuclein* gene are linked to familial Parkinson's disease, and α-synuclein accumulates in Lewy bodies and Lewy neurites. Normal and mutant forms of α-synuclein were expressed in *Drosophila* and produced adult-onset loss of dopaminergic neurons, filamentous intraneuronal inclusions containing α-synuclein and locomotor dysfunction. The *Drosophila* model thus recapitulates the essential features of the human disorder and makes possible a powerful genetic approach to Parkinson's disease.[333]

Another example is the development of a *Drosophila* model for studying the malaria parasite. Malaria is a devastating public health menace, killing over a million people every year and infecting about half a billion. It was shown that the protozoan *Plasmodium gallinaceum*, a close relative of the human malaria parasite *P. falciparum*, can develop in the fruit fly *D. melanogaster*. *P. gallinaceum* ookinetes injected into the fly developed into sporozoites infectious to the vertebrate host with similar kinetics seen in the mosquito host *Aedes aegypti*. In the fly, a component of the insect's innate immune system, the macrophage, can destroy *Plasmodia*. The experiments suggest that *Drosophila* can be used as a surrogate mosquito for defining the genetic pathways involved in both vector competence and part of the parasite sexual cycle.[334]

Biochemical and genetic networks need to be studied experimentally. The understanding of their functioning relies mainly on data collected from populations rather than from single cells. For the observation of phenotypic variability, single-cell measurements become indispensable. Understanding biology at the single-cell level requires simultaneous measurements of biochemical and genetic parameters as well as behavioral characteristics in individual cells.

The power of micro- and nanotechnologies is enhanced by an understanding of the functioning of single cells and subcellular machinery.

The output of individual flagellar motors in *E. coli* was measured as a function of the intracellular concentration of the chemotactic signaling protein. The concentration of this molecule, fused to GFP, was monitored with fluorescence correlation spectroscopy. Motors from different bacteria exhibited an identical steep input–output relation, suggesting that they actively contribute to signal amplification in chemotaxis. These experimental studies can be extended to quantitative *in vivo* studies of other biochemical networks.[335]

The studies have also been extended to single mammalian cells. A method has been demonstrated for the simultaneous measurement of the activation of key regulatory enzymes within single cells. To illustrate the capabilities of the technique, the activation of protein kinase C (PKC), protein kinase A (PKA), calcium-calmodulin-activated kinase II (CamKII) and cdc2 protein kinase (cdc2k) was measured in response to both pharmacologic or physiological stimuli. This assay strategy should be applicable to a broad range of intracellular enzymes, including phosphatases, proteases, nucleases and other kinases.

Rather than microscopic imaging, the method relies on the strengths of chemical separation technologies to identify and quantify fluorescent substrates from a cell. The determination of enzyme activation requires only that the substrate undergo a change in its electrophoretic mobility after being modified by the enzyme. The strengths of the technique include the sensitivity to use physiological to subphysiological concentrations of substrate (nanomolar to micromolar), a temporal resolution (subsecond) appropriate for the chemical reactions involved in cellular signal transduction, applicability to many different types of enzymes and the ability to perform simultaneous measurements of several enzymes within the same cell.[336]

Valuable insights into biochemical and molecular mechanisms for the understanding and identification of drug targets are provided by studies of molecular interactions at the single-molecule level. Using an optical-trap/flow-control video microscopy technique, transcription by single molecules of *E. coli* RNA polymerase was followed in real time over long template distances. These studies reveal that RNA polymerase molecules possess different intrinsic transcription rates, and different propensities to pause and stop. The experiment also shows that reversible pausing is a kinetic intermediate between normal elongation and the arrested state. The conformational metastability of RNA polymerase revealed by the single-molecule study of transcription has direct implications for the mechanisms of gene regulation in both bacteria and eukaryotes.[337]

Single-molecule studies enhance the fundamental understanding of molecular interactions and further the development of rational drug molecules. T7 DNA polymerase catalyzes DNA replication *in vitro* at rates of more than 100 bases/s and has a 3′–5′ exonuclease (nucleotide removing) activity at a separate site. This enzyme possesses a right-hand shape that is common to most polymerases with fingers, palm and thumb domains. The rate-limiting step for replication is thought to involve a conformational change between an open-fingers state in which the active site samples nucleotides and a closed state in which nucleotide incorporation occurs. DNA polymerase must function as a molecular motor converting chemical energy into mechanical focus as it moves over the template. It was shown, using a single-molecule assay based on the differential elasticity of single- and double-stranded DNA, that mechanical force is generated during the rate-limiting step and that the motor can work against a maximum template tension of about 34 pN. Estimates of the mechanical and entropic work done by the enzyme show that T7 DNA polymerase organizes two template bases in the polymerization site during each catalytic cycle. A force-induced 100-fold increase in exonucleolysis above 40 pN was observed.[338]

The suitability of *Caenorhabditis elegans* as a model organism is enhanced by the genome-wide profiling of the nematode and the analysis of gene function, gene expression, and interactions.[689] The usefulness of the zebrafish for studying metabolism is evidenced by studies of the lipid metabolism to identify genes instrumental in vertebrate digestive physiology[670].

6.3 Assay Systems

Assay systems for the identification of biologically active substances in screening programs, for the optimization of lead structures in drug discovery, for tests of efficacy, toxicity and dosage need to correspond to the biological target molecule and mechanism. Validation of the assay system with respect to the biological system for which a pharmacological intervention and therapy is sought is a crucial criterion. A suitable assay system needs to yield reproducible results and should be amenable to automation.

Signal transduction pathways, receptors and ion channels are important targets of assay systems due to their amenability to modulation by therapeutically active compounds, their widespread occurrence and the wide-ranging biological effects they produce.

Figure 6.2 shows schematic representations and flow diagrams illustrating the principals of flux assays suitable for evaluating calcium- and swelling-activated Cl⁻ channel activity plus CFTR.

Transmission of extracellular signals to the interior of the cell is dependent on receptor-mediated assembly of proteins into signaling complexes at the inner side of the plasma membrane. Protein–protein and protein–lipid interactions are essential in these processes, whereby molecular interaction and proximity provide the spatial and temporal conditions for signaling. The hydrophobic phospholipid bilayer is part of the dynamic processes of enrichment and modulation of lipid components. Utilization of fluorescent proteins interacting with lipid components allows the analysis of molecular proximity in intact cells and sheds light on the membrane dynamics of signaling processes.[456]

Research aimed at the identification of novel antiarrythmic agents is targeting Cl⁻ channels. Cl⁻ channels are involved in cardiac arrythmia, asthma, CF and diarrhea, thus representing targets for therapeutic approaches.

Voltage-gated sodium channels (VGSCs) play a central role in the generation and propagation of action potentials in neurons and other cells. Therapeutic substances modulating VGSCs are applied as local anesthetics, antiarrythmics, analgesics and antiepileptics. The identification of a multigene family of VGSCs is enabling research on selective therapeutic interventions. VGSCs are complex membrane proteins expressed inneural, neuroendocrine, skeletal muscle and cardiac cells. They respond to membrane depolarization and enable the influx of sodium ions during the rising phase of the action potential. They are indispensable for the function of the cells as evidenced by the lethal effects of their inhibition by neurotoxins. Diseases of ion channel genes are called channelopathies and are usually dominantly inherited.[458]

Table 6.6 lists voltage-gated sodium channelopathies (disease, mutations and phenotype), Table 6.7 describes the VGSC gene family and Table 6.8 describes ion-channel assay technologies, such as electrophysiology, membrane-potential measurement, flux assays, and binding assays.

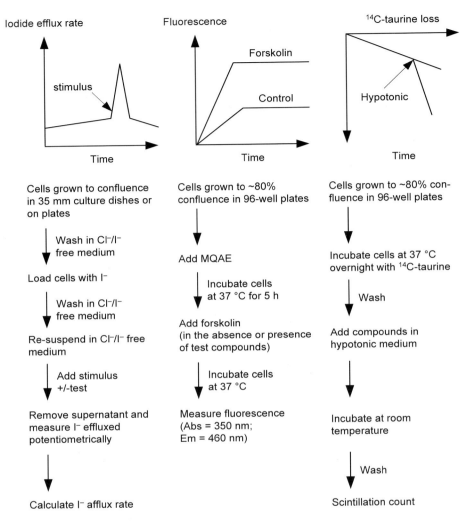

Figure 6.2. Schematic representations and flow diagrams illustrating the principles of flux assays suitable for evaluating Ca^{2+}- and swelling-activated Cl^- channel activity, as well as of the cystic fibrosis transmembrane conductance regulator (CFTR). (a) Measurement of the I^- efflux potentiometrically with an I^- specific electrode[90]. Ca^{2+}-activated Cl^- (CaCC) channels are activated by elevation of the intracellular Ca^{2+} concentration by applying an agonist that elevates the intracellular Ca^{2+} concentration in an appropriate cell type (stimulus). For example, ATP or endothelin can be used to activate such channels in smooth muscle cells by stimulating their receptors[114, 115]. I^- efflux through CFTR can be evoked using forskolin to activate adenylyl cyclase, leading to protein kinase A (PKA)-mediated activation of CFTR channels in an appropriate cell line[116] (e.g., HT29gluCl). From a practical perspective, monitoring I^- efflux through either CaCC channels or CFTR gives identical results with typical responses lasting 10–20 min and reaching a peak after 5–10 min. (b) Alternatively, Cl^- efflux can be measured using a fluorescent Cl^- indicator. In this figure, CFTR is activated by the addition of forskolin. An increase in N- (ethoxycarbomylmethyl)-

6-methoxyquinolinium bromide (MQAE) fluorescence (460 nm) indicates Cl⁻ efflux (because MQAE is quenched by Cl⁻ bindin[81]). Typically, a maximum response is seen after 5–10 min. (c) The principles of using ^{14}C-taurine efflux as a marker for monitoring swelling-activated Cl⁻ channel activity.[83, 85] The figure depicts a typical profile for ^{14}C taurine efflux from HeLa cells, a useful model for this system.[83] Typically, the response takes 10–15 min to reach a level for adequate measurement in the absence of cell volume regulation. Initial efflux occurs in an isotonic medium (red), which is then changed for a hypotonic medium (60% of initial osmolarity: blue) and an increase in ^{14}C-taurine efflux observed. Typically, adding an effective Cl⁻ channel blocker to the hypotonic medium abolishes the increased efflux.[83, 84] For ease of use, the amount of ^{14}C-taurine remaining in the cells can be also measured and related directly to the inhibition of efflux by test compounds [Refs in 457].

Table 6.6. Voltage-gated sodium channelopathies [Refs in 458].

Disease	Sodium channel	Mutations	Phenotype
Long QT syndrome 3	SCN5A (cardiac channel) 1505-1507	Deletion of KQP and length of cardiac causing	Prolonged opening action potential, persistent current
Potassium aggravated myotinia (PAM)	SCN4A (skeletal muscle channel)	6 point mutations in regions associated with inactivation	Slowed decay of transient sodium current
Paramyotonia congentia (PC)	SCN4A (skeletal muscle channel)	5 point mutations in regions associated with inactivation	Slowed decay of transient sodium current
Hyperkalemic periodic paralysis, type II (hyper PP)	SCN4A (skeletal (muscle channel)	5 point mutations in regions associated with inactivation	Slowed decay of transient sodium current
Motor end-plate disease (MED)	SCN8A (type VI channel)	Null	Complete loss of activity, chronic ataxia, dystonia, lethal paralysis
Generalized epilepsy with febrile convulsions (GEFS)	SCN1 B (β1 subunit)	Cysteine to glycine	Slowed inactivation of sodium channel á-subunits

Table 6.7. Voltage-gated sodium channel gene family[a] [Refs in 458].

Channel	Gene name/ other names	HCL	Key tissue distribution	TTX IC$_{50}$ (nM)
Pore-forming α subunit				
Brain type I	SCN1A	2q24	Brain, spinal cord	6
Brain type II	SCN2A	2q23-24	Brain, spinal cord	13
Brain type III	SCN3A	2q24	Brain (embryonic in rat)	4
Brain type VI	SCN8A/NaCH6/PN4	12q13	Brain, spinal cord, glia, DRG	3
Skeletel muscle	SCN4A/SKM1	17q23-25	Skeletal muscle	5
Cardiac	SCN5A/h1	3p21	Heart muscle	2000
PN1	SCNS9A/hNE/NaS	2q22-24	DRG, neuroendocrine cells	4
SNS	SCN10A/PN3	3p21	DRG only	31000
SNS2	SCN11A/NaN/PN5	3p21	DRG only	1500
Atypical heart/	SCN6A/nav2.3	2q21-23	Heart, uterus, lung	?
glial	SCN7A/NaG	DRG, glia		
Auxiliary β subunit				
Beta-1	SCN1B	19p13.1	Brain, muscle, DRG	–
Beta-2	SCN2B	11q22	Brain	–

[a]   Including an "atypical" voltage-gated sodium channel that has been identified in several tissues, including heart, uterus, lung, glia and dorsal root ganglia, but which does not produce sodium currents when expressed in *Xenopus* oocytes Abbreviations: DRG, dorsal root ganglion; HCL, human chromosomal localization; PN1, peripheral neuron 1; SN2, sensory neuron specific 2; SNS, sensory neuron specific; TTX, tetrodotoxin [Refs in 458].

6.4 High-throughput Screening

High-throughput screening is the description of an array of technologies to speed up the testing of compounds from various sources (natural or synthetic) for their parameters such as effects with specified test systems and ADME.

The concept is also being applied for the previous stage in the process, i.e. the attempts to discover novel gene targets by, for example, differential display.[339]

Due to the high pressure on pharmaceutical companies to reduce time-to-market and improve the success rate of new drug candidates, higher-throughput pharmaco-kinetics (HTPK) has become an integral and indispensable part of drug discovery programmes. The amalgamation of robotics, new sample preparation techniques, and highly sensitive and selective mass spectrometric detection systems has contributed to the acceleration in pharmaco-kinetic developments.[340]

Table 6.8. Ion-channel assay technologies [Refs in 458].

Assay	Detection instrument	Throughput data points per week	Gating method	Advantages	Disadvantages
Electrophysiology	Patch-clamp rig	150	Voltage/current command protocols	Gold standard assay, high information content	Low-throughput not easily amenable to automation
Bis-oxonal dye redistribution membrane potential [e.g., $DiBAC(4)_3$]	FLIPR	96-well (15000)/. 384-well (60000)	Toxin to activate channel or delay inactivation. Electrical field stimulation	High-throughput functional assay, amenable to automation	Relatively slow reporting of changes in membrane potential
^{14}C-guanidine flux assay	Scintillation counter	96-well (20000), channel or delay	Toxin to activate flux through inactivation	Measurement of ion rather than relatively high-throughput, amenable to automation	End-point assay, continuous functional recording
Aurora-FRET-based membrane-potential assay	VIPR	96-well (20000)	Toxin to activate channel or delay inactivation	High-throughput functional readout, rapid reporter of changes in membrane potential, amenable to automation, Ratiometric	Proprietary technology available through corporate alliance
Binding assay	Scintillation counter	>30000	Non-functional assay	High-throughput	Very low information content, quality membrane preparations needed, several possible binding sites for drug action

Abbreviations: FLIPR, fluorescence imaging plate reader; VIPR, voltage ion probe reader.

6.5 Automation

Automation is required at all stages of the pharmaceutical, chemical and biotechnological discovery, development and manufacturing process in order to achieve increased speed and throughput. Large central synthesis facilities exist for combinatorial chemistry programs to deliver thousands up to hundreds of thousands of compounds for mass screening. At the laboratory level, the preparation of samples, the dosing of liquids, and the feeding of sequencers, mass spectrometers, assay instruments needs to be carried out with minute volumes, and with high speed and precision around the clock. A crucial component is the efficient data management of the huge amounts of information produced, and the capability to track compounds and results. An automated laboratory, pilot plant, combinatorial synthesis facility, medical chemistry laboratory needs to be designed for automation, integration into the network, documentation, efficient control and data management.[450]

Throughput, speed and data management are combined with a keen interest in containing costs of the development and testing activities. High-throughput screening is now offering solutions to those requirements. Nanovolumes and optical tests enable the testing of the target number of 100,000 compounds per day at affordable costs.[451]

Robotics and automation are applied to synthesis, liquid handling and detection. The interface with humans provides for the flexibility and adaptation to differing tasks and routines. The delicate nature of many assays requires the inclusion of liquid handling, manual transfers, pipetting, heating and cooling, filtration, evaporation, extraction, and chromatography in integrated systems. Laboratory instruments with a low level of automation and intelligence may be attached to a core of flexible central highly automated intelligent robot stations.[452]

The development of miniaturized analytical systems with integration of sample handling, fluidics, measurement and calculations is leading to 'laboratory-on-a-chip' devices. These devices are applicable in laboratory analyses for development and also in the physician's practice. The design and manufacture involves incorporating developments from the microsystems, micro-electronics and opto-electronics industry and technologies, such as lithography, etching, placement of sensors and actuators, and the integration of the various functions. At the micro scales of these devices, there are special techniques applicable for mixing, transportation, and separation, heating and cooling, such as electrophoresis, piezoelectric actuators, diffusion and conductive heating. Microfluidics is a key technology for the design and operation of such integrated devices.[453]

The application of microsystems ('biochips') with highly integrated functions covers the analysis of DNA, protein, DNA–RNA and DNA–protein interactions, biochemical assays, and structure and expression studies.[454]

The development of laboratory-on-a-chip systems will allow routine diagnostics to be carried out at the genetic, biochemical, pharmacological and cellular level, thus enabling the researcher and the physician to study SNPs, pharmaco-genetic variation, metabolization of compounds, genetic predisposition, prenatal diagnostics and tumor cell profiling on individual patients.[455]

6.6 Combinatorial Synthesis: Chemistry, Biology, and Biotechnology

The primary use of combinatorial chemistry is for drug discovery and lead optimization. The technique has also significant potential in the fields of material science, catalyst development and studies of molecular recognition.

The basic study of intermolecular interactions is facilitated by one-bead–one-structure libraries which can be powerful tools for the discovery of ligands to synthetic receptors and vice versa. Encoded combinatorial libraries have been useful for disclosing ligands for well-designed macrocyclic host molecules and to elucidate their specificities for peptide sequences. These studies led via receptors with more flexibility to simple host molecules without elaborate design that are accessible to combinatorial synthesis. One application is the development of chemical sensors for analytes that are otherwise difficult to detect or only non-specifically detected. Such libraries have been used to find new catalysts and enzyme mimics.

Understanding the nature of non-covalent intermolecular interactions and using this knowledge for the rational design of synthetic receptor–ligand pairs is one of the most intriguing tasks in supramolecular chemistry. Combinatorial libraries offer a powerful tool to gather empirical information about host–guest interactions. This may be studied by screening of a ligand library against an individual receptor or a set of receptors, or by the screening of a library of synthetic receptors for binding studies with a set of chosen ligands.[341]

Combinatorial syntheses aim at producing large numbers of candidate substances for further tests and screening. The methods employed are chemical, biological and biotechnological. An example of biological and biotechnological methods is the generation of molecular diversity.[342]

Progress in combinatorial chemistry is aimed at the synthesis and testing of more complex molecules containing intricate ring systems, stereochemical features and elaborate presentations of functional groups – molecules that resemble real drugs and natural products. This involves the recruitment of methods and principles of complex molecule synthesis into combinatorial chemistry, thereby creating hybrid operating systems such as convergent automated parallel synthesis.

Synthetic organic chemistry is a vast, complex and expanding web of interrelated reagents, reactions, procedures, protecting groups, catalysts, purification methods and characterization methods. Chemists seek to achieve productivity gains by leveraging the existing body of knowledge and, to a relatively smaller degree, by adding to this body of knowledge. There are two strategies, i.e. the conceptual and the mechanical, to further such efforts. The conceptual levers are computer-assisted planning, retro-synthetic analysis, convergent pathways, and electronic search tools for reactions, methods and analogies. The mechanical levers and the main drivers of combinatorial chemistry are parallel processing, laboratory automation, solid-phase organic synthesis and split/combine synthesis protocols. The combination of both approaches offers the most potential for fulfilling the requirements for ever more complex molecules.[343]

These approaches and concepts of combinatorial chemistry, permeating the drug discovery and development processes, represent a paradigm shift in drug discovery and basic research. Combinatorial chemistry approaches decrease the time taken for discovery and increase the throughput of chemical screening by as much as 1000-fold. The use of mixture-based synthetic combinatorial libraries is gaining ground. Numerous mixture-based libraries of peptides, peptidomimetics and heterocycles have been synthesized and deconvoluted using the positional scanning approach. Mixture-based library approaches for drug discovery have also shown promising results in vaccine development.[344]

Combinatorial synthesis has been applied to the development of new chiral stationary phases (CSPs) for large-scale chromatographic separations, which are becoming increasingly important due to the complexity of both synthesized as well as natural products.

The technique of chromatographic enantioseparation has become the method of choice for analytical determinations of enantiopurity and the use of preparative chromatographic separation for the manufacture of enantiomerically pure forms of chiral drugs is gaining importance. The technique of simulated moving-bed chromatography is a useful tool for large-scale chromatographic separations using CSPs. Greater enantioselectivity is desirable in order to improve the economics of such separation processes.[345]

Understanding the cellular function of a protein requires either biochemical studies and/or a means to alter the function. This may be done by mutating the gene encoding the protein – the genetic approach (see Chapter 5.3 'Assay Systems and Models'). It may also be done by binding the protein directly with a small-molecule ligand – the chemical genetic approach. In fact both approaches are complementary, and may be necessary to elucidate function and to develop active pharmaceutical principles.

Such small-molecule ligands, primarily natural products or synthetic variants of them or rationally designed molecules or randomly synthesized molecules, can either inactivate or activate protein function. Efficient methods of ligand synthesis and discovery are required to provide a small-molecule partner for every gene product or target.

In addition to synthesizing and screening large numbers of small molecules, synthetic strategies are required which enable the synthesis of vast numbers of compounds with structures reminiscent of natural products and compatible with miniaturized assays. This will be necessary in order to discover non-natural compounds having the binding affinities and specificities characteristic of natural products.[346]

The ability of an enzyme to discriminate among many potential substrates is important for maintaining the fidelity of most biological functions. A method was developed for the preparation and use of fluorogenic substrates that allows for the configuration of general substrate libraries to rapidly identify the primary and extended specificity of proteases. The substrates contain the fluorogenic 7-amino-4-carboxymethylcoumarin (ACC) leaving group. Substrates incorporating the ACC leaving group show kinetic profiles comparable to those with the traditionally used 7-amino-4-methylcoumarin (AMC) leaving group. The bifunctional nature of ACC allows for the efficient production of single substrates and substrate libraries by using 9-fluorenylmethoxycarbonyl (Fmoc)-based solid-phase synthesis techniques. The approximately 3-fold increased quantum yield of ACC over AMC permits a reduction in enzyme and substrate concentrations. As a consequence, a greater number of substrates can be tolerated in a single assay, thus enabling an increase in the diversity space of the library. Soluble positional protease substrate libraries of 137, 180 and 6859 members, possessing amino acid diversity at the P4–P3–P2–P1 and P4–P3–P2 positions, respectively, were constructed. Employing this screening method, the substrate specificities of a diverse array of proteases was profiled, including the serine proteases thrombin, plasmin, Factor Xa, urokinase-type plasminogen activator, tissue plasminogen activator, granzyme B, trypsin, chymotrypsin, human neutrophil elastase, and the cysteine proteases papain and cruzain. The resulting profiles create a pharmacophoric portrayal of the proteases to aid in the design of selective substrates and potent inhibitors.[347]

Soluble polymer supports provide homogeneous reaction conditions, and facilitate product purification through selective precipitation and filtration of the polymer-bound product. The use of linear, non-cross-linked polymers in methods of liquid-phase synthesis avoids the difficulties of solution- and solid-phase methods while preserving the positive aspects of both. The solubility properties of PEG, polyacrylamide, non-cross-linked polystyrene and other soluble supports have allowed for the development of numerous molecules, including peptides, oligonucleotides, oligosaccharides and complex organic compounds. Although the choice of polymer used in liquid-phase methods may be tailored to specific applications, PEG is used most frequently due to its high solubilizing power. Its usage has resulted in the development of liquid-phase combinatorial synthesis, wherein the integration of liquid-phase synthesis with combinatorial chemistry has produced small-molecule libraries of peptides, sulfonamides and neomycin B mimics.[348]

Chemical libraries can be screened for small molecules that affect a variety of cellular processes including signal transduction, the cell cycle and cellular metabolism. In particular, it may be possible to identify molecules that induce cells to differentiate into specific cell types or, conversely, molecules that reverse differentiation or induce regenerative processes. The latter might result from either blockage of pathways that maintain that differentiated phenotype, or the induction of biochemical pathways active during development or wound healing. Molecules that affect the differentiated phenotype may ultimately find use in cellular replacement therapy or tissue engineering and regeneration.

A new microtubule-binding molecule, myoseverin, was identified from a library of 2,6,9-trisubstituted purines in a morphological differentiation screen. Myoseverin induces the reversible fission of multinucleated myotubes into mononucleated fragments. Myotube fission promotes DNA synthesis and cell proliferation after the removal of the compound and transfer of the cells to fresh growth medium. Transcriptional profiling and biochemical analysis indicate that myoseverin alone does not reverse the biochemical differentiation process. Instead, myoseverin affects the expression of a variety of growth factor, immunomodulatory, extracellular matrix-remodeling and stress response genes, consistent with the activation of pathways involved in wound healing and tissue regeneration.[349]

The sequencing of the human genome, and of numerous pathogen and microbial genomes, has resulted in an explosion of potential drug targets. These targets represent both an unprecedented opportunity and a technological challenge for the pharmaceutical industry. A novel strategy is required to implement small-molecule drug discovery with sets of incompletely characterized, disease-associated proteins. Combinatorial chemistry and other technologies serve to discover bioactive small-molecule ligands that act on candidate drug targets. Therapeutically active ligands serve to concurrently validate a target and provide lead structures for further drug development in an accelerated drug discovery process. The future of drug discovery thus holds great promise, largely because of the introduction of an array of new technologies.[350]

The next step in efficiency increase is achieved by integrating combinatorial synthesis and bioassays. Highly parallel automation in drug development requires combinatorial synthesis, parallel analysis and screening. The bottleneck has shifted from the generation of lead structures (or at least compounds) to their transformation into orally active drugs with the desired physiological properties and performance results in clinical trials. The synthesis of large and diverse libraries will have to be integrated with analysis of subsequent steps such as bioavailability, ADME and toxicity.[351]

Identification of new biological targets combined with combinatorial medical chemistry has enormous potential to facilitate the discovery of new therapeutics. These new targets require the use of different or more complex screening libraries. This complexity might be achieved through the use of new templates or the identification of new chemistries. One approach is the use of chiral libraries that have been designed using three-dimensional pharmacophores and shape descriptors to provide maximal structure–activity information after screening.[352]

Combinatorial chemistry is applying a multitude of approaches such as diversity libraries, drug-like libraries and combinatorial subset selection.[353]

Combinatorial chemistry combined with self-assembly processes studied in supramolecular chemistry, yields dynamic combinatorial chemistry (DCC). DCC is useful for the synthesis of compounds binding to specific biological and nonbiological targets by selectively binding to target molecules and being removed from a pool of interconverting compounds.[633]

The solution-phase synthesis of organic compounds offers efficiency advantages which are counter-balanced by the difficulty in separating, identifying, and purifying the resulting components. A strategy for mixture synthesis which resolves the separation problem involves the tagging of a series of organic substrates with fluorous tags of increasing fluorine content. The compounds were mixed, multistep reactions were carried out, and the resulting tagged products were demixed by fluorous chromatography and elution in order of increasing fluorine content to provide the individual pure compoonents which are detagged to obtain the final products.[569]

The increasing importance discovered for carbohydrate structures and their structure-function relationships is necessitating improved techniques in synthesis and analysis of carbohydrates, glycoproteins, and glycopeptides. Progress toward the automated synthesis has been reported and provides compunds for biological and pharmaceutical studies.[640], [570] The analytical methods range from fingerprinting by atomic force microscopy[571] to mass spectrometry determination.[641]

6.7 Genotyping: Genetic Pre-Disposition, and Heterogeneity

SNPs allow us to pinpoint the genes involved in traits and diseases such as asthma, hypertension, obesity, cancers, and diabetes. They will enable both the localization of genes and gene families as well as the development of personalized medicine.

A method for the sensitive detection of single-base mutation in DNA has been described which involves the application of a primer thiolated oligonucleotide, complementary to the target DNA as far as one base before the mutation site, on an electrode or a gold-quartz piezoelectric crystal. After hybridizing the target DNA, normal or mutant, with the sensing oligonucleotide, the resulting assembly is reacted with the biotinylated nucleotide, complementary to the mutation site, in the presence of polymerase. The labeled nucleotide is coupled only to the double-stranded assembly with the mutant site. Faradaic impedance spectroscopy and microgravimetric quartz-crystal microbalance analyses were used for the ellectronic detection of single-base mutants. The lower sensitivity limit for detection of mutant DNA is 1×10^{-14} mol/ml. The method was applied to the analysis of polymorphic blood samples with the Tay-Sachs disorder and allows the quantitaive analysis of the mutant with no PCR-pre-amplification.[560]

The International SNP Map Working Group has prepared a map which identifies and localizes 1.42 million single-nucleotide polymorphisms (SNPs) throughout te genome, most of which are in noncoding regions. This list future and expanded versions provide the foundation for the biomedical reseach focus on the systematic characterization of individual gene products and the effects of sequence variants on function.[561]

In 1999 several companies joined to form the SNP Consortium to create a database of several 100,000 SNP markers and make the available to researchers. The companies and organizations of the SNP Consortium presently are:

AstraZeneca	Glaxo Wellcome
Aventis	Motorola
Amersham Pharmacia Biotech	Novartis
Bayer	Pfizer
Bristol-Myers Squibb	Searle (Monsanto)
Hoffmann-La Roche	SmithKline Beecham
IBM	Wellcome Trust

The International SNP Map Working Group has prepared a map which identifies and localizes 1.42 million single-nucleotide polymorphisms (SNPs) throughout te genome, most of which are in noncoding regions. This list future and expanded versions provide the foundation for the biomedical reseach focus on the systematic characterization of individual gene products and the effects of sequence variants on function.[561]

SNPs and small insertions or deletions (grouped together as SNPs) are the largest set of sequence variants in most organisms. They enable the identification (respectively, localization) of human traits, QTLs and disease traits.[354] SNPs spaced at distances of 30 kb allowed the localization of several diseases, i.e. psoriasis, migraine, Alzheimer's and diabetes.

The revolution in genetics has led to the determination of the precise genetic basis of common and uncommon genetic hereditary diseases. The first fruits of this revolution are diagnostic – the ability to determine who is and who is not at risk for disease before the onset of symptoms. Such information is becoming indispensable for proper the management of patients and their families. In individuals who inherit mutant genes, simple preventive measures often can reduce morbidity and mortality, and allow more thoughtful planning for the future. The benefits of genetic testing are equally important for those family members who are found not to carry the relevant mutation; these individuals are spared unnecessary medical procedures and tremendous anxiety.

The problems of genetic testing are psychological or technical in nature. Issues related to insurance, employment discrimination and privacy have received much attention. Additional ethical concerns arise when no effective intervention is available and when prenatal testing is considered for diseases with late-onset or minimal effects.

There are significant technical challenges for genetic testing as well. In many diseases, not all of the genes capable of causing or contributing to pathogenesis are known. Even when the mutated gene is known, routine genetic testing may fail to identify mutations in many cases (25–75% or more).

There are nearly 1000 different hereditary diseases for which the causative genes are known and there are numerous ways in which genes can be mutated. The ease of detection of these mutations covers a whole spectrum from compliant mutations to refractory mutations. Compliant mutations are identifiable by using DNA sequencing or PCR. Refractory mutations involve deletions of one or more exons, insertions, translocations or alterations which affect the expression of the gene at the RNA or protein level.

Direct detection of a genetic variant is done by either DNA sequencing or micro array analysis. Even using DNA sequencing, mutations affecting only one allele may be difficult to distinguish from normal variations of sequencing baselines. Therefore, in thorough genetic testing, both strands of DNA are being sequenced.

A number of indirect methods have been developed for detecting mutations, such as single-strand conformation polymorphism (SSCP) and denaturing gradient gel electrophoresis (DGGE), which exploit the differential electrophoretic migration of nucleic acids that vary by as little as a single base. DHPLC is a related technique that detects a variation in structure between mutant and wild-type molecules but uses HPLC instead of gel electrophoresis for separation. Another group of indirect methods involves chemical or enzymatic cleavage of the DNA, which exploits the bulges or bends in the DNA duplex that are created by mismatched mutations. Alterations in protein structure resulting from certain genetic alterations may be used for analysis. Nonsense mutations, frameshifts and skipped exons are revealed by Western blot analysis or through analysis of proteins synthesized *in vitro* from PCR products.

Refractory mutations are difficult to detect due to the nature of the molecular defects. If the genomic region is deleted from the mutant allele, the PCR product from genomic DNA will lead to a false conclusion that this gene region is wild-type. Other mutations affect the expression or processing of mRNA from the affected allele, through mutations of promoter sequences, 5′ or 3′ untranslated regions or introns. These regions often comprise genomic segments 10–1000 times as large as the coding regions of the gene and may not be easily or practically examined.

Such mutations may be detected by way of their effects. Intronic mutations that affect splicing can be revealed through the analysis of RNA. Polymorphisms (benign sequence variants) within a mRNA transcript can be used to assess relative levels of expression of the two alleles of the gene. Deletions of one or a few exons can be detected by quantitative hybridization, quantitative PCR or Southern blotting. Large deletions should be detectable with FISH methods. A method called Conversion may be used to simplify the detection of refractory mutations.

Patient cells are fused with a specially designed rodent cell line, creating hybrids that stably retain a subset of the human chromosomes. About one-quarter of the derived hybrids contain a single copy of any human chromosome of interest. The diploid nature of the human genome is thereby converted to a haploid state in which mutations are easier to detect because they are not accompanied by the normal sequence of the wild-type allele. Conversion is not a substitute for the detection methods described above, but rather a technique that provides improved templates and increased sensitivity.

The benefits of genetic testing contribute to genetic counseling as well as to basic research and therapeutic development. Gene-hunting efforts may lead to the discovery of new genes that help in the elucidation of pathogenesis. The examination of families affected by the Li–Fraumeni syndrome (characterized by a marked susceptibility to cancer) without mutations in the tumor suppressor gene p53 were found to have mutations in Chk2, a checkpoint-regulating gene with protein kinase activity, thus providing insights in regulation and function of p53.[355]

Table 6.9 shows different types of mutation correlated with increasing difficulty of detection and examples of refractory mutations.

Table 6.9. Types of genetic mutations in hereditary diseases. Mutations are listed by ease of detection from compliant to refractory. Practicality, the relative ease of detecting a given mutation type will depend on the specific genetic alterations present and the detection methods used [Refs in 355].

Type of mutation	Typical ease of detection
Nonsense or frameshift mutation in coding sequence	
Missense mutation in coding sequence	Compliant
Intronic mutations affecting splicing	
Interchromosmal rearrangements	
Intrachromosomal rearrangements	
Single exon deleted	
Several contiguous exons deleted	
Entire gene deleted	
3′UTR mutations affecting transcript levels	
Intronic mutations affecting transcript levels	Refractory
Promoter mutations affecting transcript levels	

Mutation discovery and genetic profiling at the protein/peptide level has been demonstrated by MS analysis of N-terminally tagged test peptides generated in a coupled *in vitro* transcription/translation reaction from a PCR product of any continuous region of coding sequence. Truncations and amino acid substitutions in peptides coded for by the breast cancer susceptibility gene *BRCA1* were readily identified with this method.[356]

The analysis of mutations that cause genetic diseases and their diagnosis requires high-throughput methods for the identification of many different targets in an efficient, multiplexed manner. High-throughput sequence analysis is a powerful tool for both population-based genetic assessment and rapid cost-effective diagnostic tests. DNA chips (microarrays) readily adapt to the parallel format necessary to screen many samples for many mutations simultaneously.

These DNA chips are arrays of oligonucleotide probes which are spatially synthesized using either masking techniques or liquid dispersing methods. Major goals are the further reduction of overall substrate and individual feature sizes, increasing the simplicity and reducing the cost of fabrication, and elimination of target labeling. A randomly oriented fiber optic gene array for rapid, parallel detection of unlabeled DNA targets with surface immobilized molecular beacons that undergo a conformational change accompanied by a fluorescence change in the presence of a complementary DNA target has demonstrated the selective detection of genomic CF-related targets with an optical encoding scheme and an imaging fluorescence microscope system.[357]

The pharmaceutical world is realizing the fact that genetic variation is the key to identifying the biological basis behind both susceptibility to disease and response to drugs. Genetic variation within drug targets is common and genetic factors influence almost every human disease. The study of SNPs is crucial for characterizing molecular targets and also to validate the role of these targets in disease. Genetic variation as exemplified by SNPs will thus affect target, characterization and validation, and pharmaco-genetics. Examples are genetic variation affecting the response to pravastatin, SKF38393, neuroleptics, tacrine, sensitivity to chemotherapy in breast cancer and asthma treatment by interfering with 5-lipoxygenase.[358]

The study of SNPs offers a way to specifically correlate genotype with (clinical) phenotype at the individual level, thus leading to a very personalized treatment or medicine. In order to achieve that goal, gene-specific sequencing and the cataloguing of SNPs plus correlation to the phenotype is necessary.[359]

Systematic characterization of RNA splicing alterations using differential analysis of transcripts with alternative splicing (DATAS) aids in quantitative gene profiling.[360]

High-throughput SNP scoring from unamplified genomic DNA (an alternative to PCR-based essays) utilizes a structure-sAbsific 5'-nuclease (or flap nuclease) to clieve a 5'-flap sequence when two synthetic oligonucleotide probes hybridize in tandem to the target sequence[361]. Flap endonucleases (FENs) isolated from archaea are shown to recognize and cleave a structure formed when two overlapping oligonucleotides hybridize to a target DNA strand. The downstream oligonucleotide probe is cleaved and the precise site of cleavage is dependent on the amount of overlap with the upstream oligonucleotide. The use of thermostable archeal FENs was demonstrated to allow the reaction to be performed at temperatures that promote probe turnover without the need for temperature cycling. The resulting amplification of the cleavage signal enables the detection of specific DNA targets at subattomole levels within complex mixtures. Evidence was provided that this cleavage is sufficiently specific to enable discrimination of single-base differences and can differentiate homozygotes from heterozygotes in single-copy genes in genomic DNA.[362]

An improved sensitivity version of the invasive signal amplification assay was described with a limit of detection (LOD) of zeptomole (10^{-21} mol) levels of a target DNA in homogeneous format.

The invasive signal amplification reaction has been developed for quantitative detection of nucleic acids and discrimination of SNPs. The improved method couples two invasive reactions into a serial isothermal homogeneous assay using fluorescence resonance energy transfer detection. The serial version of the assay generates more than 10^7 reporter molecules for each molecule of target DNA in a 4-h reaction; this sensitivity, coupled with the exquisite specificity of the reaction, is sufficient for direct detection of less than 1000 target molecules with no prior target amplification. A kinetic analysis of the parameters affecting signal and background generation in the serial invasive signal amplification reaction and a simple kinetic model of the assay were described. The assay was able to detect as few as 600 copies of the methylene tetrahydrofolate reductase gene in samples of human genomic DNA. The assay is capable of discriminating single base differences in this gene by using 20 ng of human genomic DNA.[363]

Interleukin-1 polymorphisms have been associated with increased risk of gastric and may influence the varying susceptibility to develop gastric cancers[696] and chromosome 18q loss is connected to vascular invasion.[701]

The repertoire of technologies is expanding and includes SBE-TAGS, an array-based method for SNP genotyping,[686] detection of single-base mismatches and DNA base lesions by the electrocatalysis at DNA-modified electrodes,[687] the scanning of guanine-guanine mismatches in DNA by synthetic ligands with surface plasmon resonance,[699] and interference-based detection of nucleic acid targets on optically coated silicon.[700]

6.8 Sequencing

Gene sequencing has reached an advanced technological stage where automation allows the rapid sequencing of several 10 000 base pairs per machine per day. The future of gene sequencing may be in superfast sequencing.[364]

Large-scale sequencing efforts require efficient methods of ordering the sequences. Physical mapping has been reintroduced as an important component of large-scale sequencing projects. Restriction maps provide landmark sequences at defined intervals and high-resolution restriction maps can be assembled from ensembles of single molecules by optical means. Such optical maps can be constructed from large-insert clones and genomic DNA, and are used as a scaffold for accurately aligning sequence contigs generated by shotgun sequencing.[365]

The acceleration of sequencing is being addressed by several approaches to measuring single molecules. A variety of different DNA polymers were electrophoretically driven through the nanopore of an α-hemolysin channel in a lipid bilayer. Single-channel recording of the translocation duration and current flow during traversal of individual polynucleotides yielded a unique pattern of events for each of the several polymers tested. Statistical data derived from this pattern of events demonstrate that in several cases a nanopore can distinguish between polynucleotides of similar length and composition that differ only in sequence. Studies of temperature effects on the translocation process show that translocation duration scales as $\sim T^{-2}$. A strong correlation exists between the temperature dependence of the event characteristics and the tendency of some polymers to form secondary structure. Because nanopores can rapidly discriminate and characterize unlabeled DNA molecules at low copy number, refinements of the experimental approach demonstrated could eventually provide a low-cost high-throughput method of analyzing DNA polynucleotides.[366]

DNA and RNA molecules can be detected as they are driven through a nanopore by an applied electric field at rates ranging from several hundred microseconds to a few milliseconds per molecule. The nanopore can rapidly discriminate between pyrimidine and purine segments along a single-stranded nucleic acid molecule. Nanopore detection and characterization of single molecules represents a new method for directly reading information encoded in linear polymers. If single-nucleotide resolution can be achieved, it is possible that nucleic acid sequences can be determined at rates exceeding 1000 bases/s.[367]

A nanopore instrument, which can discriminate between individual DNA hairpins that differ by one base pair or one nucleotide, has been demonstrated. RNA and DNA strands produce ionic current signatures when driven through an alpha-hemolysin channel by an applied voltage. This nanopore detector was combined with a support vector machine (SVM) to analyze DNA hairpin molecules on the millisecond time scale. Measurable properties include duplex stem length, base pair mismatches, and loop length.[559]

Another development towards the creation of ultra-fast DNA sequencing is the progress in designing and producing micro fluidic arrays.

A nanofluidic channel device, consisting of many entropic traps, was designed and fabricated for the separation of long DNA molecules. The channel comprises narrow constrictions and wider regions that cause size-dependent trapping of DNA at the onset of a constriction. This process creates electrophoretic mobility differences, thus enabling efficient separation without the use of a gel matrix or PEF. Samples of long DNA molecules (5000–160,000 bp) were efficiently separated into bands in 15-mm-long channels. Multiple-channel devices operating in parallel were demonstrated. The efficiency, compactness and ease of fabrication of the device suggest the possibility of more practical integrated DNA analysis systems.[368]

A method for multiplexed detection of polymorphic sites and direct determination of haplotypes in 10-kb DNA fragments using single-walled carbon nanotubes (SWNT) atomic force microscopy (AFM) probes was developed. Labeled oligonucleotides are hybridized specifically to complementary target sequences in template DNA and the positions of the tagged sequences are detected by direct SWNT tip imaging. This concept was demonstrated by detecting streptavidin and IRD800 labels at two different sequences in M13mp18. The approach also permits haplotype determination from simple visual inspection of AFM images of individual DNA molecules, which was done on UGT1A7, a gene under study as a cancer risk factor. The haplotypes of individuals heterozygous at two critical loci, which together influence cancer risk, can be easily and directly distinguished from AFM images. The application of this technique to haplotyping in population-based genetic disease studies and in genomic screening studies will facilitate the correlation of genotype and phenotype.[369]

A novel sequencing approach was described that combines non-gel-based signature sequencing with *in vitro* cloning of millions of templates on separate 5-μm diameter microbeads. After constructing a microbead library of DNA templates by *in vitro* cloning, a planar array of a million template-containing microbeads in a flow cell at a density greater than 3×10^6 microbeads/cm^2 was constructed. Sequences of the free ends of the cloned templates on each microbead were then simultaneously analyzed using a fluorescence-based signature sequencing method that does not require DNA fragment separation. Signature sequences of 16–20 bases were obtained by repeated cycles of enzymatic cleavage with a type IIs restriction endonuclease, adaptor ligation and sequence interrogation by encoded hybridization probes. The approach was validated by sequencing over 269,000 signatures from two cDNA libraries constructed from a fully sequenced strain of *S. cerevisiae* and by measuring gene expression levels in the human cell line THP-1. The approach provides an unprecedented depth of analysis permitting application of powerful statistical techniques for discovery of functional relationships among genes, whether known or unknown beforehand, or whether expressed at high or very low levels.[370]

Automated DNA sequencing in 16-channel microchips was reported. A microchip prefilled with sieving matrix is aligned on a heating plate affixed to a movable platform. Samples are loaded into sample reservoirs by using an eight-tip pipetting device and the chip is docked with an array of electrodes in the focal plane of a four-color scanning detection system. Under computer control, high voltage is applied to the appropriate reservoirs in a programmed sequence that injects and separates DNA samples. An integrated four-color confocal fluorescent detector automatically scans all 16 channels. The system routinely yields more than 450 bases in 15 min in all 16 channels. In the best case using an automated base-calling program, 543 bases have been called at with an accuracy of greater than 99%. Separations, including automated chip loading and sample injection, normally are completed in less than 18 min. The advantages of DNA sequencing on capillary electrophoresis chips include uniform signal intensity and tolerance of high DNA template concentration. To understand the fundamentals of these unique features, a theoretical treatment of cross-channel chip injection called the differential concentration effect was developed. Experimental evidence demonstrating consistence with the predictions of the theory has been presented.[371]

A special sequencing approach to quickly permit DNA signatures consistently associated with phenotypes of interest has been developed and applied to dinoflagellates for rapid identification of *Pfiesteria piscicida* and related dinoflagellates from complex cultures.

The newly described heterotrophic estuarine dinoflagellate *P. piscicida* has been linked with fish kills in field and laboratory settings, and with a novel clinical syndrome of impaired cognition and memory disturbance among humans after presumptive toxin exposure. It is necessary to better characterize the organisms and these associations. Advances in *Pfiesteria* research were hampered by the absence of genomic sequence data. A sequencing strategy was employed directed by heteroduplex mobility assay to detect *P. piscicida*

18S rDNA 'signature' sequences in complex pools of DNA and those data were used as the basis for determination of the complete *P. piscicida* 18S rDNA sequence. Specific PCR assays for *P. piscicida* and other estuarine dinoflagellates were developed, permitting their detection in algal cultures and in estuarine water samples collected during fish kill and fish lesion events. These tools will enhance efforts to characterize these organisms and their ecological relationships. Heteroduplex mobility assay-directed sequence discovery is broadly applicable and may be adapted for the detection of genomic sequence data of other novel or non-culturable organisms in complex assemblages.[372]

A method involving ligation-mediated PCR (polymerase chain reaction) for quantitative *in vivo* footprinting was developed which allows high-resolution analysis of transcription-factor binding and chromatin structure as well as sequencing.[690] The repertoire of methods to sequence and identify genes was expanded by ORESTES (open reading frame ESTs) which will contribute to the construction of contigs covering full-length cDNAs.[693]

6.9   Pharmaco-Genomics

The availability of sequence data is the foundation for turning the data from modern genomics into therapies for diseases.[373] Pharmaco-genomics aims to refine the present overwhelmingly followed principle of 'one-medicine-fits-all' with the development of active principles and therapies targeting patient groups and, in the final analysis, individual patients with specific means of treatment. An ideal prior requirement is the definition of a disease at the molecular level, and a relation to the genetic make-up of the patients, defined in terms of haplotypes, gene defects, gene alleles, SNPs, genetic linkage groups and imprinting patterns. The next level is the definition with respect to protein variations, isozymes, genetic variants of receptors, and varying interactions in protein–protein, DNA–protein and RNA–protein networks, up to the cellular level of metabolic states and phenotypes.

Pharmaco-genomics may thus serve to customize pharmaceuticals and therapies for specific subgroups of patients. The direct clinical evidence of pharmaco-genomic information is demonstrated by the different susceptibility of patients to hydralazine, an arterial vasodilator, in correlation with different expression of α- and β-myosin heavy chains. The pharmaco-genetic approach contributes to understanding the causes of phamaco-kinetics and pharmaco-dynamics based on isozymes of cytochrome P450, dihydropyrimidine dehydrogenase, thiopurine methyltransferases, cholesteryl ester transfer protein, angiotensin-converting enzyme, and the serotonin transporter. The different responses may be due to different interactions with the active principle on any (or several) level(s) of the biochemical and cellular machinery.

Pharmaco-genomics may also contribute to disease correction by providing disease-associated markers which are not themselves causally related to the variation in response.

Table 6.10 lists a selection of companies and their focus area, thus illustrating the range of applications of pharmaco-genomics.[474]

Differential gene expression profiling has been added to the repertoire of pharmaco-genomics for drug discovery and development. Differential gene expression technologies

Table 6.10. Pharmacogenomics companies [Refs in 474].

Company	Area
Aeiveos Sciences Group, LLC (Seattle, WA)	Aging-related genes and gene responses.
diaDexus, LLC (Palo Alto, CA), Joint venture of Incyte (Palo Alto, CA), and SmithKline Beecham (Philadelphia, A)	Diagnostic and pharmacogenomic kits based on leads from Incyte's, SmithKlein Beecham's, and Human Genome Science's (Rockville, MD) databases
Eurona Medical, AB (Upsala, Sweden)	CRO-retrospective correlations of drug response and genetic profiling
Gemini Research, Ltd. (Cambridge, UK)	Phenotype-based gene discovery; dizygotic twin studies
Genaissance Pharmaceuticals, Inc. (New Haven)	Genetic polymorphism correlation; isogene discovery; breast cancer; vascular lesions
Genome Therapeutics Corp. (Waltham, MA)	Human high-resolution polymorphism database
Genset, SA (Paris, France)	High-density biallelic maps, 60.000 markers
Hexagen, Plc (Cambridge, UK)	Single-strand conformational polymorphism detection methodology
Lion Bioscience, AG (Heidelberg, Germany)	Proprietary sequencing and analysis software for drug target identification and gene expression data under varying conditions
MitoKor, Inc. (San Diego, CA)	Mitochondrial genome analysis
Nova Molecular, Inc. (Montreal, Canada)	CNS disease genetic profiling
Predictive Medicine (Cambridge, MA)	Millennium Pharmaceuticals spinoff
Rosetta Inpharmatics (Kirkland, WA)	"Ink-jet" technology-based oligonucleotide array studies
Sequana Therapeutics (La Jolla, CA)	High-throughput genotyping
Variagenics, Inc. (Cambridge, MA)	Anticancer program based on loss of heterozygosity

Source: *In Vivo;* ReCap; BioVista; company materials

aid in the elucidation of drug efficacy and candidate genes for the pharmaco-genetic assessment of individual variability to drug response. The underlying principle is that drug efficacy, toxicity, metabolism, and excretion are linked to specific differential expression of target genes. Table 6.11 gives a comparison of representative open and closed differential gene expression architecture systems and their sensitivities.

Pharmaco-genomics may be defined as the quantification of differential gene expression induced by an active agent either *in vitro* or an *in vivo* model. Molecular mechanisms driving individual drug pharmacodynamics are derived from the resulting data and provide a subset of gene-based markers correlative with and predictive of drug efficacy, toxicity

Table 6.11. A comparison of representative open and closed DGE architecture systems [Refs in 475].

Platform	Differential gene expression technology	Resolution (variant gene detection)	Sensitivity[a]	Coverage[b] %	Clarifying comments	Refs
Closed						
	DNA microarray	No	1:300,000	Variable		3, 4, 16
	TaqMan®/RT-PCR	NA	<1:300,000	100		8, 9
Open						
	Differential display	Yes	1:100,000	96	240 reactions	10, 13
	SAGE	Yes	<1:10,000	92	300,000 tags or 10,000– 15,000 sequencing reactions	13 14, 16
	RDA	Yes	<1:300,000	NA		13, 15
	Gene Calling®	Yes	1:125,000	95	96 restriction enzyme pair reactions	11, 13
	TOGA	Yes	<1:100,000	60 to 98	60% coverage using one enzyme, 98% using four (4 × 256 reactions)	12, 13

[a]   Sensitivity, defined as *n*-fold difference detection limit, is directly correlated to the number of iterations of analysis performed.

[b]   For many methodologies, the reported percentage coverage is not correlated to the abundance of mRNA. Coverage of rare transcripts could be significantly less.

Abbreviation: NA, no information availabe.

and metabolism. Pharmaco-genomics enables the correlation of an active molecule with a set of differentially expressed and modulated genes in target cells, organs or tissues.

Differential gene expression technologies may either be 'closed-architecture' systems [which require *a priori* knowledge of gene sequence(s) and measure only a limited set of known genes] or 'open-architecture' systems (which do not assume prior knowledge of sequences and can measure all transcripts and identify novel genes). Consequently, open-architecture systems are particularly suitable for gene discovery and pharmaco-genomic expression profiling.[475]

Open-architecture systems include:
- Differential display
- Serial analysis of gene expression (SAGE)
- Representational differential analysis (RDA)
- Subtractive hybridization
- GeneCalling®
- Total gene expression analysis (TOGA)

A strategy for disease gene identification through nonsense-mediated mRNA decay inhibition has been developed. Premature termination codons (PTCs) initiate degradation of mutant transcripts via the nonsense-mediated messenger RNA decay (NMD) pathway which is involved in an estimated one-third of mutations underlying human disorders. Targeting this mechanism allows the identification of disease genes and was validated by testing colon cancer and Sandhoff disease cell lines.[664]

The study of functional relationships between RNA expression and chemotherapeutic susceptibility allows the identification of relevance networks thus enabling identification of biochemical associations.[695]

The importance of RNA in regulatory networks is evidenced by the phenomenon of gene silencing by RNA interference (RNAi), which is the process of sequence-specific, post-transcriptional gene silencing in animals and plants by double-stranded RNA (dsRNA) homo-logous to the silenced gene. Understanding this regulatory mechanisms is essential for the control and modification of the concomitant genetic and biochemical as well as cellular processes in fundamental studies and for pharmaceutical intervention. Duplexes of 21-nucleotide RNAs were demonstrated to mediate RNA interference in cultured mammalian cells, thus representing a tool for studying gene function and offering use as gene-specific therapeutics.[675]

An RNA-directed nuclease was found to mediate post-transcriptional gene silencing in cultured *Drosophila* cells by analysis of loss-of-function phenotypes obtained upon trans-fection with double-stranded RNAs.[681] Post-transcriptional gene-silencing was analyzed in *Arabidopsis* mutants and several related proteins were identified which supports the hypo-thesis of a common ancestral mechanism of gene expression control in plants, fungi, and animals.[682] In the unicellular green alga *Chlamydomonas reinhardtii* a gene *Mut6* was cloned that is required for the silencing of a transgene and two transposon families and codes for a protein homologous to RNA helicases of the DEAH-box family, again supporting the notion of an ancestral mechanism[678]. Similar proteins are used in the nematode *Caenorhabditis elegans* and the fungus *Neurospora crassa* to effect gene silencing.[683]

Functional genomic studies in *Caenorhabditis elegans* by systematic RNAi (RNA interference) help to elucidate the interaction between genes, gene function, and the environment. They also help to understand the functions of genes in other organisms with homo-logous genes resp. proteins[676] [677] and thus will advance the development of therapeutics.

Regulatory RNAs are also prominent in bacteria controlling at the posttranscriptional level and represent targets for regulation of metabolism and of infectivity.[679]

Epigenetic mechanism acting at the level of genes without affecting gene sequence form another layer of gene regulation, exemplified by genomic imprinting which acts by several mechanism such as methylation, promoters, enhancers, antisense RNA transcripts, silencers, and chromatin boundaries.[684] Epigenetic modifications change transcription patterns in multicellular organism to effect tissue-specific gene expression. DNA methylation is involved in animals and plants for the heritability and flexibility of epigenetic states. In *Arabidopsis* a gene (*MOM*) was identified the disruption of which releases transcriptional silencing of methylated genes while not affecting methylation patterns[680]. X chromosome inactivation is an example of genetic regulation in which thousands of genes on one homologue become silenced during female embryogenesis by spreading the inactivation signal along the chromosome based on structural elements acting as DNA signals[685].

Pharmaco-genomics provides an additional means of testing drug efficacy, and can be defined in terms of the selected metabolic and/or signal transduction pathway activated by drug application. The special advantages of differential gene expression in pharmaco-genomics include:

- Comprehensive characterization of drug efficacy for incompletely characterized drugs where causal or comprehensive biochemical understanding may be lacking
- Identification of clinical surrogate markers to monitor drug efficacy
- Indication of potential drug–drug interactions
- Elucidation of novel indications and interactions.

Table 6.12 lists a selection of single-gene mutations and rare diseases with Mendelian inheritance plus some gene polymorphisms and common disease susceptibilities of polygenic origin with complex inheritance. Figure 6.3 depicts genetics and genomics for the discovery of drug targets, Figure 6.4 describes the development of a pharmaco-genetic medicine response profile and Figure 6.5 depicts the scope of genetic testing.[476]

The full contribution of pharmacogenomics can be achieved by integrating all pertinent innovative technologies from genomics via proteomics to structural and functional studies including model organisms and integrative tools like bioinformatics and modeling to productivity-enhancing technologies such as laboratory automation and combinatorial synthesis.

Table 6.12. Genes and Diseases [Refs in 476].

Single-gene mutations and rare diseases (mendelian inheritance)	Gene polymorphisms and common disease susceptibility (polygenic; complex inheritance)
Causally related to rare inherited diseases (high penetrance)	Examples of susceptibility loci:
Examples:	Common late-onset Alzheimer's disease
Cystic fibrosis (CFTR gene)	Susceptibility gene *ApoE*) on chromosome 19 (19q13)
Inheritance: autosomal recessive	Susceptibility gene loci on chromosome 12 (12q)
Location: chromosome 7 (7q31)	Migraine
Mutation: deletion of 3 bp at codon508 accounts for 70% of mutations	Susceptibility gene loci on chromosome 19 (19p13); (chromosome X (Xq24)
Huntington disease:	Non-insulin-dependent diabetes mellitus
Inheritance: autosomal dominant	Susceptibility gene loci on chromosome 12 (12q); chromosome 2 (2q)
Location: chromosome 4 (4pl 6.3)	Psoriasis
Mutation: cytosine/adenine/ guanine repeat >35 times	Susceptibility gene locus on chromosome 3 (3q21)

Discovery genetics

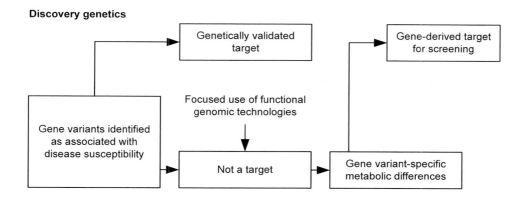

Discovery genomics

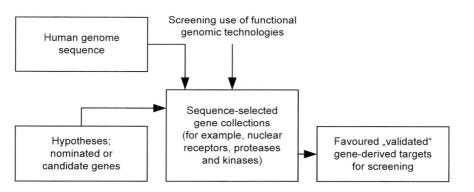

Figure 6.3. Genetics and genomics to identify drug targets. Two general strategies are used to identify genes and find new targets for drugs: genetics and genomics. Each approach shares technologies, like functional genomics, but as a part of different experimental designs. Genetics identifies disease-related susceptibility genes and genomics identifies genes that belong to similar families based on their sequence homologies. The goal of most genomic strategies is to collect genes that may be expressed and used for high-throughput screening targets. Any one of the identified genes may or may not have a connection to any disease process, with a high probability that it does not. Focused uses of functional genomic technologies include, for example, study of lines of transgenic mice that differ only in the specific polymorphisms defined in the susceptibility gene that relates to disease expression in humans. Understanding isoform-specific metabolic functions can lead to the identification of new metabolic targets for drug screening. Screening use of functional genomic technologies are used to imply validation for targets derived from discovery genomics, such as higher gene expression in a tissue of a subset of genes or the expression of protein observed in disease tissues but not seen in comparable tissue from controls [Refs in 476].

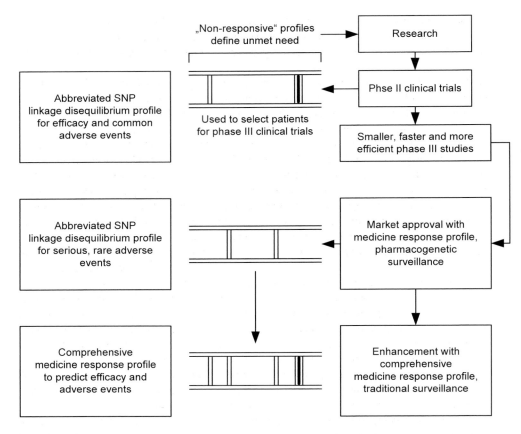

Figure 6.4. The development of a pharmacogenetic medicine response profile. An abbreviated SNP profile to predict efficacy could be identified in phase II clinical trials by detecting those SNPs along the genome that are in linkage disequilibrium when patients with efficacy are compared with patients who did not respond to the drug candidate. An abbreviated profile of these small regions of linkage disequilibrium that differentiate efficacy can then be used to select patients for larger phase studies. This could make many of these phase III studies smaller and therefore more efficient. Pharmacogenetics could also be used during the initial post-marketing surveillance period to identify SNP markers associated with serious but rare adverse events. These markers could be added to the SNP markers for efficacy and common adverse events identified during development to produce a comprehensive medicine response profile, and to identify which patients respond to the drug and which patients will be at high risk for an adverse event [Refs in 476].

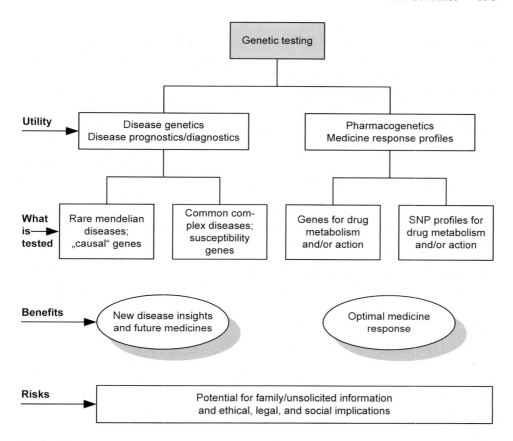

Figure 6.5. "Genetic testing" needs to be defined carefully. The magnitude of the ethical, legal and social implications of genetic testing is dependent on the information derived from the test. Genetic tests for mutations in single genes that are causally related to rare diseases and are inherited in a simple Mendelian fashion can have profound implications for the individual and family members. Genetic tests for disease-susceptibility gene polymorphisms – which are risk factors for the disease – have the added complication of uncertainty. In both cases the lack of effective intervention drives many of the issues. Pharmacogenetic profiles, on the other hand, will predict if an individual patient is likely to benefit from a medicine and be free of serious side effects. These profiles will not be designed to provide any other information, as the profile data are derived from the patients who respond with efficacy or adverse event when taking the drug, compared with patients who did not respond. It does not differentiate disease. Should a polymorphism that is found to be related to disease association be included in a profile, it can be removed and replaced by another SNP that is in linkage disequilibrium, thus avoiding any disease-specific association, even if inadvertent. This would be similar to replacing the *ApoE4SNP* by one or more of the others in linkage disequilibrium with ApoE4 hut not specifically associated with Alzheimer's disease. The ethical, legal and social implications of pharmacogenetic profiles are therefore of a lower magnitude of societal concern compared with specific genetic tests for disease [Refs in 476].

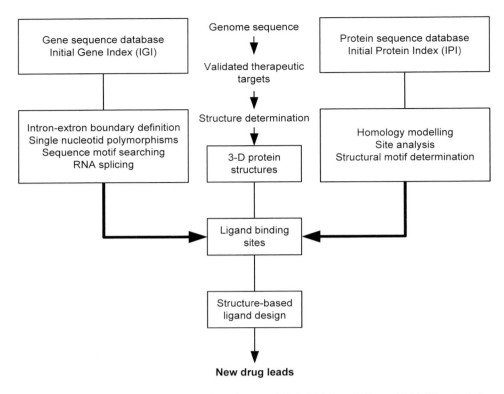

Figure 6.6. Genomic information-driven drug discovery. The Initial Gene Index and Initial Protein Index are mined in a variety of ways to identify ligand-binding sites. These in turn can act as templates for structure-based design [Refs in 567].

Biomedical development and drug discovery are increasingly driven by genomic and proteomic information. Gene sequences, intron-exon boundary definition, SNPs, sequence motifs, and RNA splicing complemented by protein sequences, and structural motifs lead to validated therapeutic targets, ligand binding sites, and drug leads. Figure 6.6 illustrates the interplay between the various contributing technologies and knowledge domains.

Table 6.13 depicts the sequential waves of technology information, from the genomics focus via the proteomics wave towards the molecular design wave integrating structural information.

Table 6.13. Three waves of innovation in drug discovery [Refs in 567].

Innovation	Tools/technologies	Representative companies
First wave		
Genomics (target discovery, antisense therapeutics, and ultimately gene therapy)	Gene/genome sequencing and expressed sequence databases Full-length cDNAs/expression	Incyte (Palo Alto, CA), Celera (Rockville, MD), Genset (Paris) Genome Therapeutics (Waltham, MA), Clontech (Palo Alto, CA), Stratagene (La Jolla, CA), Life Technologies (Rockeville, MD)
	Functional genomics	Curagen (New Haven, CT), Millennium (Cambridge, MA), Pharmagene (Cambridge, UK)
	RNA expression profiling	Affymetrix (Santa Clara, CA), Gene Logic (Gaithersburg, MD), Rosetta Inpharmatics (Kirkland, WA)
	Transgenesis/directed mutagenesis	Lexicon Genetics, (Woodlands, TX) Genome Systems (St. Louis, MO)
	Antisense Technology	Isis Pharmaceuticals (Carlsbad, CA), Hybridon (Cambridge, MA)
	SNP databases	Variagenics (Cambridge, MA), Genaissance (New Haven, CT), Celera Oxagen (Boulder, CO),
	Genetic mapping/disease genes	Myriad (Salt Lake City, UT), Decode (Reykjavik, Iceland)
	Genetic diagnostics	Quest Terterboro, NJ), Roche (Nutley, NJ), Diadexus (Palo Alto, CA), Affymetrix (Santa Clara, CA)
Second wave		
Proteomics (target validation, drug screening, antibody therapeutics, and protein therapeutics)	Protein databases	Oxford glycosciences (Oxford, UK), Large Scale Biology (Vacaville, CA)
	and protein expression analysis	Proteome (Cambridge, MA)
	Protein expression technologies and protein therapeutics	Amgen (Thousand Oaks, CA), Genentech (S. San Francisco, CA) Human Genome Sciences (Rockville, MD), Chiron (Emeryville, CA) Genetic Institute (Cambridge, MA) Lonza (Slough, UK)
	Directed evolution	Maxygen (Redwood City, CA) Phylos (Lexington, MA),
	Antibody engineering	Celltech/Medarex (Leatherhead UK), Abgenix (Fremont, CA) Cambridge Antibody Technology (Cambridge, UK)
	High-throughput screening	Aurora Biosciences (La Jolla, CA), Cambridge Drug Discovery (Cambridge, UK)

Table 6.13. Three waves of innovation in drug discovery (Cont'd).

Innovation	Tools/technologies	Representative companies
	Protein interaction databases	Proteome, MDS-Proteomics (Blainville, QC, Canada), Celera
	Affinity selection	Neogenesis (Cambridge, MA), MDS-Proteomics
	Protein pathways/protein chips	Zyomyx (Hayward, CA), Combimatrix (Seattle, WA), Ciphergen (Palo Alto, CA), Sense Proteomics (Cambridge, UK)
Third wave		
Molecular design (protein therapeutics, antibodies, and small molecules)	Protein structure determination	Structural Genomix/Syrrx (San Diego, CA), Astax (Cambridge, UK)
	Protein homology modeling	Structural Bioinformatics (San Diego, CA), Geneformatix (San Diego, CA)
	Protein engineering	Sunesis (Redwood City, CA), Sangamo (Richmond, CA) Cambridge Antibody Technology (Cambridge, UK)
	Structure-based small molecule design	Vertex Pharmaceuticals (Cambridge, MA), De Novo (Cambridge, UK)
	Molecular design tools	Molecular Simulations (San Diego, CA), Tripos (St. Louis, MO)

7 Research and Development

The R & D process is benefitting from the expanding knowledge base, and from the increasing contributions from robotics and automation, informatics and bioinformatics, equipment development, and the outsourcing of key processes and consolidation of the regulatory environment.

7.1 Biology, Medicine, and Genetics

The recognition of similarities and differences between species and the use of this knowledge for the elucidation of biochemical mechanisms, genetic lesions and relationships, disease mechanisms, identification of molecular targets is facilitating the drug discovery process.

Integration of experts from various fields of medicine, biology and genetics will bring results from fundamental research quickly into the realm of therapy and medicine.

7.2 Pre-clinical and Clinical Development

The pre-clinical development stage is accelerated by additional models of disease and testing as well as high-throughput technologies in the laboratory.

Clinical development benefits from genetic and biochemical knowledge, and the specific targeting of therapies to genetic (sub-)groups with concomitant stratification of patients.

7.3 Processes

The classical methods of chemical and fermentative synthesis and purification are augmented by the powerful genetically modified cellular systems of production, by transgenic animals and plants, as well as by sophisticated techniques of synthesis, analysis, purification, such as chiral catalysts, micro-reactors, and process intensification.

7.4   Pilot Plants

The inceasing efficiency and potency of APIs as well as the fragmentation of patient groups generally decreases the amounts of substances required for the pre-clinical, clinical and production phases. Versatile pilot plants are capable of providing such amounts, whereas large dedicated factories were often required for lower potency compounds.

7.5   Engineering

The engineering capabilities which need to be deployed are extending from classical chemical or fermentation formulation engineering towards the inclusion of the additional technologies in cell culture, metabolic engineering, micro- and nano-technology engineering, DNA or protein chip design, *in silico* biology, simulation, and bioinformatics.

7.6   Fermentation Process Development

These topics in fermentation process development include metabolic flux, metabolic engineering, process development, and downstream processing. Recent advances in metabolic engineering have led to the development of new methods for the synthesis of novel molecules, improved production of existing compounds and improved degradation of recalcitrant environmental contaminates. Increasing the flux through an existing pathway and introducing a new pathway into a host organism demand coordinated expression of the genes that encode the enzymes, tight control over gene expression and consistent expression in all cells. Several tools have been developed for the overproduction of specific proteins. Pathway redirection (metabolic engineering) requires certain specific characteristics of gene-expression tools such as cloning vectors, plasmids, and promoters for the expression of multiple genes.[374]

Metabolic engineering has been intensively studied in the eukaryotic organism *Saccharomyces cerevisiae*, which is a model for the biology of eukaryotic organisms. *S. cerevisiae* is susceptible to recombinant DNA technology and metabolic engineering is facilitated by the availability of the complete genome sequence.

The importance of bacteria and eukaryotic microorganisms is illustrated in Table 7.1, which lists metabolite production by various microorganisms, such as amino acid biosynthesis, ethanol production, lactic acid, and xylitol production, 7-ACA and 7-ADCA-production.

In *Saccharomyces cerevisiae*, substrate extension has been achieved to increase the versatility and improve the economics of *S. cerevisiae* fermentations. Examples illustrating the extension of substrate range to include Lactose, Malate, Starch and Dextrins, and Xylose are shown in Table 7.2.

Table 7.1. Metabolite production by various microrganism[α] [Refs in 502].

Organism	Property or product	Achievements	Refs
Aspergillus nidulans	Enhanced growth rate	Overproduction of glyceraldehyde-3-phosphate dehydrogenase increased the specific growth rate in both acetate and glucose in comparison with the parental strain	41
Clostridium acetobutylicum	Acetone and butanol production	Overexpression of acetoacetate decarboxylase (*adc*) and phosphotransbutyrylase *(ptb)* by introducing a *Bacillus subtilis/C acetobutylicum* shuttle vector into *C. acetobutylicum* by an improved electrotransformation protocol, which resulted in acetone and butanol formation	88
Coryne-bacterium	Amino acid bio-synthesis	Isolation of amino acid biosynthetic genes that enables enhanced enzyme activities or removal of feedback regulation in order to improve the production strains	54
Escherichia coli	Ethanol production	Integration of pyruvate dearboxylase and alco-hol dehydrogenase II from *Zymomonas mobilis* onto the chromosome of *E. coli* improve d the stability of the genes over the plasmid-based system, and the ethanol yield was near the maximum theoretical yield on 10% glucose and 8% xylose	48, 98
Escherichia coli	1,3-Propanediol production	Introduction of genes from the *Klebsiella pneumoniae dha* regulon into *E. coli* revealed 1,3-butanediol production	145
Penicillium chrysogenum	7-ACA and 7-ADCA	Production 7-ACA and 7-ADCA from new fermentation processes was obtained by trans-formation of the expandase of *Cephalosporin acremonium* into *P. chrysogenum*	19
Saccharomyces cerevisiae	Lactic acid production	A muscle bovine lactate dehydrogenase gene (LDH-A) was expressed in *S. cerevisiae*, a nd lactic acid was produced in titers of 20 g/liter with productivities of 11 g/liter/h; due to the acid tolerance of *S. cerevisiae* this organism may serve as an alternative for substitution of bacteria for lactic acid production	108
Saccharomyces cerevisiae	Xylitol production	95% xylitol conversion from xylose was ob-tained by transforming the XYL1 gene of *Pichia stipitis* ncoding e a xylose reductase into *S. cerevisiae,* making this organism an efficient host for the production of xylitol, which serves as an attractive sweetener in the food industry	39

[α]   7-ACA, 7-aminocephalosporanic acid; 7-ADCA, 7-aminodeacetoxycephalosporanic acid.

Table 7.2. Examples illustrating an extension of the substrate range in *S. cerevisiae* [Refs in 502].

Substrate	Comment	Refs
β-Glutans	An extended substrate range was obtained by introduction of a β-glucanase gene of *Trichodemia reesei* into brewer's yeast. The recombinant strain was able to utilize β-glucans, which resulted in improved filterability.	105
Lactose	Genes encoding lactose permease and β-galactosidase of *Kluyvero- myces marxianus* were introduced into *S. cerevisiae*. The recombinant strain efficiently converted lactose into ethanol when fermented in a continuous biore-actor setup.	24, 25
Lactose	A thermostable β-galactosidase encoded by *lacA* from *Aspergillus niger* was expressed from a *ADH1* promoter in β-galactosidase activity, 40% was se-creted to the cellular medium, and the recombinant strain grew on whey per-meate with a specific growth rate of 0.43 h^{-1}	75
Malate	Malate degradation is essential for deacidification of grapes when producing wine. Due to the lack of malate permease and the low malate affinity of the malic anzyme of *S cerevisiae,* this organism was metabo- lically engineered for efficient malate utilization. The *Schizosaccharo- myces pombe* malate permease (*MAE1*) and the gene encoding malic enzyme of *S. pombe* (*MAE2*) or malic enzyme of *Lactococcus lactis (mleS)* were expressed in *S. cere-visiae*, and malate degradation was successfully achieved.	154
Melibiose	Melibiase-producing baker's yeast strains were successfully constructed by introduction of *MEL1* gene into *S. cerevisiae* by genetic engineering or by classical breeding and mating. To overcome glucose repression exerted on the galactose metabolism, partial glucose derepressed strains were con-structed.	102, 117, 153
Starch and dextrins	Starch assimilation of 99% was achieved by coexpression of the *STA2* gene of *Saccharomyces diastaticus*, the *AMY1* gene of *Bacillus amyloliquefaciens,* and the *pulA* gene of *Klebsiella pneumoniae* encoding a glucoamylase, an α-amylase, and apullulanase, respectively.	50
Starch and dextrins	Improvements of baker's and brewer's yeasts were obtained by expressing the α-amylase (*AMY1*) and the glucoamylase (*GAM1*) genes of the yeast *Schwanniomyces occidentalis* in *S. cerevisiae*.	46
Xylose	The *XYL1* gene and the *XYL2* gene of *Pichia stipitis* encoding a xylose reduc-tase and a xylolit dehydrogenase, respectively, were transformed into *S. cere-visiae*. When the *XKS1* gene encoding the *S. cerevisiae* xylolukinase was overexpressed in this transformant strain, xylose was successfully utilized and converted into ethanol, with an overall yield of 63% of the maximum theoretical yield.	45

Metabolic engineering requires an integrated approach of analysis and synthesis. Studies involve expression analysis, metabolite levels, pathway analysis, fermentation experiments, and protein characterization. Synthesis incorporates recombinant DNA techniques such as cloning, transformation, construction of vectors and artificial chromosomes.

The substrate extension to include the utilization of xylose by *S. cerevisiae* involves the introduction of *XYL1* and *XYL2* from yeasts capable of utilizing xylose.

In order to maximize ethanol production in relation to glycerol and pyruvate, the metabolic pathways need to be manipulated. Figure 7.1 shows the biochemical routes to glycerol, pyruvate, and ethanol, illustrating targets for redirecting carbon flux to ethanol.

To overcome lag phases during fermentation of sugar mixtures containing glucose, sucrose, amd maltose, the pathways for sugar and saccharide utilization need to be alleviated from glucose control. Improving maltose consumption was achieved by targeting regulatory genes such as *MIG1* and *MIG2* as well as by constitutive expression of the structural *MAL* genes.

The ultimate goal of the various genome-sequencing projects is to gain sufficient understanding of the physiology of the organisms under study to enable exploitation of the genetic information to advance human health, agricultural production and industrial fermentation. A list of genes alone, however, is unlikely to provide sufficient information to manipulate the metabolism or pathophysiology of an organism in a predictable way. Metabolic flux analysis is designed to convert a list of putative enzymes into a set of metabolic pathways. This approach involves a rigorous definition of the constitutive activities of a given pathway as the sum of elementary flux modes, each of which represents a minimal sequence of metabolic steps that can cooperate independently of each other. Applications include more efficient identification of important drug targets and assigning and/or corroborating specific activities to orphan gene sequences.[375]

Applications include maximizing product yield in amino acid and antibiotic synthesis, reconstruction and consistency checks of metabolism from genome data, analysis of enzyme deficiencies, and drug target identification in metabolic networks.

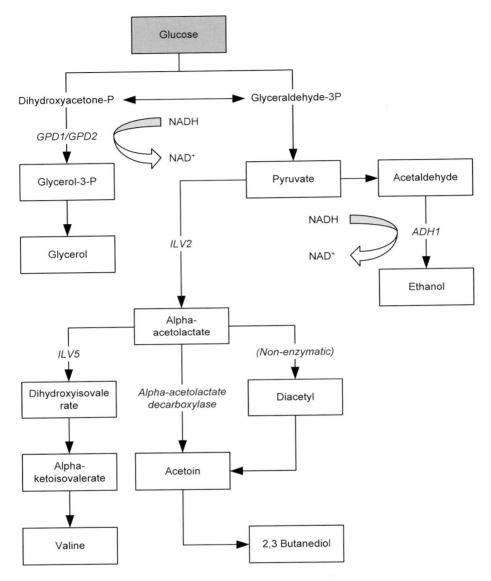

Figure 7.1. Overview of the biochemical routes leading to glycerol, pyruvate, and ethanol. Furthermore, valine biosynthesis and diacetyl formation are shown, which may be bypassed by introduction of a heterologous á-acetolactate decarboxylase that directly converts á-acetolactate to acetoin. *GPD1 and GPD2*, glycerol dehydrogenases 1 and 2; *ADH1*, alcohol dehydrogenase 1; *ILV2*, acetolactate synthetase; *ILV5*, acetolactate reductoisomerase [Refs in 502].

8 Pharmaceutical Production

Nature was the first source of active principles, albeit in crude, dilute, mixed and sometimes toxic forms, an early example being quinine (present in Cinchona bark).

Microbial biotechnology, first exemplified by penicillin, yielded industrial amounts of biosynthetic compounds, followed (after the application of molecular biology) by significant numbers of microbial metabolites, engineered compounds and recombinant proteins.[442]

The 'scale-up' of chemical and biotechnological synthesis (from automated synthesis, laboratory synthesis to pilot production and full-scale commercial production) is gaining increasing importance due to the required speed of development, the increasing complexity of the molecules (APIs) and the expanding technological repertoire being applied.

The industrial operation requires efficient design of production and optimized utilization of multi-product facilities.[376]

The increasing demand of the chemical and pharmaceutical industries for enantiomerically pure compounds has spurred the development of a range of so-called 'chiral technologies', which aim to exert the ultimate control over a chemical reaction by directing its enantioselectivity. Heterogeneous enantioselective catalysis is particularly attractive because it allows the production and ready separation of large quantities of chiral product while using only small quantities of catalyst. Heterogeneous enantioselectivity is usually induced by adsorbing chiral molecules onto catalytically active surfaces. A mimic of one such catalyst is formed by adsorbing (*R,R*)-tartaric acid molecules on Cu(110) surfaces: this generates a variety of surface phases, of which only one is potentially catalytically active and leaves the question of how adsorbed chiral molecules give rise to enantioselectivity. It was shown that the active phase consists of extended supramolecular assemblies of adsorbed (*R,R*)-tartaric acid, which destroy existing symmetry elements of the underlying metal and directly bestow chirality to the modified surfaces. The adsorbed assemblies create chiral 'channels' exposing bare metal atoms and it is these chiral spaces that are believed to be responsible for imparting enantioselectivity, by forcing the orientation of reactant molecules docking onto catalytically active metal sites. The findings demonstrate that it is possible to sustain a single chiral domain across an extended surface – provided that reflection domains of opposite handedness are removed by a rigid and chiral local adsorption geometry, and that inequivalent rotation domains are removed by successful matching of the rotational symmetry of the adsorbed molecule with that of the underlying metal surface.[377]

Using metabolic engineering, the carotenoid biosynthesis pathway in tobacco (*Nicotiana tabacum*) was modified to produce astaxanthin, a red pigment of considerable economic value. To alter the carotenoid pathway in chromoplasts of higher plants, the cDNA of the gene *CrtO* from the alga *Haematococcus pluvialis*, encoding β-carotene ketolase, was transferred to tobacco under the regulation of the tomato *Pds* (phytoene desaturase) promoter. The transit peptide of *Pds* from tomato was used to target the CRTO polypeptide to the plastids. Chromoplasts in the nectary tissue of transgenic plants accumulated $(3S,3'S)$ astaxanthin and other ketocarotenoids, changing the color of the nectary from yellow to red. This accomplishment demonstrates that plants can be used as a source of novel carotenoid pigments such as astaxanthin. The procedures described can serve as a platform technology for future genetic manipulations of pigmentation of fruits and flowers of horticultural and floricultural importance.[378]

Combinatorial biosynthesis was used to synthesize novel lipophilic carotenoids that are powerful cellular antioxidants. By co-expressing three different carotenoid desaturases in combination with a carotenoid hydratase, a cyclase and a hydroxylase on compatible plasmids in *E. coli*, four novel carotenoids were synthesized that were not previously detected in biological material or chemically synthesized. Their identification was based on their relative retention times on HPLC, spectroscopic properties, molecular weights, number of hydroxy groups and ^{1}H-NMR spectra. The carotenoids were designated as 1-HO-3',4'-didehydrolycopene, 3,1'-(H)2-γ-carotene, 1,1'-(HO)$_2$-3,4,3',4'-tetradehydrolycopene and 1,1'-(HO)$_2$-3,4-didehydrolycopene These novel acyclic derivatives differ from structurally related compounds by extension of the conjugated polyene chain as well as additional hydroxy groups at position C-1'. Their antioxidative activity was determined in a liposome-membrane model system, which showed that their ability to protect against photo-oxidation and radical-mediated peroxidation reactions was linked to the length of the conjugated double-bond system and the presence of a single hydroxy group. The protection of membrane degradation was superior to the related 1-HO and 1,1'-(HO)$_2$ lycopene derivatives, making them interesting pharmaceutical candidates.[379]

The burgeoning demand for complex, biologically active molecules for medicine, materials science, consumer products and agrochemicals is driving efforts to engineer new biosynthetic pathways into microorganisms and plants. Principles of breeding, including mixing genes and modifying catalytic functions by *in vitro* evolution, were applied to create new metabolic pathways for biosynthesis of natural products in *E. coli*. Shuffled phytoene desaturases were expressed in the context of a carotenoid biosynthetic pathway assembled from different bacterial species and the resulting library screened for novel carotenoids. One desaturase chimera efficiently introduced six rather than four double bonds into phytoene, to favor production of the fully conjugated carotenoid, 3,4,3',4'-tetradehydrolycopene. This new pathway was extended with a second library of shuffled lyopene cyclases to produce a variety of colored products. This combined approach of rational pathway assembly and molecular breeding may allow the discovery and production, in simple laboratory organisms, of new compounds that are essentially inaccessible from natural sources or by synthetic chemistry.[380]

Metabolic engineering in *S. cerevisiae* was carried out to increase the flux through central carbon metabolism. Regulatory network manipulation achieved a balanced increase in the activity of all enzymes in the pathway and augmented the metabolic flux through that pathway. Manipulation of the tightly regulated *GAL* gene regulatory network of *S. cerevisiae* produced prototroph mutant strains with increased flux through the galactose utilization pathway by eliminating three negative regulators of the *GAL* system, Gal6, Gal80 and Mig1, with a resulting 41% increase in flux through the galactose utilization pathway of the mutant as compared with the wild-type strain. The improved consumption of galactose, which is a constituent of many industrial fermentation media, in the *gal* mutants resulted not in augmented biomass formation but in excessive respiro-fermentative metabolism, whereby the ethanol production rate increased linearly with glycolytic flux.[443]

Genetic and metabolic profiling aids in the identification of suitable plant species and variants, in selecting desired mutants, and in identifying networks and the effects of gene mutations or the influence of small molecules. Metabolic profiling in particular helps in elucidating links and relationships at the metabolic regulation level.

Multiparallel analyses of mRNA and proteins are complemented by metabolic profiling as a tool for comparative display of gene function. It provides insight into complex regulatory processes and gives a characteristic description of different phenotypes. Using GC/MS, 326 distinct compounds from *A. thaliana* leaf extracts were automatically quantified. A chemical structure was assigned to approximately half of these compounds. Comparison of four *Arabidopsis* genotypes (two homozygous ecotypes and a mutant of each ecotype) demonstrated a distinct metabolic profile for each genotype. Data mining tools such as principal component analysis enabled the assignment of metabolic phenotypes. The metabolic phenotypes of the two ecotypes were more divergent than were the metabolic phenotypes of the single-loci mutant and their parental ecotypes.[448]

8.1   GenePharming (Animals and Plants)

The application of gene-transfer technology to domestic animals offers a way for the introduction of genes encoding (human) proteins, thus creating biochemical pathways that are not originally present or non-functional in these animals.

This provides mechanisms for producing therapeutic proteins and for the modification of production characteristics. Efforts to introduce specific characteristics which involve the integration of additional with existing biochemical pathways and homeostasis require higher levels of transcriptional and translational control.[381]

Farm animal genomics as a method of identifying genes controlling commercially important traits is developing by constructing maps of informative markers, scanning those maps to locate quantitative trait loci, identifying the trait genes themselves, and finally bridging the gap between the gene(s) and the ultimate trait.[382]

Transgenic plants have become attractive systems for production of human therapeutic proteins because of the reduced risk of mammalian viral contaminants in animal cell cultures and animals, the ability to do large scale-up at low cost, and the low maintenance requirements. A feasibility study was reported for the production of a human therapeutic protein through transplastomic transformation technology, which has the additional advantage of increased biological containment by apparent elimination of the transmission of transgenes through pollen. It is shown that chloroplasts can express a secretory protein, human somatotropin, in a soluble, biologically active, disulfide-bonded form. High concentrations of recombinant protein accumulation are observed (greater than 7% total soluble protein), more than 300-fold higher than a similar gene expressed using a nuclear transgenic approach. The plastid-expressed somatotropin is nearly devoid of complex post-translational modifications, effectively increasing the amount of usable recombinant protein. Approaches have been described to obtain a somatotropin with a non-methionine N-terminus, similar to the native human protein. The results indicate that chloroplasts are a highly efficient vehicle for the potential production of pharmaceutical proteins in plants.[383]

Table 8.1. The production of vaccines in transgenic plants.

Potential application/ indication	Plant	Protein	Expression systems	Refs in [446]
Hepatitis B	Tobacco	Recombinant HBsAg	AMT	15,40
	Tobacco	Murine hepatitis epitope	TMV	36
Dental caries	Tobacco	*Streptococcus mutans* surface protein Sp.	AMT	15
Autoimmune diabetes	Potato	*Vibrio cholerae* toxin B subunit-human insulin fusion	AMT	44
	Potato	Glutamic acid decarboxylase	AMT	29
Cholera and *E. coli* diarrhea	Tobacco/potato	*E. coli* heat-labile enterotoxin LT-B	AMT	15,45
Oral vaccine against cholera	Potato	*V cholerae* toxin CtoxA and CtoxB subunits	AMT	49
Mucosal vaccines not requiring adjuvants	cow-pea	D2 peptide of fibronectin -binding protein B of *Staphylococcus aureus*	CPMV	38
Diarrhea due to Norwalk virus	Tobacco/potato	Coat protein of Norwalk virus	AMT	30
Rabies	Tobacco/spinach	Rabies virus glycoprotein	AMT	22
HIV	Tobacco/ blackeyed bean	HIV epitope (gpl2o)	CPMVIAMT	14, 35
	cow-pea	HIV epitope (gp4l)	CPMV	38
Rhinovirus	blackeyed bean	Human rhinovirus epitope (HR14)	CPMV	15
Foot & mouth	blackeyed bean	Foot & mouth virus epitope (VP1)	CPMV	15,36
Resistance of mink to mink enteritis virus, dogs to canine parvovirus, and cats to feline panleukopenia virus	blackeyed bean	Mink enteritis virus epitope (VP2)	CPMV	24
Malaria	Tobacco	Malarial B-cell epitope	TMV	15,37
Influenza	Tobacco	Hemagglutinin	TMV	36
Cancer	Tobacco	c-Myc	TMV	36

AMT, Agrobacterium mediated transformation; TMV, tobacco mosaic virus; CPMV, cow-pea mosaic virus.

Plants have a significant potential for serving as production machinery for biopharmaceutical proteins, peptides and chemicals since they are easily manipulated and constitute a cheap source of biomass, protein and synthesized chemicals. The expected rise in demand for biopharmaceuticals recommends the increased attention given to the study of plant genetics and metabolism, construction of transgenics, recovery and purification of chemicals, biopharmaceuticals, and intermediates and metabolites from plant sources. The existing agricultural and agro-industrial infrastructure and processing industries offer effective support for large-scale manufacturing.

Plants are a potentially cheap source of biopharmaceuticals. It has been estimated that the manufacturing costs for recombinant proteins in plants could be 10- to 50-fold lower than the same product obtained by fermentation with *E. coli* and subsequent recovery.[445]

The types of potential products include vaccines, antibodies, enzymes, metabolites and recombinant proteins.

Table 8.1 lists examples of vaccines produced in transgenic plants, Table 8.2 shows several possibilities for antibody production in transgenic plants and Table 8.3 demonstrates the range of biopharmaceuticals under development from transgenic plants.

Table 8.2. The production of antibodies in transgenic plants.

Goal	Plant	Protein	Expression systems	Refs in [446]
Immunoglobulins				
Synthesis of secretory immunoglobin for treatment of dental caries	Tobacco	Hybrid sigA-G specific for *S. mutans* antigen II	AMT	9, 40
Synthesis of full-length IgG1	Tobacco	IgG (Guy's 13) specific for *S. mutans* surface protein (SA I/II)	AMT	17, 40
IgG assembly and secretion	Tobacco	IgG specific for human creatine kinase	AMT	11
Comparison of glycosylation in plant- and animal-derived IgG1	Tobacco	IgG (Guy's 13) specific for *S. mutants* surface protein (SA I/II)	AMT	10
Single-chain Fv fragments				
Accumulation and storage of protein in tubers	Potato	Phytochrome binding scFv	AMT	11, 21
Treatment of non Hodgkin's lymphoma	Tobacco	scFv of IgG from mouse B-cell lymphoma	AMT	19
Production of tumor-associated marker antigen	Cereals	scFvT84.66 against carcinoembryogenic antigen	Particle bombardment	33

AMT, Agrobacterium mediated transformation.

Table 8.3. The production of biopharmaceuticals in transgenic plants.

Potential application/ indication	Plant	Protein	Expression systems	Refs in [446]
Anticoagulants				
Protein C pathway	Tobacco	Human protein C (serum protease)	AMT	13
Indirect thrombin inhibitors	Tobacco, oilseed Ethiopian mustard	Human hirudin variant 2	AMT	8, 20
Recombinant hormones/proteins				
Neutropenia	Tobacco	Human granulocyte-macrophage colony-stimulating factor	AMT	2, 25
Anemia	Tobacco	Human erythropoietin	AMT	2, 25
Antihyperanalgesic by opiate activity	Thale cress, oilseed	Human enkephalins	AMT	25, 26
Wound repair/control of cell proliferation	Tobacco	Human epidermal growth factor	AMT	2,26
Hepatitis C and B treatment	Rice, turnip,	Human interferon-á	AMT	2,25
Liver cirrhosis	Potato, tobacco	Human serum albumin	AMT	2, 12, 25
Blood substitute,	Tobacco	Human hemoglobin	AMT	50
Collagen	Tobacco	Human homotrimeric collagen 1	AMT	50
Protein/peptide inhibitors				
Cystic fibrosis, liver disease, and hemorrhage	Rice	Human á-1-aminotrypsin	Particle bombardment	–
Trypsin inhibitor for trans-plantation surgery	Maize	Human aprotinin	Particle bombardment	34
Hypertension	Tobacco/ tomato	Angiotensin-1-converting enzyme	AMT	21
HIV therapies	*Nicotiana bethamiana-*	á-trichosanthin from TMV-U1 subgenomic coat protein	AMT	18
Recombinant enzymes				
Gaucher's disease	Tobacco	Glucocerebrosidase	AMT	13, 41
Neutraceuticals				
Provitamin A deficiency	Rice	Daffodil phytoene synthase	Particle bombardment	32
Amino acid deficiency	Potato	*Amaranthus hypochondriacus* Ama1 seed albumin	AMT	53

AMT, *Agrobacterium* mediated transformation.

Specific products from transgenic plants include enkephalins, IFN-α, human serum albumin, glucocerebrosidase, granulocyte macrophage colony-stimulating factor, α_1-antitrypsin inhibitor (for the treatment of CF, liver diseases, hemorrhages) and hirudin (an anticoagulant for the treatment of thrombosis) from the leech *Hirudo medicinalis.*[446]

Hepatitis B oral immunization with transgenic potato was demonstrated in preclinical animal trials with recombinant hepatitis B surface antigen (HbsAg). Mice fed transgenic HbsAg potato tubers showed an immune response, i.e. increased HbsAg-specific serum antibody, which was boosted by intraperitoneal application of a single sub-immunogenic dose of commercially available HbsAg vaccine. In order to be effective and applicable as an oral vaccine, higher resulting titers will have to be achieved by delivering larger doses of antigen. Studies of the factors influencing the accumulation of HbsAg in transgenic potato include 5′ and 3′ flanking sequences and protein targeting within the plant cells. The most significant improvements resulted from alternative polyadenylation signals and fusion proteins containing targeting signals designed to enhance the integration or retention of HbsAg in the endoplasmic reticulum of plant cells.[447]

The grain of self-pollinating diploid barley species offers two modes of producing recombinant enzymes or other proteins. One uses the promoters of genes with aleurone-specific expression duringgermination and the signal peptide code for export of the protein into the endosperm. The other uses promoters of the structural genes for storage proteins deposited in the developing endosperm. Production of a protein-engineered thermotolerant (1,3–1,4)-β-glucanase with the D hordein gene (*Hor3-1*) promoter during endosperm development was analyzed in transgenic plants with four different constructs. High expression of the enzyme and it activity in the endosperm of the mature grain required codon optimization to a C + G content of 63% and synthesis as a precursor with a signal peptide for transport through the endoplasmic reticulum and targeting into the storage vacuoles. Synthesis of the recombinant enzyme in the aleurone of germinating transgenic grain with an α-amylase promoter and the code for the export signal peptide yielded about 1 μg/mg soluble protein, whereas 54 μg/mg soluble protein was produced on average in the maturing grain of 10 transgenic lines with the vector containing the gene for the (1,3–1,4)-β-glucanase under the control of the *Hor3-1* promoter.[384]

Important chemicals and intermediates are produced by plants, in particular in glandular tissues. Glandular tissues are an anatomical plant feature widely observable in plants. Many products produced in these tissues play biologically important roles by, for example, attracting pollinators, protecting against herbivores, and defining flavor and aroma. Glandular tissues comprise trichomes which accumulate essential oils and resins, secretory cavities which accumulate aromatic oils in the epidermis of citrus fruits, ducts and cavities accumulating oleoresinsin conifers, and osmopheres in floral tissues secreting volatile compounds attracting pollinators. The compounds synthesized include monoterpenes and sesquiterpenes, diterpene resin acids, sucrose, and glucose esters of fatty acids, indole, nicotine and phenolic substances, fatty acids, nectar, proteins, and polysaccharides.[444]

Bioengineering may aim at increasing the production or concentration of such indigenous compounds or introduce the production of novel substances.

The metabolism of secretory tissues is divided into non-photosynthetic and photosynthetic glands. Oil-producing glandular tissues are often non-photosynthetic whereas crops like tomato, potato, sunflower and geranium have photosynthetic glandular trichomes. This distinction is pertinent for the manipulation of the metabolism since photosynthetic trichomes assimilate carbon dioxide and convert tissue into secretory products, whereas non-photosynthetic trichomes need to import carbon from the leaf cells for their metabolism.

The elucidation of the gene sequence of *A. thaliana* will facilitate the manipulation of regulatory and metabolic pathways since knowledge about the developmental genetics of trichome formation originates primarily from research on the non-glandular trichomes of *Arabidopsis* and numerous mutants of *Arabidopsis* have been described.

Bioengineering the development and metabolism of glandular tissues will have to take into account the commercial applications, the price of the chemicals to be produced, the agro-industrial production techniques and the genetic, regulatory and metabolic susceptibility for manipulation, as well as the potential of plant metabolism to achieve the desired increased energy and synthetic load.

Transgenic mice expressing bacterial phytase in the digestive tract (salivary glands) were developed which showed reduced fecal phosphorous. This demonstrates a biological approach to reducing phosphorous pollution from animal agriculture[674].

The expression of lysostaphin protein in mammary glands of transgenic mice confers protection against staphylococcal infection and shows the potential of genetic engineering to combat the disease in dairy cattle[702].

8.2   Vitamins

Vitamins can be produced by microbial and fungal fermentation (e.g., vitamin B_2, ribose and vitamin B_{12}).

Advanced genetic engineering has introduced a desired nutritional trait into agriculturally important rice.[385]

8.3   Amino Acids

Using directed evolution, the hydantoinase process for the production of L-methionine (L-Met) in *E. coli* was improved. This was accomplished by inverting the enantioselectivity and increasing the total activity of a key enzyme in a whole-cell catalyst. The selectivity of all known hydantoinases for D-5-(2-methylthioethyl) hydantoin (D-MTEH) over the L-enantiomer leads to the accumulation of intermediates and reduced productivity for the L-amino acid. By using random mutagenesis, saturation mutagenesis and screening it, was possible to convert the D-selective hydantoinase from *Arthrobacter* sp. DSM 9771 into an L-selective enzyme and increase its total activity 5-fold. Whole *E. coli* cells expressing the evolved L-hydantoinase, an L-*N*-carbamoylase and a hydantoin racemase, produced 91 mM L-Met from 100 mM L,D-MTEH in less than 2 h. The improved hydantoinase increased productivity 5-fold for >90% conversion of the substrate. The accumulation of the unwanted intermediate D-carbamoyl-methionine was reduced 4-fold compared to cells with the wild-type pathway. Highly D-selective hydantoinase mutants were also discovered. Enantioselective enzymes rapidly optimized by directed evolution and introduced into multienzyme pathways may lead to improved whole-cell catalysts for efficient production of chiral compounds.[386]

8.4 Proteins

In addition to the classical production of naturally occurring proteins, such as insulin, IFNs, growth hormone, growth factors and blood-clotting proteins, the techniques of molecular biotechnology enable the synthesis of superior proteins by synthesis of artificial genes or by directed evolution.

In biological systems, enzymes catalyze the efficient synthesis of complex molecules under benign conditions, but widespread industrial use of these biocatalysts depends crucially on the development of new enzymes with useful catalytic functions. The evolution of enzymes in biological systems often involves the acquisition of new catalytic or binding properties by an existing protein scaffold. This strategy was experimentally mimicked using the most common fold in enzymes, the α/β barrel, as the scaffold. By combining an existing binding site for structural elements of phosphoribosylanthranilate with catalytic template required for isomerase activity, phosphoribosylanthranilate isomerase activity evolved from the scaffold of indole-3-glycerolphosphate synthetase. Targeting the catalytic template for *in vitro* mutagenesis and recombination, followed by *in vivo* selection, results in a new phosphoribosylanthranilate isomerase that has catalytic properties similar to those of the natural enzyme, with an even higher specificity constant. The demonstration of divergent evolution and the widespread occurrence of the α/β barrel suggests that this scaffold may be a fold of choice for the directed evolution of new biocatalysts.[387]

Enzymes offer an environmentally benign alternative to chemical catalysts in many commercial and industrial applications. The expansion of their use depends on the development of new protein catalysts. This will imply the construction or identification of enzymes that have been genetically altered to improve their performance under defined, application-specific conditions. Directed evolution offers a fast and effective way of creating improved enzymes from relatively ineffective catalysts to commercially viable products by a variety of directed evolution techniques.[388]

The *Coprinus cinereus* (CiP) heme peroxidase was subjected to multiple rounds of directed evolution in an effort to produce a mutant suitable for use as a dye-transfer inhibitor in laundry detergent. The wild-type peroxidase is rapidly inactivated under laundry conditions due to the high pH (10.5), high temperature (50 °C) and high peroxidase concentration (5–10 µm). Peroxidase mutants were initially generated using two parallel approaches: site-directed mutagenesis based on structure–function considerations and error-prone PCR to create random mutations. Mutants were expressed in *S. cerevisiae* and screened for improved stability by measuring residual activity after incubation under conditions mimicking those in a washing machine. Manually combining mutations from the site-directed and random approaches led to a mutant with 110 times the thermal stability and 2.8 times the oxidative stability of wild-type CiP. In the final two rounds, mutants were randomly recombined by using the efficient yeast homologous recombination system to shuffle point mutations among a large number of parents. This *in vivo* shuffling led to the most dramatic improvements in oxidative stability, yielding a mutant with 174 times the thermal stability and 100 times the oxidative stability of wild-type CiP.[389]

The extraordinary success of breeding as a technology for modifying eukaryotic organisms is reflected in the rapid generation of diversity in domestic animals and in the combination of production-enhancing traits in crop plants. Permutation of existing variations by breeding also avoids the accumulation of deleterious mutations that accompanies an asexual evolutionary strategy. Individual proteins have traditionally been modified by choosing a single 'parent' protein and then either engineering changes based on structure and function considerations or by making random mutations and screening for improvement.

This approach has the difficulty of making mutations that improve one property without compromising others. A solution to this problem is the 'breeding' of proteins with the appropriate individual properties and then screening for 'progeny' with the desired combination. Simulating meiotic recombination at the level of single proteins is possible using the techniques of molecular breeding, such as DNA family shuffling. Thereby, multiple related genes are used as parental sequences which are combined by random fragmentation and reassembly to generate libraries of chimeric genes.

DNA family shuffling of 26 protease genes was used to generate a library of chimeric proteases that was screened for four distinct enzymatic properties. Multiple clones were identified that were significantly improved over any of the parental enzymes for each individual property. Family shuffling, also known as molecular breeding, efficiently created all of the combinations of parental properties, producing a great diversity of property combinations in the progeny enzymes. Molecular breeding, like classical breeding, is a powerful tool for recombining existing diversity to tailor biological systems for multiple functional parameters.[390]

Enzymes generated by directed evolution may play a significant role in therapies. Gene therapy for correction of genetic diseases such as sickle cell anemia involves substituting a mutant gene with its normal counterpart. This approach in gene therapy is unlikely to be effective in cancer therapy because human cancer cells contain mutations at multiple loci and have been postulated to exhibit a mutator phenotype that accelerates accumulation of further mutations. Rather than address the many mutant genes and variable combinations of mutant genes in cancer cells, a logical approach is to introduce genes encoding prodrug metabolizing enzymes that have the potential to kill cancer cells. Another approach involves protection of normal host tissues (e.g., bone marrow) against the toxicity of many chemotherapeutic agents. Chemoprotective genes can be transfected into bone marrow *ex vivo* and only a small percentage of the cells need to be transfected. Cells expressing the drug-resistance gene should have a selective proliferative advantage in the presence of chemotherapeutic drugs. The success of gene therapy might be increased by using mutated versions of enzymes for both tumor ablation and bone marrow protection.

New techniques of directed evolution make it feasible to tailor enzymes for cancer gene therapy. Novel enzymes with desired properties can be created and selected from vast libraries of mutants containing random substitutions within catalytic domains. Examples are genes for the ablation of tumors, i.e. genes that have been or can be mutated to afford enhanced activation of prodrugs and increased sensitization of tumors to specific chemotherapeutic agents. Other genes may be mutated to provide better protection of normal host tissues, such as bone marrow, against the toxicity of specific chemotherapeutic agents. Expression of the mutant enzyme could render sensitive tissues more resistant to specific cytotoxic agents.[391]

Table 8.4 gives a summary of enzymes that could be used for gene therapy together with their applications.

About 40% of commercially available enzymes are derived from filamentous fungi. These enzymes are usually produced by species of the genera *Aspergillus* and *Trichoderma*. Because they secrete large amounts of protein into the medium, they can be grown in large-scale fermentation and they are generally accepted as safe for the food industry. Improving protein production in molds is usually accomplished by introducing multiple copies of a gene encoding a desired protein into random sites on a chromosome. The market acceptance of products from such a genetically modified organism may be reduced because of growing concern about the safety of antibiotic-resistance genes and of other foreign DNA, such as remnants of bacterial vector DNA. Together, these factors make it important to

Table 8.4. Summary of enzymes that could be used by gene therapy and their applications [Refs in 391].

Tumor ablation		Protection of bone marrow	
Enzyme	Prodrug	Enzyme	Therapeutic agent
HSVTK	GCV, ACV	AGT	alkylating agents, BG
P450	procarbazine,	TS	5-FU, folate analog
	dacarbazine,	DHFR	MTX, TMTX
	CPA,	GST	mechlorethamine
	IFA	Pgp	anthracyclines, vinca
ThdPase	dFUr		alkaloids, taxol
CK	ara-C	AAG	alkylating agents
CD	5-FC	Topo I	camptothecin
E. coli gpt	6TX	Topo II	anthracyclines
E. coli nitroreductase	CB 1954	ADH	cyclophosphamide
C. acetobutylicum	metronidazol	RR	hydroxyurea
electron transport system		metallothionein	cisplatin, chlorambucil
		tubulin	Vinca alkaloids, taxol

Abbreviations not defined in text: GCV, ganciclovir, ACV, acyclovir; CPA, cyclophosphamide; IFA, ifosfamide; ThdPase, thymidine phosphorylase; dFUr, 5'-deoxy-5-fluorouridine; CK, deoxycytidine kinase; ara-C, cytosine arabinoside; CD, cytosine deaminase; 5-FC, 5-fluorocytosine; gpt, guanine phosphoribosyl transferase; 6TX, 6-thioxanthine; CB 1954, 5(-aziridine-1-yl)-2,4-dinitrobenzamide; 5-FU, 6-fluorouracil; MTX, methotrexate; TMTX, trimetrexate; Pgp, P-gylcoprotein; AAG, 3-methyladenine DNA glycosylase; Topo I, topoisomerase I; Topo II, topoisomerase II; ADH, aldehyde dehydrogenase; BR, ribonucleotide reductase.

develop methods of producing food-grade mold strains that contain multiple gene copies of a gene integrated at a predetermined locus in the genome. Ideally, the recombinant micro-organism should contain only the heterologous gene encoding the desired protein and should be free of unwanted foreign DNA. Although targeted integration of a single gene copy has been described, it is difficult and sometimes impossible to obtain mold strains that contain multiple gene copies integrated at a predetermined site in the genome.

Whereas transformation techniques for bacteria and yeasts are well established, they are less so for fungi. An alternative transformation procedure for molds based on *Agrobacterium tumefaciens* tumor-inducing DNA (T-DNA) transfer and genome integration by illegitimate recombination has been achieved, that is 100- to 1000-fold more efficient than previous methods. T-DNA can also integrate via homologous recombination into *Aspergillus awamori*. Combining these two findings allows the efficient generation of food-grade recombinant strains by targeted integration of multiple gene copies.

A. tumefaciens is known to transfer part of its tumor-inducing (Ti) plasmid to the filamentous fungus *A. awamori* by illegitimate recombination with the fungal genome. When this Ti DNA shares homology with the *A. awamori* genome, integration can also occur by homologous recombination. An efficient method was thus developed for constructing recombinant mold strains free from bacterial DNA by *A. tumefaciens*-mediated transformation. Multiple copies of a gene can be integrated rapidly at a predetermined locus in the genome, yielding transformants free of bacterial antibiotic resistance genes or other foreign DNA. Recombinant *A. awamori* strains were constructed containing up to nine copies of a *Fusarium solani pisi* cutinase expression cassette integrated in tandem at the pyrG locus. This allowed the study of how mRNA and protein levels are affected by gene copy number, without the influence of chromosomal environmental effects. Cutinase mRNA and protein were maximal with four gene copies, indicating a limitation at the transcriptional level. This transformation system will potentially stimulate market acceptance of derived products by avoiding introduction of bacterial and other foreign DNA into the fungi.[392]

8.5 Antibiotics

Antibiotics are increasingly being sought (respectively, developed) by the application of genomics and combinatorial approaches. This has been made possible by developing techniques to manipulate genes and gene clusters, and thus the metabolism of the important providers of antibiotics.

Bacteria belonging to the order *Actinomycetales* produce most microbial metabolites thus far described, several of which have found applications in medicine and agriculture. Most strains were discovered by their ability to produce a given molecule and are therefore poorly characterized physiologically and genetically. Methodologies for genetic manipulation of actinomycetes are not available and efficient tools have been developed for just a few strains. This constitutes a serious limitation to applying molecular genetics approaches to strain development and structural manipulation of microbial metabolites. To overcome this hurdle, artificial bacterial chromosomes (BAC) were developed that can be shuttled among *E. coli*, where they replicate autonomously, and a suitable *Streptomyces* host, where they integrate site-specifically into the chromosome. The existence of gene clusters and of genetically amenable host strains, such as *S. coelicolor* or *S. lividans*, makes this a sensible approach. Segments of 100 kb of actinomycete DNA can be cloned into these vectors and introduced into genetically accessible *S. lividans*, where they are stably maintained in integrated form in its chromosomes.[393]

Whereas the science of genomics has largely been driven by the desire to understand the organization and function of the human genome, the determination and characterization of smaller, less complex genomes, notably bacteria and yeast, has preceded that of the human genome, providing a testing ground for high-throughput methods in synthesis, screening, etc.

In previous decades, the search for antibiotics has been largely restricted to well-known compound classes active against a standard set of drug tests. Although many effective compounds have been discovered, insufficient chemical variability (and lack of novel targets and target mechanisms) has been generated to prevent a serious escalation in clinical resistance. Recent advances in genomics have provided an opportunity to expand the range of potential drug targets and have facilitated a fundamental shift from direct antimicrobial screening programs toward rational target-based strategies. The application of genome-based technologies such as expression profiling and proteomics will lead to further changes in the drug discovery paradigm by combining the strengths and advantages of both screening strategies in a single program.[394]

Medically useful semisynthetic cephalosporins are made from 7-aminodeacetoxy-cephalosporanic acid (7-ADCA) or 7-aminocephalosporanic acid (7-ACA). A new industrially amenable bioprocess for the production of the important intermediate 7-ADCA that can replace the expensive and environmentally unfriendly chemical method classically used has been developed. The method is based on the disruption and one-step replacement of the *cefEF* gene, encoding the bifunctional expandase/hydroxylase activity, of an actual industrial cephalosporin C production strain of *Acremonium chrysogenum*. Subsequent cloning and expression of the *cefEF* gene from *Streptomyces clavuligerus* in *A. chrysogenum* yield recombinant strains producing high titers of deacetoxycephalosporin C (DAOC). The production level of DAOC is nearly equivalent (75–80%) to the total β-lactams biosynthesized by the parental overproducing strain. DAOC deacylation is carried out by two final enzymatic bioconversions catalyzed by D-amino acid oxidase (DAO) and glutaryl acylase (GLA) yielding 7-ADCA. In contrast to the data reported for recombinant strains of *Penicillium chrysogenum* expressing ring expansion activity, no detectable contamination with other cephalosporin intermediates occurred.[395] Figure 8.1 illustrates the four different strategies for 7-ADCA production.

Identification of genes that encode essential products provides a promising approach to validation of new antibacterial drug targets. A mariner-based transposon was developed, TnAraOut, that allows efficient identification and characterization of essential genes by transcriptionally fusing them to an outward-facing, arabinose-inducible promoter, located at one end of the transposon. In the absence of arabinose, such TnAraOut fusion strains display pronounced growth defects. Of a total of 16 arabinose-dependent TnAraOut mu-

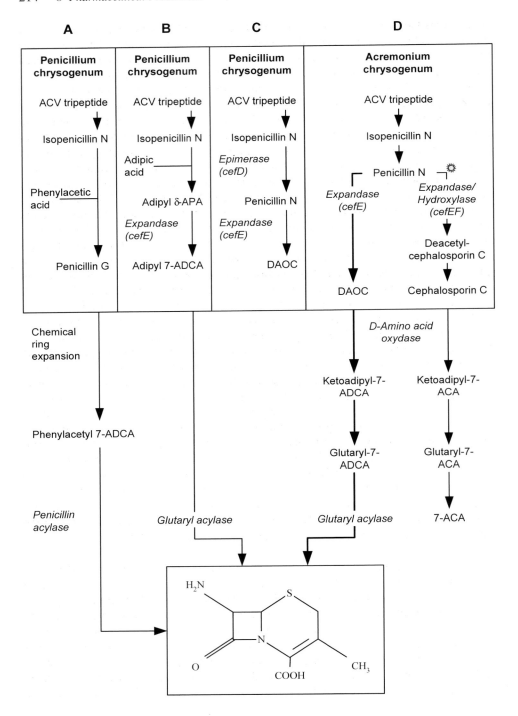

Figure 8.1. Four different strategies for 7-ADCA production. (A) Current industrial process for producing 7-ADCA involving the chemical ring expansion of penicillin G to phenylacetyl-7-ADCA. The aromatic side chain is removed using a penicillin acylase[1]. (B) Production of 7-ADCA by biosynthesis of adipyl-cephalosporins in recombinant strains of *P chrysogenum* harboring the *cefE* gene of 8. *clavuligerus* and feeding adipic acid as side chain precursor[7]. (C) Process using transformants of *P chrysogenum* carrying the *cefD* and *cefE* genes from *S. clavuligerus*. Some transformants produced by levels of DAOC[3][4]. (D) Industrial bioprocess involving the *A. chrysogenum ΔcefEF-cefE* transformants described in this work. Heavy arrows indicate the new process to 7-ADCA. The inactivated biosynthetic step *(cefEF,* expandase-hydroxylase) is indicated by a sun (I). DAOC constitutes the starting material for 7-ADCA production by a process analogous to the conversion of cephalosporin C to 7-ACA [Refs in 395].

---

tants characterized in *Vibrio cholerae*, four were found to carry insertions upstream of known and hypothetical genes not previously shown to encode essential gene products. One of the essential genes identified by this analysis appears to be unique to *V. cholerae* and thus may represent an example of a species-specific drug target.

More than 50 prokaryotic genome-sequencing projects are currently under way. Based on the results of completed genome sequence projects, these efforts will lead to the identification of many hypothetical ORFs of unknown function. One class of genes of high interest are those that are essential for bacterial growth and survival. The identification and characterization of such 'essential' genes has played an important role in the ability to understand basic cellular processes. The fact that a large fraction (10–30% depending on the organism) of the potential ORFs in sequenced bacterial genomes are highly conserved, yet we cannot assign a function to them, suggests that essential biological processes remain to be discovered in the post-genomic era. Since most antibiotics target essential cellular processes, essential gene products of microbial cells represent promising new targets for antibiotics.[396]

Polyketides represent a large and diverse group of natural products with an impressive wealth of anticancer, immunosuppressive, antiparasitic and antimicrobial activities. These natural products are assembled by polyketide synthases in a common pathway that resembles fatty acid biosynthesis. Simple units such as malonyl-CoA and methylmalonyl-CoA are used in decarboxylative condensations to grow a polyketide chain from a starter unit such as propionyl-CoA. Variations in the number of extensions, choice of starter units and extender units, and the amount and stereochemistry of reduction of the polyketide chain after each condensation, all contribute to the structural diversity of these natural products. More structural diversity is accomplished by glycosylases, methyltransferase and oxidases, which act on these polyketide structures.

The development of combinatorial biosynthetic strategies, in which components of different polyketide pathways are combined to generate new polyketides, has attracted interest as a new drug discovery tool. Extension of previous approaches focused on combinations of different polyketide synthase components to inclusion of pathways that generate different polyketide starter units provides an additional element of structural diversity.

The side chain of the antifungal antibiotic ansatrienin A from *Streptomyces collinus* contains a cylohexanecarboxylic acid (CHC)-derived moiety. This moiety is also observed in trace amounts of ω-cyclohexyl fatty acids (typically less than 1% of total fatty acids) produced by *S. collinus*. Coenzyme A-activated CHC (CHC-CoA) is derived from shikimic acid through a reductive pathway involving a minimum of nine catalytic steps. Five putative CHC-CoA biosynthetic genes in the ansatrienin biosynthetic gene cluster of *S. collinus* have been identified. Plasmid-based heterologous expression of these five genes in *S. avermitilis* or *S. lividans* allows for production of significant amounts of ω-cyclohexyl fatty acids (as high as 49% of total fatty acids). In the absence of the plasmid these organisms are dependent on exogenously supplied CHC for ω-cyclohexyl fatty acid production. Doramectin is a commercial antiparasitic avermectin analog produced by fermenting a *bkd* mutant of *S. avermitilis* in the presence of CHC. Introduction of *the S. collinus* CHC-CoA biosynthetic gene cassette into this organism resulted in an engineered strain able to produce doramectin without CHC supplementation. The CHC-CoA biosynthetic gene cluster represents an important genetic tool for precursor-directed biosynthesis of doramectin and has potential for directed biosynthesis in other important polyketide-producing organisms.[397]

An enormous range of medically important polyketide and peptide natural products assembled by modular polyketide synthases (PKSs), non-ribosomal peptide synthases (NRPSs) and mixed PKS/NRPS systems have macrocyclic structures, including the antibiotics erythromycin (PKS) and daptomycin (NRPS), the immunosuppressants cyclosporin (NRPS) and rapamycin (PKS/NRPS), and the antitumor agent epothilone (PKS/NRPS). PKSs and NRPSs are large, multifunctional proteins that are organized into sets of functional domains termed modules. The order of modules corresponds directly to the sequence of monomers in the product. Synthetic intermediates are covalently tethered by thioester linkages to a carrier protein domain in each module. The thiol tether on each carrier domain is phosphopantetheine, which is attached to a conserved serine residue in the carrier protein in a post-translational priming reaction catalyzed by a phosphopantetheinyltransferase.

In the biosynthesis of many macrocyclic natural products by multidomain megasynthases, a C-terminal thioesterase (TE) domain is involved in cyclization and product release. It has not been determined whether TE domains can catalyze macrocyclization (and elongation in the case of symmetric cyclic peptides) independently of upstream domains. The inability to decouple the TE cyclization step from earlier chain assembly steps has precluded determination of TE substrate specificity, which is important for the engineered biosynthesis of new compounds. The excised TE domain from tyrocidine synthetase efficiently catalyzes cyclization of a decapeptide-thioester to form the antibiotic tyrocidine A and can catalyze pentapeptide-thioester dimerization followed by cyclization to form the antibiotic gramicidin S. By systematically varying the decapeptide-thioester substrate and comparing cyclization rates, it was shown that only two residues (one near each end of the decapeptide) are critical for cyclization. This specificity profile indicates that the tyrocidine synthetase TE and, by analogy many other TE domains, are able to cyclize and release a broad range of new substrates and products produced by engineered enzymatic assembly lines.[398]

Peptide antibiotics such as penicillins and cephalosporins, vancomycins, bacitracin, actinomycin D, the antitumor peptide bleomycin, and the immunosuppressant cyclosporin A are all assembled on NRPSs. The identity and sequence of amino acid residues in these non-ribosomal peptides is dictated by the organization of sets of iterated modules in the megasynthetases.

In non-ribosomal biosynthesis of peptide antibiotics by multimodular synthetases, amino acid monomers are activated by the adenylation domains of the synthetase and loaded onto the adjacent carrier protein domains as thioesters, then the formation of peptide bonds and translocation of the growing chain are effected by the synthetase's condensation domains. Whether the condensation domains have any editing function has been unknown. Synthesis of aminoacyl-CoA molecules and direct enzymatic transfer of aminoacyl-phospho-pantetheine to the carrier domains allow the adenylation domain editing function to be by-passed. This method was used to demonstrate that the first condensation domain of tyroci-dine synthetase shows low selectivity at the donor residue (D-phenylalanine) and higher selectivity at the acceptor residue (L-proline) in the formation of the chain-initiating D-Phe–L-Pro dipeptidyl-enzyme intermediate.[399]

Modular polyketide synthases catalyze the biosynthesis of medically important natural products through an assembly-line mechanism. Although these megasynthases display very precise overall selectivity, their constituent modules are remarkably tolerant toward diverse incoming acyl chains. By appropriate engineering of linkers, which exist within and between polypeptides, it is possible to exploit this tolerance to facilitate the transfer of biosynthetic intermediates between unnaturally linked modules. This protein engineering strategy provides a new strategy for combinatorial biosynthesis, in which modules rather than individual enzymatic domains, are the building blocks for genetic manipulation, and furthermore gives insights into the evolution of modular polyketide synthases.[400]

Lovastatin is an inhibitor of the enzyme (3S)-hydrooxymethylglutaryl-CoA (HMG-CoA) reductase that catalyzes the reduction of HMG-CoA to mevalonate, a key step in cholesterol biosynthesis. This activity confers on lovastatin its medically important anti-hypercholesterolemic activity and other potentially important uses. It is a secondary metabolite from the filamentous fungus *Aspergillus terreus* and has been shown to be derived from acetate via a polyketide pathway.

Polyketides, the ubiquitous products of secondary metabolism in microorganisms, are made by a process resembling fatty acid biosynthesis that allows the suppression of reduction or dehydration reactions at specific biosynthetic steps, giving rise to a wide range of often medically useful products. The lovastatin biosynthesis cluster contains two type I polyketide synthase genes. Synthesis of the main nonaketide-derived skeleton was found to require the previously known iterative lovastatin nonaketide synthase (LNKS), plus at least one additional protein (LovC) that interacts with LNKS and is necessary for the correct processing of the growing polyketide chain and the production of dihydromonacolin L. The non-iterative lovastatin diketide synthase (LDKS) enzyme specifies formation of 2-methylbutyrate and interacts closely with an additional transesterase (LovD) responsible for assembling lovastatin from this polyketide and monacolin J.[401]

Additional enzymes may add to the versatility of gene clusters and novel compounds synthesized by combining previously unrelated genes from different organisms.

Chalcone synthases, which biosynthesize chalcones (the starting materials for many flavonoids), were believed to be specific to plants. The *rppA* gene from the Gram-positive, soil-living filamentous bacterium *Streptomyces griseus* encodes a 372-amino-acid protein that shows significant similarity to chalcone synthases. Several *rrpA*-like genes are known, but their functions and catalytic properties have not been described. A homodimer of RppA catalyzes polyketide synthesis: it selects malonyl-CoA as the starter, carries out four successive extensions and releases the resulting pentaketide to cyclize to 1,3,6,8-tetrahydroxynaphthalene (THN). Site-directed mutagenesis showed that, as in other chalcone synthases, a cysteine residue is essential for enzyme activity. Disruption of the chromosomal *rppA* gene in S. griseus abolished melanin production in hyphae, resulting in 'albino' mycelium. THN was readily oxidized to form 2,5,7-trihydroxy-1,4-naphtoquinone (flaviolin), which then randomly polymerized to form various colored compounds. THN formed by RppA appears to be an intermediate in the biosynthetic pathways for not only melanins but also various secondary metabolites containing a naphthoquinone ring. RppA is a chalcone-synthase-related synthase that synthesizes polyketides and is found in the *Streptomyces* and other bacteria.[402]

The elucidation of molecular structures of enzymes in biosynthetic pathways aids in the design of proteins with modified specificities.

Isopenicillin N synthase (IPNS), a non-heme iron-dependent oxidase, catalyzes the biosynthesis of isopenicillin N (IPN), the precursor of all penicillins and cephalosporins. The key steps in this reaction are the two iron-dioxygen-mediated ring closures of the tripeptide δ-(L-α-aminoadipoyl)–L-cysteinyl–D-valine (ACV). The four-membered β-lactam ring forms initially, associated with a highly oxidized iron(IV)-oxo-(ferryl) moiety, which subsequently mediates closure of the five-membered thiazolidine ring. The observation of the IPNS reaction in crystals by X-ray crystallography has been described. IPNS-Fe^{2+} substrate crystals were grown anaerobically, exposed to high pressures of oxygen to promote reaction and frozen. Their structures were subsequently elucidated by X-ray crystallography. Using the natural substrate ACV, this resulted in the IPNS–Fe^{2+}–IPN product complex. With the substrate analog, δ-(L-α-aminoadipoyl)–L-cysteinyl–L-S-methyl-cysteine (AcmC) in the crystal, the reaction cycle was interrupted at the monocyclic stage. These mono- and bicyclic structures support the hypothesis of a two-stage reaction sequence leading to penicillin. The formation of a monocyclic sulfoxide product from AcmC is best explained by the interception of a high-valency iron-oxo species.[403]

Escherichia coli was metabolically engineered to produce a crucial intermediate for complex polyketides synthesized by the bacterium Saccharopolyspora through the action of a multifunctional polyketide synthase (PKS). Genetically engineered E.coli convertes exogenous propionate into the macrocyclic core of the antibiotic erythromycon, 6-deoxyerythronolide B (6dEB) with a productivity comparable to an industrially used high-producing mutant of S. erythraea.[503]

The sequential coordinated action of the modular PKSs is illustrated in Figure 8.10, where (A) shows the 6-deoxyerythronolide B synthase with catalytic domains, (B) depicts the production of triketide lactone, and (C) illustrates the rifamycin synthetase, a polyketide synthase naturally primed by a nonribosomal PKS loading module which may be engineered to utiliize exogenous acids for the sythesis of substituted macrocycles.

8.6   Biocatalysis

The chemical processes and transformations underpinning all biological phenomena are executed by enzymes that transform small molecules, macromolecular substrates and transfer energy. The numerous chemical reactions are enabled by hundreds to thousands of proteins, enzymes, and by RNAs and ribozymes, which have catalytic activity for the conversion of specific substrates to products. The location and lifetime of proteins is controlled by the family of related proteases that hydrolyze peptide bonds in a controlled way. The degree of specificity ranges from highly specific actions on particular substrate bonds to a general non-specific hydrolysis action that attacks a broad set of substrates. The high specificity is usually found in signaling pathways, whereas the broad reactions apply to protein degradation. The signaling pathways are initiated by catalytic action of cascades of protein kinases that are related proteins and catalyze phosphoryl transfer from ATP to the side-chain hydroxyl of serine, threonine or tyrosine residues. The selectivity is achieved by specific protein–protein interactions between kinase and protein substrate acting in cascades and ultimately producing changes in activity and location of proteins as well as selective gene activation.

Enzymes may also catalyze transformations of a highly selective and unique nature such as in transformations of small molecules. Examples are the fragmentation of 1-amino-cyclopropane-1-carboxylate to the fruit ripening hormone ethylene, the cycloreversion of thymine dimers in DNA repair, the synthesis of isopenicillin and the reduction of molecular nitrogen (N_2) to ammonia (NH_3).[488]

Enzymes demonstrate both high specificities and significant reaction rate accelerations. The relative values of enzymic over non-enzymic reactions may be from 10^{10} to 10^{23} (orotidine decarboxylase) and the turnover numbers range from one catalytic event per minute to 10^5 per second (hydration of CO_2 to HCO_3^- by carbonic anhydrase). The molecular entities of enzymes cover proteins, ribozymes and catalytic antibodies.

The sequential reactions in elongating acyl transfers in the synthesis of polyketide natural products and non-ribosomal peptide antibiotics such as erythromycin, rapamycin, epothilone, lovastatin, penicillins, cyclosporin and vancomycin resemble molecular solid-state assembly lines. Such multimodular enzymes may be utilized in combinatorial biosynthesis by way of reprogramming for the manufacture of unnatural analogs of natural products.

IPNS belongs to a family of iron-containing enzymes which use Fe^{2+} to activate O_2 and simultaneously the specific co-substrate for redox reactions whereby both atoms of O_2 are reduced to water and the tripeptide ACV undergoes oxidation with C–S and C–C bond formation to generate β-lactam molecules.

A related enzyme, the expandase enzyme, is deployed by cephalosporin-producing organisms to expand the five-membered penicillin ring to the six-membered cephalosporin ring.

The increasingly stringent environmental requirements demand new catalytic synthetic processes as does the necessity to produce the novel types of organic compounds generated in biomolecular research as active molecules or targets. Enzyme-catalyzed chemical transformations are providing useful improvements over non-biological organic syntheses and offer solutions to synthetic challenges. In organic synthesis the enzymes may be used under non-natural conditions of solvent (other than water), with non-biological reagents and on unnatural substrates. The chiral nature of enzymes enables the creation of stereochemically defined products with significant reaction rate accelerations of 10^5 to 10^8. The specificities, the activity and the stability of enzyme catalysts may be altered by genetic engineering. Enzymes can be produced on a large scale by using recombinant DNA technologies and thus offer possibilities for new industrial chemical and pharmaceutical synthetic processes.[489]

Table 8.5 lists the enzymes commonly used in organic synthesis and the type of reactions they are involved in.

The synthesis of enantiomerically pure intermediates and active products is a major requirement for the pharmaceutical industry. Hydrolytic biocatalysts such as esterases, lipases and proteases are employed for the preparation of enantiopure compounds from racemic precursors, prochiral compounds, and diastereomeric mixtures. Hydrolytic enzymes alsocatalyze reverse hydrolysis and thus offer access to both enantiomers of a specific compound. Examples are the use of enol esters as *trans*-esterification reagents and the combination of hydrolytic enzymes with racemization catalysts.

Regeneration of the cofactor is crucial for an economic industrial process and it drives the reaction towards the desired product, thereby reducing the accumulation of intermediates, facilitating the recovery of product and increasing the enantioselectivity.

The usefulness of enzymes is being enhanced by deploying them in organic solvents rather than aqueous media. In such solvent or solvent–water mixtures enzymes can catalyze reactions impossible in water, display greater stability and show behavior such as molecular memory. Enzymatic selectivity, and substrate-, stereo-, regio- and chemo-selectivity are affected and occasionally modified by the solvent. Enzyme reactions of industrial applicability have been demonstrated in organic solvents, supercritical fluids and the gaseous phase.[490]

Table 8.5. Enzymes commonly used in organic synthesis [Refs in 489].

Enzymes	Reactions
Esterase, lipases	Ester hydrolysis
Amidases (proteases, acylases)	Amid hydrolysis, formation
Dehydrogenases	Oxidoreduction of alcohols and ketones
Oxidases (mono- and dioxygenases)	Oxidation
Peroxidases	Oxidation, epoxidation, halohydratation
Kinases	Phosphorylation (ATP-dependent)
Aldolases, transketolases	Aldol reaction (C-C bond)
Glycosidases, glycosyltransferases	Glycosidic bond formation
Phosphorylases, phosphatases	Formation and hydrolysis of phosphate
Sulphotransferases	Formation of sulfate esters
Transaminases	Amino acid synthesis (C-N bond)
Hydrolases	Hydrolysis
Isomerases, lyases, hydratases	Isomerization, addition, elimination, replacement

Modular enzymes fall into three categories of modular catalysts:
- Enzymes where catalysis and substrate specificity reside in separate domains
- Multisubstrate enzymes wherein the binding sites for individual substrates are modular
- Multienzyme systems which catalyze programmable metabolic pathways

The ribosomal protein biosynthetic machinery encompasses all three types of modularity. The ribosome catalyst can be separated from the element determining the substrate specificity, i.e. the mRNA template. The two acylated tRNA substrates for the peptide-bond formation bind to different ribosomal sites, the A and P sites, and the multistep pathway catalyzed by the modular system may be reprogrammed by codon choice.

The individual modules of polyketide synthases (PKSs) and of NRPSs represent another example of modularity. An electrophile and a nucleophile are covalently attached to two different subunits of a multidomain enzyme. The selectivity for the electrophile resides in the domain catalyzing bond formation between two substrates and the selectivity for the nucleophile is controlled by a transfer domain which attaches the nucleophile onto a carrier domain.[491]

Multienzyme modular assemblies such as PKSs and NRPSs have flexible swinging tethers which channel covalently bound intermediates between successive active sites. Swinging arms plus specific protein–protein interactions offer mechanisms for the transfer of substrates between modules and offer concepts for the development of one-pot multi-reaction biocatalytic processes.

The quest for purpose-designed biocatalysts relies on the rational redesign of natural biocatalysts and on evolutionary procedures to obtain novel catalysts. To rationally design biocatalysts requires knowledge of the connection between structure and function, for proteins as well as for ribozymes. The capabilities for manipulations of novel structures, organisms, bio-electronic hybrid devices will be targeted at process biocatalysts, novel therapeutic enzymes, specifically designed cellular machinery and metabolic networks.[492]

Rational design, directed evolutionary methods and combinatorial synthesis are the available techniques for the development of novel biocatalysts. Evolutionary protein design methods utilize random mutagenesis, gene recombination and high-throughput screening, and resemble breeding procedures. Molecular breeding can be extended to involve complex synthetic tasks involving several interacting enzymes and even the creation of new pathways in microorganisms, plants and animals. Genes encoding the enzymes for synthesizing in a series of reactions may be combined in suitable organisms. An example of engineering a biosynthetic pathway for carotenoid synthesis is the transfer of genes from *Erwinia* sp. into *E. coli* together with a gene library of shuffled versions of a *Erwinia*-saturase led to novel carotenoids as well as novel synthetic pathways for known carotenoids.[380]

De novo catalyst design by rational principles is emerging as a tool for the design of catalysts. Iron and oxygen binding sites were introduced into thioredoxin by computational design. The rudimentary active sites demonstrate enzyme activities selecting between different oxygen chemistries and different control mechanisms are observed. The reactivity of the active sites is determined by the microenvironment and the interactions of the protein matrix with the metal center.[493]

Table 8.6. Observed catalytic activities for designed proteins [Refs in 493].

Site	Fenton* $k_{cat}\,s^{-1}$	Fenton* $K_m\,mM$	SOD $k\,10^6 \cdot M^{-1}\,s^{-1}$	Udenfriend[‡] $k\,s^{-1}$
G1	0.80	1.70	0.75	9.70
G2	0.70	1.60	0.10	0.01
G3	0.50	0.60	0.10	16.00
S1	1.50	6.50	2.30	0.01
S2	0.50	1.30	6.40	0.01
D1	0.03	0.14	3.30	30.00

*   Absence of ascorbate reduces rates 100- to 1000-fold, indicating that intrinsic H_2O_2 disproportio-na-tion is not a major contributing factor.
‡   Pseudo first-order rate constant: M catechol, M^{-1}; 1:1 Fe/protein, s^{-1}.

Models of Fe.His3.O2 sites were designed in *E. coli* thioredoxin and Table 8.6 lists the observed catalytic activities for designed proteins.

The correlation of structure with function was studied in the important enzyme and biocatalyst class of the proteases. Consistent structural similarities among unrelated proteases were observed, such as more helices and loops, smaller surface areas, smaller radii of gyration and higher Cα densities, leading to tighter packing structures. A neural network predicts protease function from structural data with over 86% accuracy. Such structural–functional relationship studies yield the predictive designs for constructing novel proteases.[494]

Fundamental studies of enzyme catalysis decipher the mechanisms of enzymatic reactions at the molecularquantum mechanical level for the refined design of novel catalysts. Hybrid quantum mechanics/molecular mechanics calculation with Austin Model 1 system-specific parameters in studies of the S_N2 displacement reaction of chloride from 1,2-dichloroethane (DCE) by nucleophilic attack of the carboxylate of acetate in the gas phase and by Asp124 in the active site of haloalkane dehalogenase from *Xanthobacter autotrophicus* GJ10 showed that the activation barrier for nucleophilic attack of acetate on DCE depends on the geometry of reactants. This haloalkane dehalogenase lowers the activation barrier for dehalogenation of DCE by 2–4 kcal/mol relative to the single point energies of the enzyme's quantum mechanics atoms in the gas phase. S_N2 displacements of this type are infinitely slower in water than in the gas phase.[495]

Novel protein domains may be generated by combinatorial shuffling of polypeptide segments to generate specific three-dimensional structures with desired functions. The architecture of protein domains has evolutionarily evolved by combinatorial assembly and exchange of smaller polypeptide segments (respectively, the pertinent gene sequences). DNA encoding the N-terminal half of a β-barrel domain from the cold-shock protein CspA was fused with fragmented genomic DNA from *E. coli*. The resulting repertoire of chimeric genes coding for chimeric polypeptides was cloned for display on filamentous bacteriophage. Phages displaying folded polypeptides were selected by proteolysis and the resulting protease-resistant chimeric polypeptides consisted of genomic sequences in their original reading frames. One of the obtained proteins had a fold similar to CspA without showing sequence homology. Several chimeric proteins were soluble polypeptides forming monomers and exhibiting cooperative unfolding. One chimeric protein actually contained several very slowly exchanging amides and was more stable than CspA itself. Combinatorial assembly of segments from non-homologous proteins and diverse sources can be used to generate novel proteins, domains and architectures.[496]

Evolutionary methodology can be used to generate catalytic antibodies for chemical process steps. The reactive immunization procedure employed uses reactive chemicals as immunogens. The immune system selects chemical steps such as the formation of a covalent bond between the antibody and the antigen rather than a simple recognition and binding process. Catalytic aldolase antibodies were generated by immunization with two different but structurally related β-diketone haptens and the resulting antibodies were sequenced to allow study of similarities, differences, and structural determinants of properties of independently evolved catalysts. Kinetic and sequence analysis, mutagenesis, and structural and modeling studies demonstrate a somatic mutation placing a lysine residue into a deep hydrophobic pocket as the crucial event of the catalyst evolution. In addition to the non-covalent interactions which have been the focus of immunochemistry, covalent chemistries can be procured from the immune repertoire.[497]

Table 8.7 lists the kinetic parameters for antibody-catalyzed aldol and *retro*-aldol reactions.

Biocatalysis on an industrial scale covers a range of reactions, solvents and products, thus extending from the manufacture of large-volume chemicals such as in acrylamide production to numerous syntheses of optically active alcohols, amines, nitriles, peptides and semisynthetic antibiotics, to numerous intermediates.[498]

Several recently developed biocatalytic systems at several chemical companies are listed in Table 8.8, illustrating the range of products (amides, alcohols, acids, amino acids, penicillins, N-heterocyclic compounds, and non-proteinogenic amino acids).

The factors determining the industrial and economic feasibility of biocatalytic processes are depicted in Figure 8.2, showing the key aspects of the synthetic process (and its economics), biocatalyst selection and characterization, biocatalyst engineering, its application in industrial use, and product recovery.

For reasons of stability, handling, separation and recovery, the industrial biocatalyst is preferably applied as an entrained molecule.

Lipases are an important class of enzymes for biocatalysts. They serve to synthesize enantiomerically pure alcohols, chiral amines, and nitrilases for amino- and hydroxy-carboxylic acids.

The bioconversion of steroids, the products of which are used in steroidal hormone pharmaceuticals, is carried out at a scale of more than 1000 tons/year starting from natural phytosterols from soya, conifers and from rape seed (β-sitosterol, campesterol, stigmasterol, brassicasterol, and β-sitostanol).

Ribozymes cleave RNA molecules at specific sites. In order to modify and extend the number of cleavable target sequences on the mRNA, an allosteric version of a ribozyme, a maxizyme, *in vivo* activity and specificity was developed several custom-designed maxizymes showed sensor functions, thus indicating the applicability of the technology for the design of maxizymes with use in analytics and therapy.[499]

Table 8.7. Kinetic parameters for antibody-catalyzed aldol and *retro*-aldol reactions [Refs in 497].

Substrate	Antibody	$K_{cat} \text{min}^{-1}$ *	$K_m \mu M$ *	K_{cat}/K_{un} [†]	$(K_{cat}/K_m)/K_{un}$ [‡]
	40F12	0.0054	33	2.4×10^4	7.2×10^8
	42F1	0.0026	41	1.1×10^4	2.8×10^8
	38C2	0.0067	17	2.9×10^4	1.7×10^9
	40F12	0.071	203	6.5×10^5	3.2×10^9
	42F1	0.048	271	4.3×10^5	1.6×10^9
	38C2	0.21	123	1.9×10^6	1.5×10^7
	40F12	0.21	317	8.8×10^5	2.8×10^9
	42F1	0.14	266	5.8×10^5	2.2×10^9
	38C2	1.1	184	4.6×10^6	2.5×10^{10}
	40F12	0.11	130	6.1×10^2	4.7×10^6
	33F12	0.21	317	1.2×10^3	3.8×10^6
	38C2	4.8	20	2.7×10^4	1.4×10^9
	40F12	0.11	130	1.1×10^5	8.5×10^8
	33F12	0.11	43	1.1×10^5	2.6×10^9
	38C2	1.0	14	1.0×10^6	7.1×10^{10}

* The kinetic data of K_{cat} and K_m (per antibody active site) were obtained in PBS at pH 7.4 by fitting experimental data to nonlinear regression analysis using GRAFIT software.
[†] Aldol reactions with unit (M);. [‡] *Retro*-aldol reactions with a unit (M^{-1}).

The term maxizyme is chosen for the designed ribozymes because they are highly active, minimized dimeric ribozymes [minimized, active, X-shaped (functioning as a dimer), intelligent (allosterically controllable) ribozymes

Biocatalytic processes for the manufacturing of complex or sensitive molecules require highly selective separation and purification methods. The *in situ* separation of inhibitory or toxic byproducts or the shifting of unfavourable equilibria are additional aims of bioprocess control technologies. *In situ* product removal advances both the control of biocatalytic processes and the recovery of target molecules.[587]

The sensitivity of biological molecules to detrimental forces during purification and recovery may also be addressed by means of modifying the target molecules. The manufacturing of retroviral vectors for gene therapy is complicated by the sensitivity of these viruses during processing. To isolate viruses resistant to these manufacturing processes, breeding of ecotropic murine leukemia virus strains was performed by DNA shuffling. The envelope regions were shuffled to generate a recombinant library and subjected to three consecutive concentration processes with amplification of the surviving virus after each cycle.

Table 8.8. Recently developed biocatalytic systems at several chemical companies [Refs in 498].

Figure	Product	Substrate	Reaction	Bio-catalyst	Enzyme	Organ-ism*	Scale Tons yr^{-1}	Yield	Applica-bility
A. Amides, alcohols, acids (source BASF)									
5a	Enantio-pure alcohols	Racemic alcohols	Resolu-tion	Enzymes	Lipases		Thousand	Excellent	Broad spectrum of alcohols
5b	R-amide S-amine	Racemic amines	Resolu-tion	Enzymes	Lipases		Several hundred	Excellent	Broad spectrum of amines
5c	R-mandelic acid	Racemic mandelo-nitrile	Hydroly-sis	Enzymes	Nitrilases		Several	Excellent	
B. Amino acids, penicillins (source DSM)									
6a	Non-proteo-genic L-amino acids	Racemic amino acid, amides	Kinetic resolution	Enzymes	Amidases	*P. putida, M. neo-aurum O. an-tropii* rec. *E. coli*	Few to several hundred		Broad spectrum of L- + D-amino acids
6b	L-Aspar-tic acid	Fumaric acid	Addition of ammo-nia	Enzymes	Aspartic acid ammonia lyase	*E. coli*	Thousand		L-Aspartic acid
6b	Aspar-tame (L-α aspartyl-L-phenyl-alanine methyl ester)	N-pro-tected L-aspartic acid, D/L-phenyl-alanine methyl ester	Selective coupling	Enzymes	Thermo-lysine	*B. subtilis*	Thousand		Aspartame
6c	6-APA$^{\#}$	Penicillin G/V	Hydro-lysis	Enzymes	Penicillin acylase	*E. coli*	Thousand		6-APA$^{\#}$
6c	Semi syn-thetic penicil-lins	6-APA$^{\#}$	Selective coupling	Enzymes	Acylases	*E. coli P. putida*	Few to several hundred		Semi syn-thetic penicillins and cepha-losporins

Table 8.8. Recently developed biocatalytic systems at several chemical companies (Cont'd).

Figure	Product	Substrate	Reaction	Bio-catalyst	Enzyme	Organism*	Scale Tons yr^{-1}	Yield	Applicability
C. N-Heterocyclic compounds (source Lonza)									
7a	6-hydroxy-nicotinic acid	Niacin	Addition of water	Whole cells	Niacin Hydroxy-lase	A. xy-loso-oxidans LKI	Few	65 gl^{-1}	Insecticides
7b	5-Hydroxy-pyrazine-carboxylic acid	2-Cyano-pyrazine	Addition of water	Whole cells	Nitrilase/hydroxy-lase	Agro-bacterium DSM 6336	Development product	40 gl^{-1}	Anti-tuberculosis drugs
7c	6-Hydroxy-S-nicotine	(S)-nicotine	Addition of water	Whole cells	Hydroxy-lase	oxydans NRRL-B3603	Development product	30 gl^{-1}	
7c	4-[6-Hy-droxy-pyridin-3-yl]-4-oxobu-tyrate	(S)-nicotine	Complex reaction	Whole cells	Several	Pseudo-monas sp. DSM865 3	Development product	15 gl^{-1}	Analogues of Epi-batine
D. Non-proteinigenic amino acids (source Lonza)									
7d	S-Pipe-razine-2-carboxylic acid	(R,S)-Pipera-zine-2-carboxy-lic acid	Selective amidase	Whole cells	Stereo specific amidases	*K. terri-gena DSM 9174*	Development product	9 gl^{-1}	Various bio-active compounds
7e	5-Methyl pyrazine-2-carboxylic acid	2,5-Di-methyl-pyrazine	Selective oxidation of a methyl substituent on an aromatic N-hetero-cycle	Whole cells	Xylene oxidation pathway	*P. putida ATCC330 15*	Several	20 gl^{-1}	carboxylic acid, Synton for anti-lipolytic drug

*   Organism: *A. oxydans, Arthrobacter oxydans; A. xylosoxydans, Achromobacter xylosoxydans; B. subtilis, Bacillus subtilis; E. coli, Escherichia coli; K. terrigena, Klebsiella terrigena; M. neoaurum, Mycobacter neoaurum; O.antropii, Ochrobactrum antropii; P. putida, Pseudomonas putida.*

\#   6-APA, 6-Aminopenicillanic acid

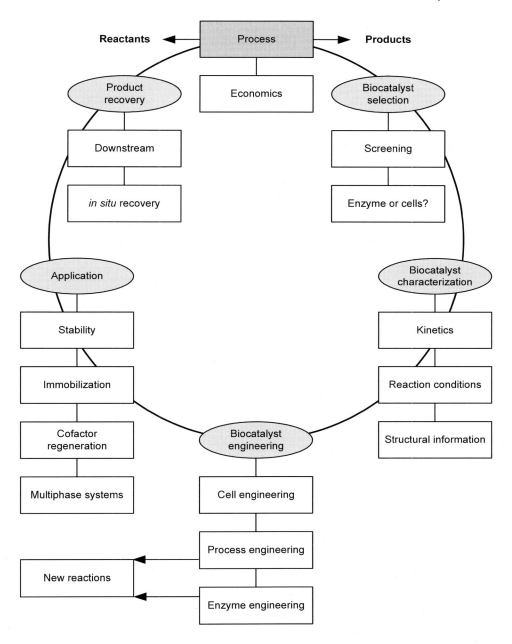

Figure 8.2. The biocatalysis cycle [Refs in 498].

Several viral clones with greatly improved stability were isolated. The envelopes of the resistant viruses differed in DNA and protein sequence and were complex chimeras from multiple parents. DNA shuffling is useful in breeding viral strains with improved characteristics for gene therapy and other manufacturing or biomedical applications.[597]

A high-throughput quantitative method was developed for the isolation of enzymes with novel substrate specificities from large libraries of protein variants. Protein variants are displayed on the surface of microorganisms and incubated with a synthetic substrate comprising a fluorescent dye, a positively charged moiety, the target fissile bond, and a fluorescence resonance energy transfer (FRET) quenching moiety. Enzymatic cleavage of the scissile bond effects the release of the FRET quenching partner while the fluorescent product remains on the cell surface, allowing isolation of catalytically active clones by fluorescence-activated cell sorting (FACS). *Escherichia coli* expressing the serine protease OmpT was enriched over 5,000 fold in a single round from cells expressing an inactive OmpT variant. Variant proteases with catalytic activities enhanced 60-fold were isolated. The method allows the high-throughput screening of large catalytic libraries on the principle of catalytic turnover.[630]

The efforts to construct an allosteric enzyme which functions *in vivo* have thus led to a maxizyme, a construct, which includes a sensor function and thus allows to specifically cleave abnormal chimeric mRNA targets without affecting the wild-type mRNA, a task that cannot be achieved with conventional hammerhead ribozymes. Maxizymes can be designed to attack and cleave specific sequences of interest, such as in the expression of an abnormal chimeric gene lacking a NUX cleavage site, a target generated from reciprocal chromosomal trans-locations observed in several leukemia diseases such as acute lymphoblastic leukemia (ALL), acute promyelocytic leukemia (APL), and chronic myelogenous leukemia (CML). Specific cleavage of abnormal mRNA without affecting normal transcripts was also observed in studies of Alzheimer's disease.

The utilization of short ribozymes functioning as dimers enables the design of a maxizyme with one substrate-binding region for the unique junction sequence thus acting as the sensor and the second binding region for an efficient cleavage site thus operating as the scissors.

Ribozymes are also potential targets for pharmaceutical intervention with bio-active molecules, small molecules, acting as modulators, agonists or antagonists. A screening procedure has been developed for identifying hammerhead-ribozyme inhibitors and was used to detect inhibitory activities in extracts from actinomycete strains.[697]

Biotransformations, bioconversions impact on the whole range of biological, biotechnological and biomedical applications, such as in the analysis of genes and transcripts, the therapeutic treatment of abnormal molecules in diseases, the construction of specific enzymes and proteins for therapy, analysis, production, the assembly and construction of synthetic bio-synthetic pathways in organisms for the synthesis of intermediates, antibiotics, vitamins, enzymes, flavors.

8.7 Natural Compounds

The supply of natural products can be increased from rare original sources by identification of the relevant gene (gene cluster) and its transfer to a more suitable host(s).[404] This also allows the facile generation of novel compounds by reshuffling the genes, or introducing or deleting genes or functions.

Organic compounds such as carotenoids can be produced by aqua-farming of microalgae. Microalgae are a potential source of single-cell protein, as a means of sequestering carbon dioxide from effluent gases, and can be used in effluent purification and for producing biofuels. High production costs preclude the use of microalgae as commercial sources of commodity products that can be produced inexpensively on a vast scale by plants.

Astaxanthin is a high-value carotenoid produced from microalgae that is produced commercially. Astaxanthin is ubiquitous in nature, especially in the marine environment, and is responsible for eliciting the pinkish-red hue to the flesh of salmonids, shrimp, lobster and crayfish. Cultivation methods have been developed to produce *Haematococcus* containing 1.5–3.0% astaxanthin by dry weight, with potential applications as a pigment source in aquaculture, poultry feeds and in the nutraceutical area.[405]

8.8 Recovery/(Bio-) Processing

The delicate nature of biomolecules such as (glycosylated) proteins, polysaccharides and metabolites necessitates appropriate methods for purification and separation, such as chromatography, lyophilization and crystallization.

The chiral nature of biomolecules and, increasingly, of APIs requires the production of enantiomerically pure compounds. The difference in biological activity of the two enantiomers of a chiral drug has raised the demand for enantiomerically pure products, especially in the pharmaceutical and veterinary industry. Simulated moving-bed chromatography is developing into an efficient tool for the separation of the two isomers of a chiral molecule, at all production scales, from laboratory to pilot plant to production plant.[406]

The cell possesses a complex protein machinery to help some proteins to fold after biosynthesis or heat shock. It consists of molecular chaperones, thought to prevent misfolding and aggregation by binding to polypeptide chains that are not fully folded, and some enzymes able to catalyze specific interactions in polypeptide chains.

Proteins that contain disulfide bonds often fold slowly *in vitro* because the oxidation and correct pairing of the cysteine residues becomes the rate-limiting step and the bonds formed are not always the correct ones. Many proteins, especially those that are secreted by eukaryotes, are stabilized by disulfide bonds. Examples of such proteins include those used for medical or biotechnological purposes, such as interleukins, IFNs, antibodies and their fragments, insulin, TGF, and many toxins and proteases. Expression of recombinant proteins as inclusion bodies in bacteria can be a very efficient way to produce cloned proteins, as long as the inclusion body protein can be successfully refolded.

Aggregation is the leading cause of decreased refolding yield. Refolding of non-covalently immobilized denatured protein on gel beads has been used to avoid aggregation. Another approach is the *in vitro* method to cope with the aggregation by use of immobilized minichaperones. This technique was extended to include reagents for the oxidative catalysis of disulfide bond formation and for the isomerization of *cis–trans* peptidyl–prolyl linkages.

An immobilized and reusable molecular chaperone system for oxidative refolding chromatography was constructed. Its three components – GroEl minichaperone (191–345) (which can prevent protein aggregation), DsbA (which catalyzes the shuffling and oxidative formation of disulfide bonds) and peptidyl–prolyl isomerase – were immobilized on an agarose gel. The gel was applied to the refolding of denatured and reduced scorpion toxin Cn5. The 66-residue toxin, which has four disulfide bridges and a *cis* peptidyl–proline bond, had not previously been refolded in reasonable yield. The protein was recovered in 87% yield with 100% biological activity.[407]

Inteins are self-splicing proteins that occur as in-frame insertions in specific host proteins. In a self-splicing reaction, inteins excise themselves from a precursor protein, while the flanking regions, the exteins, become joined to restore host gene function. The ability to construct intein fusions to proteins of interest has broad potential applications. One of these is affinity fusion-based protein purification, where an intein is used in conjunction with an affinity group to purify a desired protein. Self-cleavage, rather than splicing of the intein, releases the desired protein, thereby eliminating the need for protease addition and simplifying overall processing. For this application, the intein must be altered to yield controllable cleavage rather than splicing, and the intein should be relatively small for scale-up considerations.

A self-cleaving element for use in bio separations has been derived from a naturally occurring, 43-kDa protein splicing element (intein) through a combination of protein engineering and random mutagenesis. A mini-intein (18 kDa) previously engineered for reduced size had compromised activity and was therefore subjected to random mutagenesis and genetic selection. In one selection a mini-intein was isolated with restored splicing activity, while in another, a mutant was isolated with enhanced, pH-sensitive C-terminal cleavage activity. The enhanced-cleavage mutant has utility in affinity fusion-based protein purification. These mutants also provide new insights into the structural and functional roles of some conserved residues in protein splicing.[408]

Chirality is a determinative property of most biological molecules. A particular stereoisomer is generally preferred in physiologic processes such as metabolism, intra- and intercellular communication, and as building blocks for the formation of macromolecules. Living organisms discriminate between the enantiomers of exogenous compounds (e.g., drugs) at virtually all levels of interaction and respond differently to them. Even minor enantiomeric impurities may cause severe pharmacologic and toxicologic side effects. The development of refined analytical techniques for the precise determination of biorelevant chiral molecules is therefore of great interest. Various sensor technologies have been developed for chiral compounds. Chiral selectors for enantiomer discrimination have largely built upon empiric chromatographic phases such as metal complexes, cycodextrin derivatives and Chirasil-Val derivatives.

Based on the stereoselectivity of immunoglobulins, a new chiral sensor for the detection of low-molecular-weight analytes was developed. Using surface plasmon resonance detection, enantiomers of free, underivatized α-amino acids can be monitored in a competitive assay by their interaction with antibodies specific for the chiral center of this class of substances. The sensitivity to the minor enantiomer in non-racemic mixtures exceeds currently available methods; therefore, such immunosensors can readily detect traces of enantiomeric impurities and are attractive for a range of applications in science and industry.[409]

8.9 Chemical–Biotechnological Syntheses

Membranes and biotechnological tools can be used for improving traditional production systems to maintain the sustainable growth of society and industry. Typical examples include new and improved foodstuffs, in which the desired nutrients are maintained during thermal treatment, novel pharmaceutical products with defined enantiomeric compositions and the treatment of wastewater, wherein pollution by traditional processes is problematic.

Table 8.9, 8.10 and 8.11 show the industrial application of enzymes, the application of biocatalytic membrane reactors in the agro-food industry, and the application of biocatalytic membrane reactors in pharmaceutical and biomedical treatments, respectively.

Biocatalytic membrane reactors combine selective mass transport with chemical reactions and the selective removal of products from the reaction site increases the conversion of product-inhibited or thermodynamically unfavorable reactions. Membrane reactors using biological catalysts can be used in production, processing and treatment operations. Recent advances towards environmentally friendly technologies make these membrane reactors particularly attractive because they do not require additives, are able to function at moderate temperatures and pressure, and reduce the formation of by-products. The catalytic action of enzymes is extremely efficient and selective compared with chemical catalysts. These enzymes demonstrate higher reaction rates, milder reaction conditions and greater stereospecificity.

The potential advantages of membrane reactor technology over more conventional approaches include its higher efficiency and reduced costs due to the integration of bioconversion and product purification, thus reducing equipment costs and the number of processing steps. Enzymatic membranes will also contribute to the growth of new research areas, such as non-aqueous enzymology, the use of non-biological catalysts (such as cyclodextrins) and the development of novel biosensors for diagnostic purposes.[410]

Solid-state fermentation (SSF) using inert supports impregnated with chemically defined liquid media has several potential applications in both scientific studies and the industrial production of high-value products, such as metabolites, biological control agents and enzymes. As a consequence of its more defined system, SSF on inert supports offers numerous advantages, such as improved process control and monitoring, and enhanced process consistency, compared with cultivation on natural solid substrates. SSF fermentation processes involve the growth of microorganisms (typically fungi) on moist solid substrates in the absence of free-flowing water and have considerable economical potential in producing products for the food, feed, pharmaceutical, and agricultural industries. See Table 8.12.

Table 8.9. Industrial applications of enzymes [Refs in 410].

Type of industry	Enzyme	Application
Detergent	Proteases	To remove organic stains
	Lipases	To remove greasy stains
	Amylases	To remove residues of starchy foods
	Cellulases	To restore a smooth surface to the fiber and restore the garment to its original colors
Food	Proteases and lipases	To intensify flavor and accelerate the aging process
	Lactases	To produce low-lactose milk and related products for special dietary requirements
Wine	β-Glucanase	To help the clarification process
	Cellulase	To aid the breakdown of cell walls
	Cellulase and pectinase	To improve clarification and storage stability
Fruit juices	Pectinases	To improve fruit-juice extraction and reduce juice viscosity
	Cellulase	To improve juice yield and color of juice
Oils and fats	Lipases	The industrial hydrolysis of fats and oils or the production of fatty acids, glycerin, poly-unsaturated fatty acids used to produce pharmaceuticals, flavors, fragrances and cosmetics
Alcohol	α-Amylases	Liquefaction of starch or fragmentation of gelatinized starch
	Amiloglucosidase	Saccharification or complete degradation of starch and dextrins into glucose
Starch and sugar	α-Amylases	Enzymatic conversion of starch to fructose: liquefaction, saccharification and isomerization Liquefaction of starch
	Glucoamylase and pullulanase	Saccharification
	Glucose isomerase	Isomerization of glucose
Animal feed	β-Glucanases	The reduction of β-glucans
Brewing industry	β-Glucanases	The reduction of β-glucans and pentosans
Fine chemical	Lipases, amidases and nitrilases	Enantiomeric intermediates for drugs and agrochemicals; hydrolysis of esters, amides, nitriles or esterification reactions
Leather	Lipases	To remove fats in the de-greasing process
Textiles	Amylases and cellulases	To produce fibers from less-valuable raw materials
Pulp and paper	Xylanases	Used as a bleaching catalyst during pre-treatment for the manufacture of bleached pulp for paper

Table 8.10. Applications of biocatalytic membrane reactors in the agro-food industry [Refs in 410].

Reaction	Membrane bioreactor	Purpose
Hydrolysis of lactose to glucose and β-galactose (β-galactosidase)	Axial-annular flow reactor	Delactosization of milk or whey for human consumption
Hydrolysis of high-molecular-weight protein in milk (trypsin and chymotrypsin)	Asymmetric hollow fiber with gelified enzyme	Production of baby food
Hydrolysis of raffinose (á-galactosidase and invertase)	Hollow fiber reactor with segregated enzyme	Production of monomeric sugars
Hydrolysis of starch to maltose (á-amylase, β-amylase, pullulanase)	CSTR with UF membrane	Production of syrups
Fermentation of sugars (yeast)	CSTR with UF membrane	Brewing industry
Anaerobic fermentation (yeast)	CSTR with UF membrane	Production of alcohol
Hydrolysis of pectines (pectinase)	CSTR with UF membrane	Production of bitterness and clarification of fruit juice and wine
Fermentation of *Lactobacillus bulgaricus*	CSTR with UF membrane	Production of carboxylic acids
Removal of limonene and naringin (β-cyclodextrin)	CSTR with UF membrane	Production of bitterness and clarification of fruit juice
Hydrolysis of K-casein (endopeptidase)	CSTR with UF membrane	Milk coagulation for dairy products
Hydrolysis of collagen and muscle proteins (protease, papain)	CSTR with UF membrane	Meat tenderization
Conversion of glucose to gluconic acid (glucose oxidase and catalase)	Packed bed reactor	Prevention of discoloration and off-flavor of egg products during storage
Hydrolysis of triglycerides to fatty acids and glycerol (lipase)	UF capillary membrane reactor	Production of foods, cosmetics and emulsificants
Hydrolysis of cellulose to cellobiose and glucose (cellulase and β-glucosidase)	Asymmetric hollow fiber reactor	Production of ethanol and protein
Hydrolysis of malic acid to lactic acid (*Lactobacillus oenos*)	MF capillary membranes with entrapped cells	Improve taste in white wine
Hydrolysis of fumaric acid to L-malic acid (fumarase)	UF capillary membrane reactor	Production of food additives
Hydrolysis of olive oil triglycerides (lipase)	Hydrophobic plate-and-frame membrane reactor	Treatment of oils
Hydrolysis of soybean oil (lipase)	Hydrophilic hollow fiber membrane reactor	Treatment of oils
Hydrolysis of butteroil glycerides (lipase)	Hydrophobic flat-sheet membrane reactor	Treatment of oils and products for the cosmetics industry
Hydrolysis of milk fat (lipase)	Spiral-wound polypropylene membrane reactor	Treatment of fats and oils

Abbreviations: CSTR, continuous stirred tank reactor; UF, ultra filtration; MF, micro filtration.

Table 8.11. Applications of biocatalytic membrane reactors in pharmaceutical and biomedical treatments [Refs in 410].

Reaction	Membrane reactor	Purpose
Conversion of fumaric acid to L-aspartic acid (*Escherichia coli* with aspartase)	Entrapment in polyacrylamide gel	Pharmaceuticals and feed additives
Conversion of L-aspartic acid to L-alanine(*Pseudomonas dacunhae*)	Entrapment in polyacrylamide gel	Pharmaceuticals
Conversion of cortexolone to hydrocortisone and prednisolone (*Curvularia lunata/Candida simplex*)	Entrapment in polyacrylamide gel	Production of steroids
Conversion of acetyl-D,L-amino acid to L-amino acid (aminoacylase)	Ionic binding to DEAE-sephadex	Production of L-amino acids for pharmaceutical use
Synthesis of tyrosine from phenol, pyruvate and ammonia (tyrosinase)	Entrapment in cellulose triacetate membrane	Production of L-amino acids for pharmaceutical use
Hydrolysis of a cyano-ester to ibuprofen (lipase)	Entrapment in biphasic hollow fiber reactor	Production of anti-inflarmmatories
Production of ampicillin and amoxycillin (penicillin amidase)	Entrapment in cellulose triacetate fibers	Production of antibiotics
Hydrolysis of a diltiazem precursor (lipase) blocker	Entrapment in biphasic hollow fiber reactor	Production of calcium-channel
Hydrolysis of 5-*p*-HP-hydantoine to D-*p*-HP-glycine (hydantoinase and carbamylase)	Entrapment in UF polysulfone membrane	Intermediate for the production of cephalosporin
Dehydrogenation reactions [NAD(P)H-dependent enzyme systems]	Confination with UF-charged membrane	Production of enantiomeric amino acids
Hydrolysis of DNA to oligo-nucleotides (DNase)	Gelification on UF capillary membrane	Production of pharmaceutical substances
Hydrolysis of hydrogen peroxide (bovine liver catalase)	Entrapment in cellulose triacetate membrane	Treatment in liver failure
Hydrolysis of whey proteins (trypsin, chymotrypsin)	Polysulfone UF membrane	Production of peptides for medical use
Hydrolysis of arginine and asparagines (arginase and asparaginase)	Entrapment in polyurethane membrane	Care and prevention of leukemia and cancer

Abbreviation: UF, ultra filtration.

Table 8.12. Examples of solid-state fermentation using impregnated inert supports.

Microorganism	Product	Support	Refs in [411]
Colletotrichum truncatum	Spores	Vermiculite, perlite	11, 34
		Rice hulls	34
Beauveria bassiana	Spores	Clay granules	35
Penicillium roquefortii	Spores	Pozolano	17
Coniothyrium minitans	Spores	Hemp, bagasse, perlite	8
Gibberella fujikuroi	Gibberellic acid	Bagasse, PUF	36
Brevibacterium sp.	L-Glutamic acid	Bagasse	37
Aspergillus niger	Citric acid, polyols	Amberlite	19
	Citric acid	Bagasse	38
Penicillium chrysogenum	Penicillin	Bagasse	39
Rhizopus delemar	Lipase	Amberlite	21
Penicillium citrinum	Nuclease P1	PUF	20
Vibrio costicola	L-Glutaminase	Polystyrene	18
Aspergillus oryzae	Protease	PUF	40
	Amylase	PUF	41

Abbreviation: PUF, polyurethane foam.

8.10 Gene Therapy Vectors/Systems

Strategies in genetic engineering require permanent directed modification of the target genome by safe vectors or modes of transfer. Site-specific integration is an important tool in achieving non-deleterious integration. An efficient, site-specific, unidirectional integration in mammalian cells has been achieved by application of the integrase from a *Streptomyces* phage[657].

Cancer immunotherapy by genetically engineered dendritic cells that have been transfected with genes encoding tumor-associated antigens provides an intriguing combination of gene therapy and immunotherapy[658].

An important target for gene therapy is the mitochondrial genome, mutations of which may be involved in many rare diseases. Such mitochondriopathies may be amenable to treat-ment by genetic modifications [659].

8.11 Production:
Safety, Efficacy, Consistency, and Specificity

Consumer and patient safety are the prerequisites for (bio-)pharmaceutical product development, production and marketing. The ability to provide an effective, pure and safe product is the primary factor determining the product's success. With an ever-increasing number of national and international regulations, quality assurance' has become a complex task for project managers. Good manufacturing practices (GMPs) are an asset.

GMP is aimed at assuring the quality of the product by assuring the quality of the process. GMP should also be part of the process development (e.g., development reports and approval requirements), and proceed through validation, manufacturing, controls and end-product testing, as well as reaching into the distribution network of the product.[412]

The biopharmaceutical products manufactured by processes that use mammalian cell cultures have gained increasing importance in recent years. A strong awareness of the considerations for safety and quality of such products has also emerged and led to improvements in cultivation and production technology, validation procedures and process organization. Cultivation techniques have developed towards the use of chemically defined culture media, characterization of cell lines has intensified and the downstream processing as well as product analyses have become more rigorous, including process organizations incorporating GMP.[413]

The development of validated manufacturing processes is a prerequisite for pharmaceutical application of the newer biotechnologicals, such as DNA for plasmid-based genes in vaccines and gene therapy. Using bioprocess-design information it is possible to create efficient and consistent processes for these materials. Key issues are the required purity, the sensitivity of the chromosomal DNA and larger plasmids to hydrodynamic forces, and the impact of the various characteristics of plasmids on the recovery and purification of DNA for pharmaceutical purposes.[414]

Interest in producing large quantities of supercoiled plasmid DNA has increased as a result of gene therapy and DNA vaccines. Due to the commercial interests in these approaches, the development of production and purification strategies for gene therapy vectors has been performed in pharmaceutical companies within a confidential environment. It is thus important to describe the downstream operations for the large-scale purification of plasmid DNA to attain a final product that meets specifications and safety requirements.[415]

Table 8.13. The principal approval specifications and recommended assays for assessing the purity, safety and potency of DNA preparations for gene therapy and DNA vaccines[a] [Refs in 415].

Impurity	Recommended assay	Approval specification
Proteins	BCA protein assay	Undetectable
RNA	Agarose-gel electrophoresis	Undetectable
gDNA	Agarose-gel electrophoresis	Undetectable
	Southern blot	<0.01 μg (μg plasmid)$^{-1}$
Endotoxins	LAL assay	<0.1 EU (μg plasmid)$^{-1}$
Plasmid isoforms (linear, relaxed, denatured)	Agarose-gel electrophoresis	$<5\%$
Biological activity and identity	Restriction endonucleases	Coherent fragments with the plasmid restriction map
	Agarose-gel electrophoresis	Expected migration from size and supercoiling
	Transformation efficiency	Comparable with plasmid standards

[a]   Abbreviations: BKA, bicinchoninic acid; LAL, *Lymulus* amebocyte lysate; EU, endotoxin units.

Table 8.13 shows the principal approval specifications of DNA preparations for gene therapy and DNA vaccines.

Figure 8.3 shows the process flow sheet for large-scale purification of plasmid DNA.

Table 8.14 presents a comparison of laboratory methods and large-scale pharmaceutical processes for plasmid DNA purification.

Table 8.15 outlines the purification of clinical-grade plasmid DNA.

Table 8.14. A comparison of laboratory methods and large-scale pharmaceutical processes for purifying super coiled plasmid DNA.

Process step	Laboratory method	Large-scale process
Cell lysis	RNase, lysozyme	No enzymes, only GRAS reagents
Removal of cell debris	Centrifugation	Filtration, centrifugation or expanded bed chromatography
Removal of host impurities (RNA, gDNA, proteins and endotoxins)	RNase, proteinase K, organic solvents (phenol and chloroform)	Salting out PEG precipitation
Concentration	Alcohol precipitation	Alcohol precipitation, PEG precipitation
Plasmid purification	Ultra centrifugation (mutagenic reagents and ethidium bromide) IEC (gravity flow columns provided in commercial kits) RPC (organic, toxic solvents)	IEC and/or SEC (use only GRAS reagents)

[a] Abbreviations: GRAS, generally regarded as safe; IEC, ion-exchange chromatography; PEG, polyethylene glycol; RPC, reverse-phase chromatography; SEC, size-exclusion chromatography.

Table 8.15. The purification of clinical-grade plasmid DNA [a,b]. Purification strategy [Refs in 415].

Target disease	Clarification and concentration	Purification	Refs
Cancer	None	Expanded bed IEC SEC	24
Melanoma	Isopropanol precipitation Ammonium acetate precipitation PEG precipitation	SEC	26
Cystic fibrosis	Isopropanol precipitation Ammonium sulfate precipitation	HIC Desalting	29
Melanoma	Isopropanol precipitation Ammonium acetate precipitation PEG precipitation	Fixed bed IEC SEC	48

[a] Abbreviations: HIC, hydrophobic interaction chromatography; IEC, ion-exchange chromato-graphy; PEG, polyethylene glycol; SEC, size-exclusion chromatography.

[b] All described strategies are based on the alkaline-lysis method.

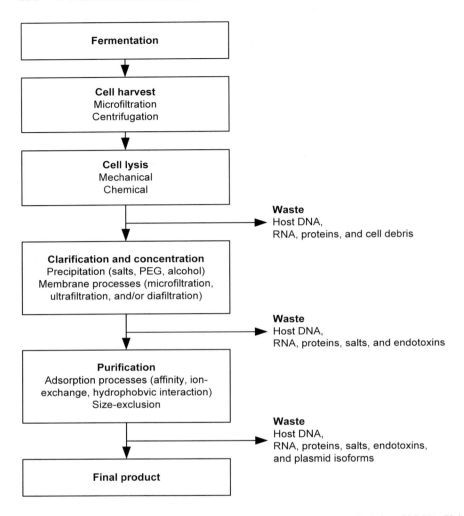

Figure 8.3. Process flow sheet for the large-scale purification of supercoiled plasmid DNA. Unit operations to be considered during process development are indicated together with with the eliminated impurities [Refs in 415].

8.12 Registration

The registration process is converging in the major regions of the world towards mutual recognition of the highest standards, with the US Food and Drug Administration (FDA) and European Medicines Evaluation Agency (EMEA) getting close to similar standards for trials, documentation, and validation.

9 Safety

Safety is the uppermost concern for patients, physicians, producers, the population at large, the environment, the biosphere and (due to genetic modifications through, for example, gene therapies) also for future generations.

9.1 Medical Safety

Medical safety essentially involves ensuring the safety of the patients from unwanted and undesirable side effects, whereby side effects are assessed in the context of severity of disease and acceptable side effects versus therapeutic effect.

The use of transplanted organs from tissue engineering or from donor species provides new and additional aspects of safety, e.g., dealing with unwanted gene transfer. Somatic gene therapy needs to consider undesirable integration in germ cells, and subsequent biological effects and perpetuation.

9.2 Biological Safety

Biological safety concerns the effect of APIs, therapies, gene therapy vectors as well as biological manufacturing systems (e.g., bacteria, cells and transgenics) on the humans involved and the biological systems potentially affected. Examples are considerations of antibiotic-resistance genes, virulence determinants and specific genetic traits.

9.3 Chemical Safety

Chemical safety is the acute toxicity of compounds which needs to be considered thoroughly in view of the increasing potency and the potentially different effects on genetically different subtypes of humans or other entities of the biosphere.

9.4 Equipment Safety

The safety of equipment considers primarily the function and effects of malfunction of production equipment, diagnostic equipment, and the increasing numbers and types of medical devices, ranging from implanted pumps to prosthetics to the conceived nano-robots carrying out essential supervisory or remedial functions in patients' bodies.

10 Environment

Biotechnology is playing an increasing role in the environmental realm, in environmental control and in the remediation field by providing genetically modified plants, e.g., for carbon dioxide fixation, by accumulating metal ions in plants, as a source of renewable raw materials and as an alternative for fossil fuels[502] More specifically, antibodies offer alternatives to chemical and physical technologies in the multitude of possible topologies and the resulting sensitive detection and specific interactions for removal of organic pollutants.[416]

A revealing study of the environmental impact of biotechnological production of PHA (polyhydroxy acid) in agricultural plants has been published with a comparative view on the sustainability of biological processes. This study involves a thorough 'cradle-to-grave' analysis and reveals that the total amount of fossil fuel required to produce 1 kg of PHAs exceeds that required to produce an equal amount of polystyrene. The replacement of conventional polymers with fermentation-derived PHAs does not yield benefits with respect to fossil fuel consumption. Future assessments of biological processes must incorporate the use of raw materials (which are renewable) and also address the indirect consumption of non-renewable energy sources required for the process.[417]

Genetically engineered microorganisms or plants may be used for the remediation of soils contaminated with metals[707] or degradable organic compounds.

10.1 Pharmaceuticals and the Environment

Novel APIs, medical devices and gene therapy vectors/systems are increasingly being discovered and designed based on their interaction with genetic and biological networks. The similarity of the genomes of many different species and the conservation of protein structure and function necessitates consideration of the effects on the environment and the biosphere.

10.2 Biological Containment

Biological containment involves the use of cells, vectors, genes, and APIs which, upon release from the targeted or optimal environment, will not interact with the biological systems they encounter. This means that high-performance strains are auxotrophic for rare compounds unlikely to be found in nature or are unable to function at the temperatures/oxygen levels found in the environment.

10.3 Physical/Chemical Containment

Physical/chemical containment takes into account the toxicity, virulence, and potential for change and evolution of biological agents, and uses barriers, closed systems, and physical destruction by irradiation, chemicals and temperature to prevent the escape of viable agents.

10.4 Process-Integrated Environmental Protection

The manufacturing processes need to take a holistic view of the total life cycle of products including their manufacture and disposal (from factory and from patient). The choice of chiral syntheses, metabolically engineered cells, microreactors for optimized processes, and medical devices designed for function and recycling (and manufacturability) are key considerations to be applied early in the development process.

10.5 Waste/Effluent Treatment and Recycling

Molecular/genetic techniques involve the direct modification of genetic material and require special consideration of the waste streams produced. The adequacy of existing treatment, disposal and recycling processes of waste streams from biotechnological laboratories and industrial processes, especially those using genetically modified organisms, is focused on the DNA content of these wastes, the properties of extracellular (or 'naked') DNA and the ability to transfer genetic information between bacteria (e.g., antibiotic resistance) or into higher organisms.[418]

Particle-based biofilm reactors provide the potential to develop compact and high-rate processes. In these reactors, a large biomass content can be maintained (up to 30 g/l) and the large specific surface area (up to 3000 m^2/m^3) ensures that the conversions are not limited by the biofilm liquid mass-transfer rate. Engineered design and control of particle-based biofilm reactors have been established and reliable correlations exist for the estimation of the design parameters. Thus, a new generation of high-load, efficient biofilm reactors are operating throughout the world, with several full-scale applications for industrial and municipal waste-water treatment.

Biofilm reactors are used in situations where the reactor capacity obtained by freely suspended organisms is limited by the biomass concentration and hydraulic residence time.

The main reactor types are upflow sludge blankets, biofilm fluidized beds, expanded granular sludge blankets, and biofilm airlift suspension and internal circulation reactors.

Table 10.1 shows examples of design characteristics and applications of particle-based biofilm reactors.[419]

Table 10.1. Examples of design characteristics and applications of particle-based biofilm reactors.

Reactor type	Commercial names	Flow pattern	Liquid velocity	Height to diameter	Mixing	Examples of application	Refs in [419]
EGSB	BIOBED, Biothane (USA)	Upflow	10–15 m h^{-1} (upward)	4–5	Liquid, gas produced	The production of formaldehyde from methanol	17–21
IC	IC, Paques (The Netherlands)	Mixed	Bottom: 10–30 m h^{-1} Top: 4–8 m h^{-1} (circulation)	3–6	Gas produced	Inuline and fructose production from chicory beet	22, 23, 45
BAS		Mixed	0.4–0.8 m s^{-1} (circulation)	4–5	Gas produced	Brewery	11, 22, 14 15, 23

Abbreviations: EGSB, expanded granular sludge blanket; IC, internal circulation; BAS, biofilm airlift suspension.

11   Ethics

Gene therapy and ES cell research are a major focus of ethical considerations in terms of research developments and applications.

The emphasis of underlying issues differs in Europe and the USA. Whereas Europe confronts the debate from its historic plurality and with consideration of vast excesses the USA debate brings the argument of public financing stronger into the debate. The patients' views and interests, which in the case of pressure groups during the AIDS epidemic have already changed procedures of the FDA and public perceptions by bringing about fast-track transfer of pharmaceuticals into therapy, are significantly influencing the debate in USA.[420]

The creation of novel organisms as exemplified by discussions and research on creating a free-living organism with a minimal genome is also promoting ethical debate. To create a novel organism, scientists must determine which genes form the minimum set necessary for basic metabolism and replication, construct this minimum gene set, and provide or create the necessary non-genetic components for successful gene expression. There is still a large technological gap between what has been achieved to date (defining a portion of a minimum set of genes necessary for an organism to survive under permissive laboratory conditions) and actually 'creating life'.[421]

Ethical considerations across the whole spectrum of biotechnology are being discussed in Europe, dealing with environmental applications, food production, human health, animal welfare, research, industry, the 'north/south' divide, biodiversity, public perceptions and media, and stress the consideration of the interest of future generations.[422]

Research dealing with generating organisms with a minimum set of genes has so far focussed on the smallest known cellular genome, that of *Mycoplasma genitalium* (580 kb). *M. genitalium* (with 517 genes) has the smallest gene complement of any independently replicating cell identified so far. Global transposon mutagenesis was used to identify non-essential genes in an effort to identify whether the naturally occurring gene complement is a true minimum genome under laboratory growth conditions. The positions of 2209 transposon insertions in the completely sequenced genomes of *M. genitalium* and its close relative *M. pneumoniae* were determined by sequencing across the junction of the transposon and the genomic DNA. These junctions define 1354 distinct sites of insertion that are not lethal. The analysis suggests that 265–350 of the 480 protein-coding regions of *M. genitalium* are essential under laboratory growth conditions, including about 100 genes of unknown function.[423]

The ethical and practical aspects of applying biomedicine and life-extending therapies to generate an 'immortal' phenotype are questions that society needs to consider as developments unfold. Society, even in the advanced 'rich' parts of the world, will divide into those that can afford expensive therapies and those that cannot.

On a global scale, there will exist societies that can hardly sustain the minimum requirements for food, water, clothing, shelter, and medicine, while affluent individuals and societal groups will co-exist. Unprecedented mechanisms need to be developed to cope with these developments and their consequences, intra-societally as well as inter-societally (i.e. between nations or groups of nations).[424]

12 Companies, Institutes, Networks, and Organizations

Actinova Ltd	185 Cambridge Science Park, Cambridge CB4 0GA, UK
Advanced Cell Technology	One Innovation Drive, Worcester, MA 01605, USA
Aeterna Laboratories (Arthritis)	Quebec, PQ, Canada
Affymetrix Inc.	3380 Central Expressway, Santa Clara, CA 95051, USA
Agouron Pharmaceuticals Inc. (AIDS)	3565 General Atomics CT., La Jolla, CA 92037, USA
Agouron Pharmaceuticals Inc. (Arthritis)	10350 N Torrey Pines Rd., La Jolla, CA 92037-1018, USA
AGOWA	Gesellschaft fuer molekular-biologische Technologie mbH, Glienicker Weg 185, 12489 Berlin, Germany
Agracetus Campus	Monsanto Company, 8520 University Green, Middleton, WI 53562, USA
Akkadix Corp.	11099 North Torrey Pines Road, Suite 200, La Jolla, CA 92037, USA
Alexion Pharmaceuticals	25 Science Park #360, New Haven, CT 06511, USA
Alliance for Ageing Research	2021 K Street, NW, Suite 305, Washington, DC 20006, USA
Alliance for Ageing Research, Amarillo Biosciences (AIDS)	800 W 9th Ave., Amarillo, TX 79101-3206, USA
Amgen Inc.	One Amgen Center Drive, Thousand Oaks, CA 91320, USA
Amgen Staffing	P. O. Box 2569, Thousand Oaks, CA 91319-2569, USA
Amrad (Malaria)	576 Swan St., Richmond VIC3121, Australia
AnalyticCon AG	Hermannswerder Haus 17, 14473 Potsdam, Germany
AnorMed (Arthritis) (AIDS)	20353 64 Ave., Langley, BC V2Y 1N5, Canada
Antibioticos SA	Laboratorios de Biotechnologia and Bioquimica, Avenida de Antibioticos 59–61, 24009 Leon, Spain
Antigen Express	One Innovation Drive, Worcester, MA 01605, USA
Applied Biosystems Division	PE Biosystems, 850 Lincoln Center Drive, Foster City, CA 894404-1128, USA www.appliedbiosystems.com
Applied Logic Associates Inc.	5615 Kirby Drive, Houston, TX 77005, USA
Applied Molecular Evolution Inc.	3520 Dunhill Street, San Diego, CA 92121, USA

Aquila Biopharmaceuticals Inc. (Malaria)	175 Crossing Blvd. #200, Framingham, MA 01702-4473, USA
Arena Pharmaceuticals	6166 Nancy Ridge Drive, San Diego, CA 92121, USA
Arena Pharmaceuticals	6166 Nancy Ridge Drive, San Diego, CA 92121, USA
Arevia GmbH	Fennstrasse 49, 13353 Berlin, Germany
ARIAD Pharmaceuticals Inc.	26 Landsdowne Street, Cambridge, MA 02139, USA
Aronex Pharmaceuticals Inc. (Cancer) (Antibacterial and Antifungal) (AIDS)	8707 Technology Forest Pl., Spring, TX 77381-1191, USA
ArQule, Inc.	200 Boston Ave. Medford, MA 02155, USA Tel (++1) 781 395 4100 Fax: (++1) 781 395 1225
Artemis Pharmaceuticals GmbH	Neurather Ring 1, 51063 Koeln, Germany
AstraZeneca	Alderley Park, Macclesfield, Cheshire SK10 4TG, UK
Atrix Laboratoies Inc. (Antibacterial and Antifungal)	701 Central Ave., Fort Collins, CO 80526-1843, USA
Atugen AG	Robert-Roessle-Strasse 10, 13125 Berlin, Germany
Aurora Biosciences Corp.	11010 Torreyana Road, San Diego, CA 92121, USA Tel. (++1) 858 404 6600 Fax: (++1) 858 404 8485
AutoImmun (Arthritis)	Lexington, MA 781-860-0710, USA
Avanir Pharmaceuticals (AIDS)	9393 Towne Center Dr. #200, San Diego, CA 92121-3070, USA
Aventis Pharma	London Road, Homes Chapel, Crewe, Cheshire CW4 8BE, UK
Aventis Research & Technologies	Gebäude H 777, Industriepark Hoechst, GmbH & Co. KG, 65926 Frankfurt/M, Germany
Avigen Inc. (Cancer)	1201 Harbor Bay Pkwy. #1000, Alameda, CA 94502-6586, USA
Axys Pharmaceuticals	180 Kimball Way, South Francisco, CA 94080-6218, USA
BASF-LYNX Bioscience	Im Neuenheimer Feld 515, 69120 Heidelberg, Germany
BASF Bio-research	100 Research Dr., Worcester, MA 01605-4312, USA
Bayer Corp., Pharmaceutical Division	PO Box 3238, Scranton, PA 18505-0238, USA
Bayer Inc. (Antibacterial and Antifungal)	400 Morgan Ln., West Haven, CT 06516-4175, USA
Bayer AG (Arthritis)	Kaiser-Wilhelm-Allee 1, 51373 Leverkusen, Germany
B. Braun Biotech International GmbH	Schwarzenberger Weg 73–79, 34212 Melsungen, Germany
BD Biosciences	1 Becton Drive, Franklin Lakes, NJ 07417 USA www.bdbiosciences.com

Bell Laboratories	Lucent Technologies, 600 Mountain Avenue, Murray Hill, NJ 07974, USA
Biocarta, Inc	3830 Valley Center Drive 705-161, San Diego, CA 92130, USA www.biocarta.com
Biochem Technology	1190 Saratoga Ave. # 140, San Jose, CA 95129-3433, USA
Biochemie GmbH	A-6250 Kundl, Austria Fax and Tel. (++43) 5338200442
BioCryst Pharmaceuticals (AIDS)	2190 Parkway Lake Dr., Birmingham, AL 35244-1879, USA
Biogen Inc.	14 Cambridge Ctr., Cambridge, MA 02142-1481, USA
BioGenes Gesellschaft für Biopolymere mbH	Koepenicker Strasse 325, 12555 Berlin, Germany
Bioimmune Systems (AIDS)	1177 S 1680 W, Orem, UT 84058-4930, USA
BioInvent Therapeutic AB	Soelveg 41, Lund 22370, Sweden
Bioject Medical Technologies (Malaria)	7620 SW Bridgeport Rd., Portland, OR 97224-7700, USA
Biomatrix Inc. (Arthritis)	65 Railroad Ave. #2, Ridgefield, NJ 07657-2130, USA
BioSource Deutschland GmbH	Tel. 0800 100 85 74, 12555 Berlin, Germany
BioSyn (AIDS) (Antibacterial and Antifungal)	3401 Market St. #300, Philadelphia, PA 19104-3319, USA
Biotechnology, Boston Life Sciences (Arthritis)	137 Newbury Street, Boston, MA 02116, USA
Biotecon GmbH	Gesellschaft fuer Biotechnologische Entwicklung und Consulting mbH, Tegeler Weg 33, 10589 Berlin, Germany
Biozym GmbH	Steinbrinksweg 27, POB 180, 31833 Hessisch-Oldendorf, Germany
Bitop Gesellschaft für Biotechnische Optimierung mbH	Alfred-Herrhausen-Strasse 44, 58455 Witten, Germany
Boehringer Ingelheim Anml Hlth	2621 N Belt Hwy., St Joseph, MO 64506-2002, USA
Boehringer Ingelheim GmbH	Binger Strasse 173, 55216 Ingelheim am Rhein, Germany
Biotechnology Research and Information Network	Darmstaedter Strasse 34, 64673 Zwingenberg, Germany
Brain GmbH	Darmstaedter Strasse 34
Biotechnology Resesarch and Information Network	64673 Zwingenberg, Germany
Bristol-Myers-Squibb Pharmaceutical	5 Research Parkway, Wallingford, Research Institute, CT 06492, USA
Britsh Biotech	Watlington Rd., Oxford, Oxfordshire, OX4 6LY, United Kingdom

Brucker Daltonics Inc.	15 Fortune Dr. Bldg. #2, Billerica, MA 01821-3958, USA
Bruker Daltonik GmbH	Fahrenheitstrasse 4, 28359 Bremen, Germany
BTX Inc., A Division of Genetronics	11199 Sormento Valoley Road, San Diego, CA 92121, USA www.btxonline.com
Cadus Pharmaceutical Corp.	777 Old Saw Mill River Rd., Tarrytown, NY 10591, USA
Cambridge Antibody Technology (Arthritis)	The Science Park Melbourn, Royston, Hertfordshire SG8 6JJ, UK
Carl Zeiss Inc.	1 Zeiss Dr., Thornwood, NY 1054-1996, Microscopy & Imaging Systems, USA
CELERA Genomics	45 West Gude Drive, Rockville, MD 20850, USA www.celerajobs.com
Celgene (Arthritis)	7 Powderhorn Dr. #1,Warren, NJ 07059-5190, USA
Cell Genesis Inc. (Arthritis)	342 Lakeside Dr., Foster City, CA 94404-1146, USA
CellGenix Technology Transfer GmbH	Elsässer Strasse 2N, 79110 Freiburg, Germany
Cellomics Inc.	635 William Pitt Way, Pittsburgh, PA 15238, USA
Cell Signaling Technology	166B Cummings Center, Beverly, MA 01915, USA
CellTec GmbH Biotechnologie	Frohmestrasse 110, 22459 Hamburg, Germany
Cell-Tech Inc.	8982 Mansfield Rd., Shreveport, LA-71118-2128, USA
Celltech Ltd.	216 Bath Rd., Slough Berkshire SL1 4EN, UK
Cell Therapeutics (Cancer)	201 Ellrot Ave W#400, Seattle, WA 98119-4237, USA
Cellular Products GmbH	Delitzscher Strasse 141, 04129 Leipzig, Germany
Centocor Inc.	200 Great Valley Pkwy, Malvern, PA 19355-1307, USA
Charles River Laboratories, Inc	251 Ballardvale Street, Wilmington, MA 01887-1000, USA www.CRIVER.COM
Charles River Therion	Rensselaer Technology Park, 185 Jordan Road, Troy, NY 12180-7617, USA
Chemical Synthesis,Process Development, SNP Consortium, Cognetics	421 Wakara Way; Suite 201, Salt Lake City, UT 84108, USA Tel.: (++1) 801 581 0400 ext. 222 Fax: (++1) 801 581 9555
Chiron Corp.	4560 Horton Street, Room 4.4144, Emeryville, CA 94608, USA
Clontech	1020 East Meadow Circle, Palo Alto, CA 94303, USA
Clontech Laboratories GmbH	Tullastrasse 4, 69126 Heidelberg, Germany
Cobra Therapeutics	The Science Park, University of Keele, Keele, Staffordshire ST5 5SP, UK
Cold Spring Harbor Laboratory	1 Bungtown Road, Cold Spring Harbor, NY 11724, USA

Collateral Therapeutics Research (Cardiovascular Disease)	11622 El Camino Real # 300, San Diego, CA 92130-2051, USA
Columbia Laboratories Inc. (AIDS)	2875 NE 191st St., Miami, FL 33180-2831, USA
Combinature Biopharm AG, c/o GenProfile AG	Robert-Roessle-Strasse 10, 13125 Berlin, Germany
Comparative Genetics	SmithKline Beecham Pharmaceuticals, Third Avenue, Harlow, Essex CM19 5AW, UK
Computational Chemistry and Informatics Unit,	GlaxoWellcome R & D, Medicines Research Centre, Gunnels Wood Road, Stevenage, Hertfordshire SG1 2NY, UK
Connetics Corp.	3400 W Bay Shore Rd., Palo Alto, CA 94303-4227, USA
Connex Gesellschaft zur Optimierung von Forschung und Entwicklung	Am Kopferspitz 19, 82152 Martinsried, Planegg, Germany
Cortech Tec. (Arthritis)	422 N Douty St. # 422, Hanford, CA 93230-3977, USA
Corvas Intl. Inc. (Cardiovascular Disease) (Malaria)	3030 Science Park Rd. # 301,San Diego, CA 92121-1183, USA
Cosmix	Molecular Bilogicals GmbH, Mascheroder Weg 1b, 38124 Braunschweig, Germany
CRO Laboratories	89 Adams Rd, North Grafton, MA 0153-2001, USA
CSORO Commonwealth Scientific and Industrial Research Org.	Normanby Rd., Clayton, VIC 3168, Australia
Cubist Pharmaceuticals (antibacterial and antifungal)	24 Emily Street, Cambridge, MA 02139, USA
CuraGen Corp.	555 Long Wharf Drive, 11th Floor, New Haven, CT 06511, USA
CV Therapeutics	3172 Porter Drive, Palo Alto, CA 94304, USA
Cyanotech Corp.	73-4460 Queen Kaahumanu Highway, Suite 102, Kailua-Kona, HI 96740, USA
Cyclacel Ltd.	Dundee Technopole James Lindsay Place, Dundee DD1 5JJ, UK
Cypress Bioscience Inc. (Arthritis)	4350 Executive Dr. #325, San Diego, CA 92121-2118, USA
Cypress Pharmaceuticals (Cardiovascular Disease)	4350 Executive Dr. #325, San Diego, CA 92121-2118, USA
Cytomatrix	50 Cummings Park, Woburn, MA 01801-2123, USA
Cytran Inc. (AIDS)	10230 NE Points Dr., Kirkland, WA 98033-7869, USA
Cytrx (Cancer) (Malaria)	154 Technology Pkwy, Norcross, GA 30092-2911, USA
Daylight Inc.	1561 Laurel St., San Carlos, CA 94070-5114, USA
Department of Genomics, Targets and Cancer Research	Pfizer Central Research, Eastern Point Road, Groton, CT 06340, USA

Departments of Cardiovascular Research and Pathology	1 DNA Way, South San Francisco, CA 94080, USA
Departments of Obstetrics & Gynaecology National University of Singapore	5 Lower Kent Ridge Rd, Singapore119074
DeveloGen AG	Rudolf-Wissell-Strasse 28, 37079 Goettingen, Germany
Diatide Inc. (Cardiovascular Disease)	9 Delta Dr. #7, Londonderry, NH 03053-2372, USA
Digital Gene Technologies	11149 North Torrey Pines Road, La Jolla, CA 92037, USA
Discovery Partners International	11149 North Torrey Pines Road, La Jolla, CA 92037, USA
DNAVEC Research Inc.	Tsukuba-City, Ibaraki, Japan
DNX Transgenic Sciences	5, Cedar Brook Drive, Cranbury, NJ 08512, USA www.dnxsciences.com Tel.: (++1) 609 860 0806 Fax: (++1) 609 860 8515
Double Twist.com	Dr Kuebler GmbH, Siebertstrasse 6, 81675 Muenchen, Germany
Drug Discovery Ltd	Royal College Building, 204 George Street, Glasgow G1 1XW, UK
DuPont Pharmaceuticals Research Laboratories	4570 Executive Drive, Suite 400, San Diego, CA 92121, USA
DuPont Pharmaceuticals Co. (AIDS)	974 Center Rd., Wilmington, DE-19805-1269, USA
Dura Pharmaceuticals Inc.	7475 Lusk Blvd., San Diego, CA 92121-5796, USA
Elan Pharmaceuticals (AIDS)	South San Francisco, CA, USA
Eli Lilly Co.	Bldg. 3 # 31, Indianapolis, IN 46285-0001, USA
Enzyme Design	Novo Nordisk, Building 2C, 2880 Bagsvaerd, Denmark
Eppendorf Scientific Inc.	One Cantiague Road, Westbury, NY 11590-0207, USA
Eppendorf-Netherler-Hinz GmbH	Barkhausenweg 1, 22339 Hamburg, Germany
Epimmune Inc. (Malaria)	655 Nancy Ridge Dr., San Diego, CA 92121-3221, USA
Evotec Biosystems AG	Schnackenburgallee 114, 22525 Hamburg, Germany
Exelixis Inc.	P.O.Box 511, So. San Francisco, CA 94083-511, USA www.exelixis.com
Frauenhofer, Max-Planck-Society/ Fijusawa (AIDS)	Deerfield, IL 60015-4850, USA
Fraunhofer Research Center	2901 Hubbard St., Ann Arbor, MI 48105-2435, USA
GAIFAR (German–American Institute for Applied Biomedical Research GmbH)	Herrmannswerder 15–16, 14473 Potsdam, Germany
GATC GmbH	Fritz-Arnold-Strasse 23, 78467 Konstanz, Germany
G D Searle Co (Arthritis)	5200 Old Orchard Rd., Skokie, IL 60077-1034, USA

GelTex Pharmaceuticals Inc. (AIDS)	153 Second Avenue, Waltham, MA 02451-1122, USA
Gene Alliance	Max-Volmer-Strasse 4, 70724 Hilden, Germany
Gene Logic	708 Quince Orchard Road, Gaithersburg, MD 20878, USA Tel.: (++1) 301 987 1700 Fax: (++1) 301 987 1701
Genelabs Technologies Inc. (Asthma)	505 Penobscot Dr., Redwood City, CA 94063-4738, USA
Genelabs Technologies Inc. (Cancer)	505 Penobscot Dr., Redwood City, CA 94063-4738, USA
Gene Therapy Systems	10190 Telesis Court, San Diego, CA 92121, USA
Geneformatics Inc.	5830 Oberlin Drive, Suite 200, San Diego, CA 92121-3754, USA
GeneMachines	935 Washington Street, San Carlos, CA 94070, USA www.genemachines.com
Genencor	925 Page Mill Road, Palo Alto, CA 94304, USA
Genetics Institute GmbH	Lochamer Strasse 11 82152 Planegg Germany
Genetic Technologies	SmithKline Beecham Pharmaceuticals, New Frontiers Science Park, Harlow, Essex CM19 5AW, UK
Genetic Therapy Inc. (Cancer)	9 W Watkins Mill Rd., Gaithersburg, MD 20878-4021, USA
Genetix Pharmaceuticals (Cancer)	840 Memorial Drive, Cambridge, MA 02139, USA
Genom Analytik GmbH c/o Zentrum fuer Umweltforschung und Technologie Germany	Loebener Strasse, UFT, 28359 Bremen, Germany
Genomatix Software GmbH	Bioinformatics & Sequence Analysis, Karlstrasse 55, 80333 Muenchen, Germany
Genome Pharmaceuticals Corporation (GPC AG)	Lochamer Strasse 29, 82152 Martinsried, Germany
Genome Pharmaceuticals Corporation	Fraunhoferstrasse 20, 82152 Martinsried/Munich, Germany Tel.: (++49) 8985652650 Fax: (++49) 8985652610
Genomica Corp.	1745 38th Street, Boulder, CO 80301, USA
Genta (Arthritis)	San Diego, CA 92121, USA
Genta (AIDS)	99 Hayden Avenue, Lexington, MA 02173 USA
Genzyme Transgenics TSI	2 Taft CT. #202, Rockville, MD 20850-5355, USA
Genzyme Corp.	1 Kendall Sq., Cambridge, MA 02139-1562, USA
Genzyme Molecular Oncology and Genzyme Corp.	PO Box 9322, Framingham, MA 01701-9322, USA
Geron Corp.	230 Constitution Drive, Menlo Park, CA 94025, USA
Gilead Sciences Inc. (AIDS)	333 Lakeside Dr., Foster City, CA 94404-1147, USA
Glaxo Wellcome (AIDS)	Research Triangle Park, NC 27709-5428 USA

Glaxo Wellcome Research and Development	Bioanalysis and Drug Metabolism, Park Road, Ware, SG12 ODP, UK
GuraGen Corp.	555 Long Wharf Drive, New Haven, CT 06511, USA
Hemispherx Biopharma (AIDS)	1617 JFK Blvd., Philadelphia, PA 19103-1821, USA
HepaVec AG	Robert-Roessle-Strasse 10, 13122 Berlin, Germany
Hoechst Marion Roussel (now Aventis) (Asthma) (Antibacterial and Antifungal)	RT 202.206 Bridgewater, NJ 08807, USA
Hoffmann-LaRoche Inc. Preclinical Research	340 Kingsland St., Nutley, NJ 07110-1199, USA
Hoffmann-La Roche Ltd (Antibacterial and Antifungal)	Basel, Switzerland
Hollis-Eden Pharmaceuticals (Malaria) (AIDS)	9333 Genesee Ave.# 200, San Diego, CA 92121-2113, USA
Hybridon (AIDS)	620 Memorial Drive, Cambridge, MA 02139, USA
Hyseq Inc.	670 Almanor Avenue, Sunnyvale, CA 94085, USA www.hysec.com Fax: (++1) 408 524 8129
Ibis Therapeutics, Isis Pharmaceuticals Inc.	2292 Faraday Avenue, Carlsbad, CA 92008, USA Tel.: (++1) 760 603 2347 Fax: (++1) 760 431 2768
ICOS Inc. (Asthma) (Cardiovascular Disease)	22021 20 th Ave., Bothell, WA 98021-4406, USA
IDEA (Innovative Dermale Applikationen GmbH)	Frankfurter Ring 193a, 80807, Muenchen, Germany
IDEC Pharmaceuticals Corp.	11011 Torreyana Rd., San Diego, CA 92121-1104, USA
Idun Pharmaceuticals	11085 North Torrey Pines Road, Suite 300, La Jolla, CA 92307, USA
IGB Fraunhofer Institute for Interfacial Engineering and Biotechnology	Nobelstrasse 12, 70569 Stuttgart, Germany
Immune Response Corp.	5935 Darwin CT., Carlsbad, CA 92008-7399, USA
Immunex Corp. (Arthritis) (Cancer) (AIDS)	51 University St., Seattle, WA 98101-2936, USA
Incyte Genomics, Inc	4678 World Parkway Circle, St. Louis, MO 63134-3156, USA www.incite.com
Incyte Microarray Systems & Genomics	6519 Dumbarton Dr., Fremont, CA 94555-3619, USA
Inflazyme Pharmaceuticals Ltd. (Arthritis)	5600 Parkwood Way, Richmond, BCV6V 2M2, Canada
Inflazyme (Asthma)	Richmond, BC V6X 2GO, Canada

InforMax

610 Executive Boulevard,10 th Floor, North Bethesda, MD 20852, USA
www.informaxinc.com

Ingenex (Cancer)

Menlo Park, CA 94025, USA

Integrated Genomics Inc.

2201 West Campbell Park Drive, Chicago, IL 60612, USA

Integrated Protein Technologies Agracetus Campus, Monsanto Company

8520 University Green, Middleton, WI 53562, USA

Interferon Sciences Inc. (AIDS)

783 Jersey Ave., New Brunswick, NJ 08901-3660, USA

Intra Biotics Pharmaceuticals) (Antibacterial and Antifungal)

1255 Terra Bella Ave., Mountain View, CA 94043-1833, USA

Invitrogen Corp.

1600 Faraday, Carlsbad, CA 92008, USA

Isenselt (Intelligente Sensorsoftware und Bioinformatik AG iG)

Fahrenheitstrasse 9, 28359 Bremen, Germany

IVAX Corp. (AIDS)

4400 Biscayne Blvd., Miami, FL 33137-3212, USA

Ixsis (Cancer)

1903 1/2 Capitol Ave., Sacramento, CA 95814-4212, USA

Jacobus Pharmaceuticals (AIDS)

37 Cleveland Ln., Princeton, NJ 08540-3790, USA

Janssen Pharmaceiticals (antibacterial and antifungal)

1125 Trenton Harbourton Rd.,Titusville, NJ 08560-1504, USA

Juergen Drews, International Biomedicine Management Partners, Orbimed Advisors

767 Third Avenue, New York, NY 10158, USA
Tel.: (++1) 212 739 6400
Fax: (++1) 212 739 6444

Jerini Bio Tools GmbH

Rudower Chaussee 5, 12489 Berlin, Germany

Kissei Pharmaceuticals Co. Ltd. (Arthritis)

Meijiseimei Sakaimachi Bldg., 1-9-6 Sakaimachi, Kokurakita-Ku, Kitakyushi City, Fakuoka Pref., Japan

KOSAN Biosciences

3832 Bay Center Place, Hayward, CA 94545, USA

KTB Tumorforschungs GmbH

Breisacher Strasse 117, 79106 Freiburg, Germany

LaVision BioTec GmbH

Hoefeweg 74, 33619 Bielefeld, Germany

LETTI (CEA-Technologies Avancées)

DEIN/SPE/GCO, CE Saclay, 91191 Gif-sur-Yvette Cedex, France

LeukoSite Inc.

75 Sydney St., Cambridge, MA 02139-4134, USA

Lexicon Genetics

4000 Research Forest Drive, The Woodlands, TX 77381, USA

Ligand Pharmaceuticals Inc. (AIDS)

10275 Science Center Dr., San Diego, CA 92121-1117, USA

Lion Bioscience AG

Im Neuenheimer Feld 515, 69120 Heidelberg, Germany

Lorus Therapeutics Inc. (AIDS)

7100 Woodbine Ave., Markham, ON L3R 5JR, Canada

Lynx Therapeutics Inc.

25861 Industrial Boulevard, Hayward, CA 94545, USA

Max-Planck-Gesellschaft MIPS Datenbank für Proteinsequenzen und Molekulare Strukturbiologie	Hofgartenstr. 8, 80539 Muenchen, Germany
Magainin Pharmaceuticals Inc. (Antibacterial and Antifungal) (Asthma)	5110 Campus Dr., Plymouth Meeting, PA 19462-1111, USA
Maxygen Inc.	3410 Central Expressway, Santa Clara, CA 95051, USA
Medarex Inc.	707 State Rd.206, Princeton, NJ 08540-1437, USA
MEDICS (The European Center of Competence for Biomedical Microdevices)	c/o FhG IBMT, Industriestrasse 5, 66280 Sulzbach, Germany
MediGene AG	Lochhamer Strasse 11, 82152 Martinsried, Germany
MediGenomix GmbH	Lochamer Strasse 11, 82152 Martinsried Planegg, Germany
Memorec Stoffel GmbH	Medical Molecular Research Cologne, Stockheimer Weg 1, 50829 Koeln, Germany
Merck & Co Inc. (AIDS)	1 Merck Dr., White House Station, NJ 08889-3497, USA
Merck KGaA	Frankfurter Strasse 250, 64293 Darmstadt, Germany
Merck Research Laboratories	P. O. Box 2000, Mail Drop RY80Y-215, Rahway, NJ 07065, USA
Merck Research Laboratories	770 Sumneytown Pike, West Point, PA 19486, USA
Merck Sharp and Dohme Research laboratories Neuroscience Research Centre	Terlings Park, Eastwick Road, Harlow, Essex CM20 2QR, UK
Message Pharmaceuticals	30 Spring Mill Drive, Malvern, PA 19355, USA
Metanomics GmbH & CoKG	Tegeler Weg 33, 10589 Berlin, Germany
MIKROGEN GmbH	Fraunhofer Strasse 20, 82152 Martinsried, Germany
Millennium Pharmaceuticals	P. O. Box 798, Burlington, MA 01803, USA
Millennium Pharmaceuticals and Millennium Predictive Medicine	640 Memorial Drive, Cambridge, MA 02139, USA
Millenium Pharmaceuticals (Asthma)	640 Memorial Drive, Cambridge, MA 02139, USA
Miltenyi Biotec GmbH	Friedrich-Ebert-Strasse 68, 51429 Bergisch Gladbach, Germany
MIT Artificial Intelligence Lab	545 Technology Square, Cambridge, MA 02139, USA
MitoKor	11494 Sorrento Valley Road, San Diego, CA 92121, USA
Mixture Sciences	Torrey Pines Institute for Molecular Studies, 3550 General Atomics Court, San Diego, CA 92121, USA
Molecular Devices Corporation	1311 Orleans Drive, Sunnyvale, CA 94089, USA www.moleculardevices.com Tel.: (++1) 800 635 5577

Molecular Dynamics Amersham Pharmacia Biotech	928 E Arques Ave., Sunnyvale, CA 94085-4520, USA
Monsanto Company	700 Chesterfield Village Parkway North, St Louis, MO 63198, USA
MorphoSys AG	Am Klopferspitz 19, 82152 Martinsried, Germany
MPB Cologne GmbH Molecular Plant & Protein Biotechnology	Eupener Strasse 161, 50923 Koeln, Germany
MWG Biotech AG	Anzinger Strasse 7a, 85560 Ebersberg, Germany
Myriad Genetics Inc.	320 Wakara Way, Salt Lake City, UT 84108, USA
NABI Biomedical (Antibacterial and Antifungal)	1000 Chestnut St., Evansville, IN 47713-1952, USA
Nanogate GmbH	Gewerbepark, Eschberger Weg, 66121 Saarbruecken, Germany
Nanogen Inc.	10398 Pacific Center Center, San Diego, CA 92121, USA
National Institute of Health	9000 Rockville Pike, Bethesda, MD 20892, USA
National Institute on Aging	Gerontology Research Center, 5600 Nathan Shock Drive, Baltimore, MD 21224-6825, USA
NeoGenesis	840 Memorial Drive, Cambridge, MA 02139, USA
Neo Rx Corp. (cardiovascular disease)	410 W Harrison St., Seattle, WA 98119-4000, USA
Neose Technologies Inc.	102 Witmer Road, Horsham, PA 19044, USA
Neurocrine Biosciences Inc. (autoimmune disease)	10555 Science Center Dr., San Diego, CA 92121-1100, USA
New World Science and Technology Inc.	8401 Colesville Road, Suite 610, Silver Spring, MD 20910, USA
Nexell Therapeutics Inc. (AIDS)	9 Parker, Irvine, CA 92618-1605, USA
Novartis Agricultural Discovery Institute Inc.	3115 Merryfield Row, Suite 100, San Diego, CA 92121, USA
Novartis Pharmaceuticals Corp.	556 Morris Avenue, Building SEF 1029, Summit, NJ 07901-1398, USA
Novartis (Antibacterial and Antifungal)	Basel, Switzerland
Novartis Corp. (Antibacterial and Antifungal)	59 State Route 10, East Hanover, NJ 07936-1005, USA
Novo Nordisk A/S	Enzyme Design, 2CS.01, 1 Novo Alle, 2880 Bagsvaerd, Denmark
Novo Nordisk Biotech Inc.	1445 Drew Avenue, Davis, CA 95616, USA
NOXXON Pharma AG	Gustav-Meyer-Allee 25, 13355 Berlin, Germany
Oncotest GmbH	Am Flughafen 8–10, 79110 Freiburg, Germany
Ono Pharmaceuticals Co. Ltd (Arthritis)	2-1-5 Doshomachi, Chuo-Ku, Osaka-City, Osaka Pref., Japan

Ontogeny Inc.	45 Moulton Street, Cambridge, MA 02138 USA
Orchid BioSciences	303 College Road East, Princeton, NJ 08540, USA
Organon Inc.	Department of Immunology, PO Box 20, 5340 BH Oss, The Netherlands
Origene Technologies Inc.	6 Taft Court, Suite 300, Rockville, MD 20850, USA
Orpegen Pharma Gesellschaft fuer Biotechnologische Forschung, Entwicklung und Produktion mbH	Czernyring 22, 69115 Heidelberg, Germany
Pangea Systems	4040 Campbell Ave, Menlo Park, CA 94025, USA
PanVera Corp.	545 Science Drive, Madison WI 53711, USA
Paradigm Genetics	104 Alexander Drive, Building 2, PO Box 14528, Research Triangle Park, NC 27709-5428, USA
Pasteur Merieux Connaught (Malaria) (AIDS)	Discovery Dr., Swiftwater, PA 18370, USA
PathoGenesis Corp.	201 Elliott Avenue West, Seattle, WA 98119, USA
PE Biosystems	500 Old Connecticut Path, Framingham, MA 01701, USA
Penederm Inc. (Antimicrobial and Antifungal)	320 Lakeside Dr. #A, Foster City, CA 94404-1146, USA
Pfizer Inc. (Antibacterial and Antifungal) (AIDS)	235E 42 nd St., New York, NY 10017-5755, USA
Pharmabiodyn GmbH	Ferdinand-Porsche-Strasse 51, 79211 Denzlingen, Germany
Pharmacia & Upjohn	7000 Portage Road, Kalamazoo, MI 49001, USA
Pharmacia & Upjohn	103 Carnegie Ctr., Princeton, NJ 08540-6235, USA
Pharmacia Corp. (Antibacterial and Antifungal)	100 RT 206N, Peapack, NJ 07977, USA
Pharmacopoeia Inc. (Asthma)	3000 Eastpark Blvd., Cranbury, NJ 08512-3516, USA
Phytotech Inc.	1 Deer Park Drive, Suite 1, Monmouth Junction, NJ 08852, USA
Pico Rapid Technologie GmbH	Leobener Strasse, UFT, 28359 Bremen, Germany
Polydex Pharmaceuticals Inc. (AIDS)	2705 Lajoie, Trois Rivières, QC G8Z 3G4 Canada
PPL Therapeutics Inc.	1700 Kraft Drive, Blacksburg, VA 24060, USA
PPL Therapeutics Ltd	Roslin, Edingburgh, EH25 9PP, UK
Preclinical Research Hoffmann-La Roche, Inc	340 Kingsland St., Nutley, NJ 07110-1199, USA
Price Waterhouse Coopers	Harman House, 1 George Street, Uxbridge, Middlesex UB8 1QQ, UK
ProBioGen GmbH	Rudower Chaussee 5, 12489 Berlin, Germany

Progen Biotechnik GmbH	Maasstrasse 30, 69123 Heidelberg, Germany
Progenitor (Asthma)	1730 Ackerman Dr. #4, Lodi, CA 95240-6385, USA
Proligo Biochemie GmbH	Georg-Heyken-Strasse 14, 21147 Hamburg, Germany
Promega Corp.	2800 Woods Hollow Road, Madison, WI 53177-5399, USA
Promega GmbH	Hi-Tech-Park, Schildkroetstrasse 15, 68199 Mannheim, Germany
Protana A/S	Staermosegaardsvej 16, 5230 Odense M, Denmark
Protein Design Labs Inc.	34801 Campus Dr., Fremont, CA 94555-3606, USA
Qiagen, Inc.	28159 Avenue Stanford, Valencia, CA 91355, USA www.qiagen.com
Qiagen GmbH	Max-Volmer-Strasse 4, 40724 Hilden, Germany
R & D Division Smith-Kline-Beecham Pharmaceuticals	1250 South Collegeville Road, P.O.Box 5089, Collegeville, PA 19426-0989, USA
R W Johnson Pharmaceutical Research Institute (Antibacterial and Antifungal)	920 RT 202, Raritan, NJ 08869-1420, USA
Regeneron Pharmaceuticals Inc.	777 Old Saw Mill River Road, Tarrytown, NY 10591, USA
Regis Technologies Inc.	8210 Austin Avenue, Morton Grove, IL 60053-3205, USA
Rentschler Biotechnolgie GmbH	Erwin-Rentschler-Str. 21, 88471 Laupheim, Germany
Research Genetics	2130 Memorial Parkway, SW Huntsville, AL 35801, USA
Rhein Biotech GmbH	Eichsfelder-Strasse 11, 40595 Duesseldorf, Germany
RheoGene	706 Forest Street, Charlottsville, VA 22903, USA
Rhone-Poulenc Rorer	13 Quai Jules Guesdo, Vitry-Sur-Seine 94400, France
Rhone-Poulenc Rorer Ltd.	Dagenham Research Centre, Rainham Road South, Dagenham, Essex RM10 7XS, UK
Rigel Corp. (Asthma)	240 E Grand Ave., South San Francisco, CA 94080-4811, USA
Roche Bioscience	3401 Hillview Avenue, Palo Alto, CA 94304, USA
Rosetta Inpharmatics	12040 115th Ave., NE, Kirkland, WA 98034-6947, USA
Roslin Institute	Midlothian, Edinburgh EH25 9PS, UK
Santa Fe Institute	1399 Hyde Park Rd., Santa Fe, NM 87501-8943, USA
(Sarawak) MediChem Research Inc. (AIDS)	12305 New Ave., Lemont, IL 60439-3687, USA
Schering-Plough	Research Institute, 2015 Galloping Hill Road, Kenilworth, NJ 07033-1300, USA
Schering-Plough Lab	1 Giralda Farms, Madison, NJ 07940-1010, USA
Science Research Laboratory Inc.	15 Ward Street, Somerville, MA 02143-4241, USA
SCIOS Inc. (Cardiovascular Disease)	820 W Maude Ave., Sunnyvale, CA 94085-2910, USA

SensLab GmbH	Permoserstrasse 15, 04318 Leipzig, Germany
Sepracor Inc. (Asthma)	33 Locke Dr., Marlborough, MA 01752-1146, USA
Sequenom Inc.	11555 Sorrento Road, San Diego, CA 92121, USA
Serono Laboratories (AIDS)	100 Longwater Dr., Norwell, MA 02061-1616, USA
Serono Pharmaceutical Research Institute Serono International SA	14, Chemin des Aulx, 1228 Plan-les-Ouates, Geneva, Switzerland
Smith-Kline Beecham Pharmaceuticals	709 Swedeland Road, PO Box 1539, King of Prussia, PA 19106, USA
Smith-Kline-Beecham Pharmaceuticals	R & D Division, 1250 South Collegeville Road, PO Box 5089, Collegeville, PA 19426-0989, USA
Smith-Klein-Beecham (Arthritis) (AIDS) (Antibacterial and Antifungal)	1 Franklin Plz #200, Philadelphia, PA 19102-1225, USA
Squibb Institute for medical Research (Cardiovascular Disease)	RT 206, Trenton, NJ 08648, USA
Stern Cell Sciences Pty Ltd	Level 1, 28 Riddell Parade, Elsternwick Victoria 3185, Australia
Strathmann Biotech GmbH	Sellhopsweg 1, 22459 Hamburg, Germany
SUGEN Inc.	230 E Grand Avenue, South San Francisco, CA 94080-4811, USA
SUGEN Inc. (AIDS Community Research Cnsrt)	1048 El Camino Real # 8, Redwood City, CA 94063-1687, USA
Sunesis Pharmaceuticals Inc.	3696 Haven Avenue, Suite C, Redwood City, CA 94063, USA
Super Gen Inc. (Arthritis) (Cancer)	2 Annabel Ln. #220, San Ramon, CA 94583-1398, USA
Synsorb Biotech Inc. (AIDS)	1204 Kensington Rd NW, Calgary, AB T2N 3P5, Canada
Tanox Biosystems Inc.	10301 Stella Link Rd. #110, Houston, TX 77025-5497, USA
Target Genomics	c/o Pfizer Ltd, Ramsgate Rd., Sandwich, Kent CT13 9NJ, UK
Targeted Genetics Corp. (AIDS)	1100 Olive Way # 100, Seattle, WA 98101-1823, USA
Texas Biotechnology Corp. (Asthma)	7000 Fannin St.#1920, Houston, TX 77030-5500, USA
The Automation Partnership	Melbourn Science Park, Melbourn, Royston SG8 6HB, UK
The Genetics Company	Winterthurerstrasse 190, 8057 Zuerich, Switzerland www.the-genetics.com Tel. (++41) 1 635 66 20 eistetter@the-genetics. com
Therapeutic Antibodies (Malaria) (now Protherics)	Blasenwaun Ffostrasol, Llandysul Dyfed SA 44 5JT, UK
Third Wave Technologies Inc.	502 South Rosa Road, Madison, WI 53719, USA
TissUse GmbH	Delitzscher Strasse 141, 04129 Leipzig, Germany

Titan Pharmaceuticals Inc. (Cancer)	400 Oyster Point Blvd. # 505, South San Francisco, CA 94080-1920, USA
TOPLAB (angewandte Biologie)	Fraunhoferstrasse 18a, 82152 Martinsried/Muenich, Germany
TRACE Biotech AG	Mascheroder Weg 1b, 38124 Braunschweig, Germany
TranXenoGen	Fuller Building, P. O. Box 707, 222 Maple Avenue, Shrewsbury, MA 01545, USA
Trianle Pharmaceuticals Inc. (AIDS)	4611 University Dr., Durham, NC 27707-3458, USA
Tularik Inc.	2 Corporate Dr., South San Francisco, CA 94080-7047, USA
United Biomedical Lab. (AIDS)	25 David's Dr., Hauppauge, NY 11788-2037, USA
Univlever Research	Colworth House Laboratory, Colworth House, Sharnbrook, Bedford, MK44 1LQ, UK
US Bioscience (AIDS)	100 Front St. # 400, W. Conshohocken, PA 19428-2874, USA
Vanguard Medica (Asthma)	Chancellor Court Surrey Research Park, Guilford Surrey, GU2 7YG, UK
Vax Gen Inc. (AIDS)	1000 Marina Blvd. #200, Brisbane, CA 94005-1841, USA
Verex Labs. Inc. (AIDS)	14 Inverness Dr. E #D100, Englewood, CO 80112-5604, USA
Vical Inc. (Malaria)	4510 Executive Dr. #215, San Diego, CA 92121-3023, USA
Warner Lambert (Antibacterial and Antifungal)	182 Tabor Rd, Morris Plains, NJ 07950-2597, USA
Wyeth-Ayerst Labs.	31 Morehall Road, Malvern, PA 193551759, USA
Wyeth-Ayerst Research (CNS Disorders)	37 Commonwealth Ave., Boston, MA 02116-2354, USA
Xenova Ltd. (Cancer)	240 Bath Rd., Slough Berkshire SL1 4EF, UK
Xerox PARC	3333 Cayote Hill Road, Palo Alto, CA 94304, USA
Xoma Corp. (antibacterial and antifungal)	2910 7th St. #100, Berkeley, CA 94710-2743, USA
Zymark Corp.	68 Elm Street, Hopkinton, MA 01748-1668, USA

References

[1] David Eisenberg, Edward M. Marcotte, Ionannis Xenarios and Todd O. Yeates, Protein function in the post-genomic era, *Nature*, **405** (2000), 823–826.

[2] S. Beck, A. Olek and J. Walter, From genomics to epigenomics: a loftier view of life, *Nature Biotechnology*, **17** (1999), 1144.

[3] C. Klein, Validation of genomics-derived drug targets using yeast, *Drug Discovery Today*, **5** (2000), 37–38.

[4] J. Skolnick, J. S. Fetrow and A. Kolinski, Structural genomics and its importance for gene function and analysis, *Nature Biotechnology*, **18** (2000), 283–287.

[5] J. G. Sutcliffe, P. E. Foye, M. G. Erlander, B. S. Hilbush, L. J. Bodzin, J. T. Durham and K. W. Hasel, TOGA: an automated parsing technology for analyzing expression of nearly all genes, *Proceedings of the National Academy of Sciences USA*, **97** (2000), 1976–1981.

[6] Richard A. Shimkets, David G. Lowe, Julie Tsu-Ning Tai, Patricia Sehl, Hongkui Jin, Renhui Yang, Paul F. Predki, Bonnie E. G. Rothberg, Michael T. Murtha, Matthew E. Roth, Suesh G. Shenoy, Andreas Windemuth, John W. Simpson, Jan F. Simons, Michael P. Daley, Steven A. Gold, Michael P. McKenna, Kenneth Hillan, Gregory T. Went and Jonathan M. Rothberg, Gene expression analysis by transcript profiling coupled to a gene database query, *Nature Biotechnology*, **17** (1999), 798–803.

[7] Stephen L. Madden, Clarence J. Wang and Greg Landes, Serial analysis of gene expression: from gene discovery to target identification, *Drug Discovery Today*, **5** (2000), 415–425.

[8] N. Kaminski, J. D. Allard, J. F. Pittet, F. Zuo, M. J. D. Griffiths, D. Morris, X. Huang, D. Sheppard and R. A. Heller, Global analysis of gene expression in pulmonary fibrosis reveals distinct programs regulating lung inflammation and fibrosis, *Proceedings of the National Academy of Sciences USA*, **97** (2000), 1778–1783.

[9] Petra Ross-Macdonald, Paulo S. R. Coelho, Terry Roemer, Seema Agarwal, Anuj Kumar, Ronald Jansen, Kel-Hoi Cheung, Amy Sheehan, Dawn Symoniatis, Lara Umansky, Matthew Heidtman, F. Kenneth Nelson, Hiroshi Iwasaki, Karl Hagers, Mark Gerstein, Perry Miller, G. Shirleen Roeder and Michael Snyder, Large-scale analysis of the yeast genome by transposon tagging and gene disruption, *Nature*, **402** (1999), 413–418.

[10] S. Brenner, S. R. Williams, E. H. Vermaas, T. Storck, K. Moon, C. McCollum, J.-I Mao, S. Luo, J. J. Kirchner, S. Eletr, R. B. DuBridge, T. Burcham and G. Albrecht, *In vitro* cloning of complex mixtures of DNA on microbeads: physical separation of differentially expressed cDNAs, *Proceedings of the National Academy of Sciences USA*, **97** (2000), 1665–1670.

[11] David J. Lockhart and Elizabeth, A. Winzeler, Genomics, gene expression and DNA arrays, *Nature*, **405** (2000), 827–836.

[12] Mei-Ling Ting Lee, Frank C. Kuo, G. A. Whitmore and Jeffrey Sklar, Importance of replication in microarray gene expression studies: statistical methods and evidence from repetitive cDNA hybridizations, *Proceedings of the National Academy of Sciences USA*, **97** (2000), 9834–9839.

[13] San Ming Wang, Scott C. Fears, Lin Zhang, Jian-Jun Chen and Janet D. Rowley, Screening poly(dA/dT)(–) cDNAs for gene identification, *Proceedings of the National Academy of Sciences USA*, **97** (2000), 4162–4167.

Katherine J. Martin and Arthur B. Pardee, Identifying expressed genes, *Proceedings of the National Academy of Sciences USA*, **97** (2000), 3789–3791.

[14]   Tiziana Sturniolo, Jiayi Ding, Laura Raddrizzani, Oezlem Tuereci, Ugur Sahin, Michael Braxenthaler, Fabio Gallazzi, Maria Pia Protti, Francesco Sinigaglia and Juergen Hammer, Generation of tissue-specific and promiscuous HLA ligand databases using DNA microarrays and virtual HLA class II matrices, *Nature Biotechnology*, **17** (1999), 555–561.

[15]   Adel M. Talaat, Preston Hunter and Stephen Albert Johnston, Genome-directed primers for selective labeling of bacterial transcripts for DNA microarray analysis, *Nature Biotechnology*, **18** (2000), 679–682.

[16]   R. T. Simpson, *In vivo* methods to analyze chromatin structure, *Current Opinions in Genetics and Development*, **9** (1999), 225–229.

[17]   Bas van Steensel and Steven Henikoff, Identification of *in vivo* DNA targets of chromatin proteins using tethered Dam methyltransferase, *Nature Biotechnology*, **18** (2000), 424–428.

[18]   Stephen Rea, Frank Eisenhaber, Donal O'Carroll, Brian D. Strahl, Zu-Wen Sun, Manfred Schmid, Susanne Opravil, Karl Mechtler, Chris P. Ponting, C. David Allis and Thomas Jenuwein, Regulation of chromatin structure by site-specific histone H3 methyltransferases, *Nature*, **406** (2000), 593–599.

[19]   Christina M. Grozinger and Stuart, L. Schreiber, Regulation of histone deacetylase 4 and 5 and transcriptional activity by 14-3-3-dependent cellular localization, *Proceedings of the National Academy of Sciences USA*, **97** (2000), 7835–7840.

[20]   Asifa Akhtar, Daniele Zink and Peter B. Becker, Chromodomains are protein–DNA interaction modules, *Nature*, **407** (2000), 405–409.

[21]   Vincent Galy, Jean-Christophe Olivo-Marin, Harry Scherthan, Valerie Doye, Nadia Rascalou and Ulf Nehrbass, Nuclear pore complexes in the organization of silent telomeric chromatin, *Nature*, **403** (2000), 108–112.

[22]   Xuetong Shen, Gaku Mizuguchi, Ali Hamiche and Carl Wu, A chromatin remodeling complex involved in transcription and DNA processing, *Nature*, **406** (2000), 541–544.

[23]   Patrick N. Gilles, David J. Wu, Charles B. Foster, Patrick J. Dillon and Stephen J. Chanock, Single nucleotide polymorphic discrimination by an electronic dot-blot assay on semiconductor microchips, *Nature Biotechnology*, **17** (1999), 365–370.

[24]   W. Matthias Howell, Magnus Jobs, Ulf Gyllensten and Anthony J. Brooks, Dynamic allele-specific hybridization, *Nature Biotechnology*, **17** (1999), 87–88.

[25]   Kai Tang, Dong-Jing Fu, Dominique Julien, Andreas Braun, Charles R. Cantor and Hubert Köster, Chip-based genotyping by mass spectrometry, *Proceedings of the National Academy of Sciences USA*, **96** (1999), 10016–10020.

[26]   Claire M. McCallum, Luca Comai, Elizabeth A. Greene and Steven Henikoff, Targeted screening for induced mutations, *Nature Biotechnology*, **18** (2000), 455–457.

[27]   Maureen D. Megonigal, Eric F. Rappaport, Robert S. Wilson, Douglas H. Jones, James A. Whitlock, Jorge A. Ortega, Diana J. Slater, Peter C. Nowell and Carolyn A. Felix, Panhandle PCR for cDNA: a rapid method for isolation of *MLL* fusion transcripts involving unknown partner genes, *Proceedings of the National Academy of Sciences USA*, **97** (2000), 9597–9602.

[28]   Akhilesh Pandey and Matthias Mann, Proteomics to study genes and genomes, *Nature*, **405** (2000), 837–846.

[29]   Alison Abbott, A post-genomic challenge: learning to read patterns of protein synthesis, *Nature*, **402** (1999), 715–720.

[30]   M. R. Wilkins, J.-C. Sanchez, A. A. Gooley, R. D. Appel, I. Humphery-Smith, D. F. Hochstrasser and K. L. Williams, Progress with proteome projects: why all proteins expressed by a genome should be identified and how to do it, *Biotechnology and Genetic Engineering Reviews*, **13** (1995), 19–50.

[31]   M. Wilkins, Proteomic paradigms for perceiving purpose, *TIBTECH*, **18** (2000), 91–92.

[32]   Eckhard Nordhoff, Anne-M. Krogsdam, Helle F. Jorgensen, Birgitte H. Kallipolitis, Brian F. C. Clark, Peter Roepstorff and Karsten Kristiansen, A rapid identification of DNA-binding proteins by mass spectrometry, *Nature Biotechnology*, **17** (2000), 884–888.

[33] G. Marius Clore, Accurate and rapid docking of protein–protein complexes on the basis of inter-molecular nuclear Overhauser enhancement data and dipolar couplings by rigid body minimization, *Proceedings of the National Academy of Sciences USA*, **97** (2000), 9021–9025.

[34] Steven P. Gygi, Garry L. Corthals, Yanni Zhang, Yvan Rochon and Ruedi Aebersold, Evaluation of two-dimensional gel electrophoresis-based proteome analysis technology, *Proceedings of the National Academy of Sciences USA*, **97** (2000), 9390–9395.

[35] M. J. Grossel, H. Wang, B. Gadea, W. Yeung and P. H. Hinds, A yeast two-hybrid system for discerning differential interactions using multiple baits, *Nature Bio-technology*, **17** (1999), 1232–1233.

[36] Stephen McCraith, Ted Holtzman, Bernard Moss and Stanley Fields, Genome-wide analysis of vaccinia virus protein–protein interactions, *Proceedings of the National Academy of Sciences USA*, **97** (2000), 4879–4884.

[37] Fang Liu, Qi Wan, Zdenek B. Pristupa, Xian-Min Yu, Yu Tian Wang and Hyman B. Niznik, Direct protein–protein coupling enables cross-talk between dopamine D5 and γ-aminobutyric acid A receptors, *Nature*, **403** (2000), 274–280.

[38] Steven P. Gygi, Beate Rist, Scott A. Gerber, Frantisek Turecek, Michael H. Gelb and Ruedi Aebersold, A quantitative analysis of complex protein mixtures using isotope-coded affinity tags, *Nature Biotechnology*, **17** (1999), 994–999.

[39] Ruud M. T. de Wildt, Chris R. Mundy, Barbara D. Gorick and Ian M. Tomlinson, Antibody arrays for high-throughput screening of antibody–antigen interactions, *Nature Biotechnology*, **18** (2000), 989–994.

[40] Alia Qureshi Emili and Gerard Cagney, Large-scale functional analysis using peptide or protein arrays, *Nature Biotechnology*, **18** (2000), 393–397.

[41] Gavin MacBeath and Stuart L. Schreiber, Printing proteins as microarrays for high-throughput function determination, *Science*, **289** (2000), 1760–1763.

[42] T. A. Craig, L. M. Breson, A. J. Tomlinson, T. D. Veenstra, S. Naylor and R. Kumar, Analysis of transcription complexes and effects of ligands by microelectrospray ionization mass spectrometry, *Nature Biotechnology*, **17** (1999), 1214–1218.

[43] Alfred Maelicke, Proteomics, *Nachrichten aus Chemie, Technik und Laboratorium*, **47** (1999), 1433–1435.

[44] S. Müllner, Eine Bewertung des Technologie- und Marktpotentials für die Life-Science-Industrie, *transkript*, **8** (5) (1999), 40–44.
Georgius Amexis, Paul Oeth, Kenneth Abel, Anna Ivashina, François Pelloquin, Charles R. Cantor, Andreas Brau, Konstantin Chumakow, Quantitative Mutant Analysis of Viral Quasispecies by Chip-Based Matrix-Assisted Laser Desorption/Ionization Time-of-Flight Mass Spectrometry, *Proccedings of the National Academy of Sciences USA*, **98** (2001), 12097–12102.

[45] T. Ito, K. Tashiro, S. Muta, R. Ozawa, T. Chiba, M. Ishizawa, K. Yamamoto, S. Kuhara and Y. Sakaki, Toward a protein–protein interaction map of the budding yeast: a comprehensive system to examine two-hybrid interactions in all possible combinations between the yeast proteins, *Proceedings of the National Academy of Sciences USA*, **97** (2000), 1143–1147.

[46] Peter Uetz, Loic Giot, Gerard Cagney, Traci A. Mansfield, Richard S. Judson, James R. Knight, Daniel Lockshon, Vaibhav Narayan, Maithreyn Srinivasan, Pascale Pochart, Alia Qureshi-Emili, Ying Li, Brian Godwin, Diana Conover, Theodore Kalbfleisch, Govindan Vijayadamodar, Meijia Yang, Mark Johnston, Stanley Fields, Jonathan M. Rothberg, A comprehensive analysis of protein–protein interactions in *Saccharomyces cerevisiae*, *Nature*, **403** (2000), 623–627.
Benno Schwikowski, Peter Uetz, Stanley Fields, A network of protein-protein interactions in yeast, *Nature Biotechnology*, **18** (2000), 1257–1261.

[47] J. Keith Joung, Elizabeth I. Ramm and Carl O. Pabo, A bacterial two-hybrid selection system for studying protein–DNA and protein–protein interactions, *Proceedings of the National Academy of Sciences USA*, **97** (2000), 7382–7387.

[48] Carl T. Rollins, Victor M. Rivera, Derek N. Woolfson, Terence Keenan, Marcos Hatada, Susan E. Adams, Lawrence J. Andrade, David Yaeger, Marie Rose van Schravendijk, Dennis A. Holt, Michael Gilman and Tim Clackson, A ligand-reversible dimerization system for controlling protein–protein interactions, *Proceedings of the National Academy of Sciences USA*, **97** (2000), 7096–7101.

[49] L. Li, R. W. Garde and J. V. Sweedler, Single-cell MALDI: a new tool for direct peptide profiling, *TIBTECH*, **18** (2000), 151–160.

[50] F. Oesterhelt, D. Oesterhelt, M. Pfeiffer, A. Engel, H. E. Traub and D. J. Müller, Unfolding pathways of individual bacteriorhodopsins, *Science*, **288** (2000), 143–146.
      J. G. Forbes and G. H. Lorimer, Unraveling a membrane protein, *Science*, **288** (2000), 63–64.

[51] D. Baker, A surprising simplicity to protein folding, *Nature*, **405** (2000), 39–42.

[52] G. Milligan, Receptors as kissing cousins, *Science*, **288** (2000), 65–67.

[53] M. Rocheville, D. C. Lange, U. Kumar, S. C. Patel, R. C. Patel and Y. C. Patel, Receptors for dopamine and somatostatin: formation of hetero-oligomers with enhanced functional activity, *Science*, **288** (2000), 154–157.

[54] P.-A. Binz, M. Müller, D. Walther, W. V. Bienvenut, R. Gras, C. Hoogland, G. Bouchet, E. Gasteiger, R. Fabretti, S. Gay, P. Palagi, M. R. Wilkins, V. Rouge, L. Tonella, S. Paesano, G. Rossellat, A. Karmime, A. Bairoch, J.-C. Sanchez, R. D. Appel and D. F. Hochstrasser, A molecular scanner to automate proteomic research and to display proteome images, *Analytical Chemistry*, **71** (1999), 4981–4988.

[55] J. Rappsilber, S. Siniossoglou, E. C. Hurt and M. Mann, A generic strategy to analyze the spatial organization of multi-protein complexes by cross-linking and mass spectro-metry, *Analytical Chemistry*, **72** (2000), 267–275.

[56] E. Saxon and C. R. Bertozzi, Cell surface engineering by a modified Staudinger reaction, *Science*, **287** (2000), 2007–2010.

[57] B. D. Stahl and C. D. Allis, The language of covalent histone modifications, *Nature*, **403** (2000), 41–45.

[58] V. Galy, J.-C. Olivo-Marin, H. Scherthan, V. Doye, N. Rascalou and U. Nehrbass, Nuclear pore complexes in the organization of silent telomeric chromatin, *Nature*, **403** (2000), 108–112.

[59] Ulrich Schubert, Luis C. Anton, James Gibbs, Christopher C. Norbury, Jonathan W. Yewdell and Jack R. Bennink, Rapid degradation of a large fraction of newly synthesized proteins by proteasomes, *Nature*, **404** (2000), 770–774.

[60] Sangeet Singh-Gasson, Roland D. Green, Yongjian Yue, Clark Nelson, Fred Blattner, Michael R. Sussman and Franco Cerrina, Maskless fabrication of light-directed oligo-nucleotide microarrays using a digital micromirror array, *Nature Biotechnology*, **17** (1999), 974–978.

[61] R. C. Merckle, Biotechnology as a route to nanotechnology, *TIBTECH*, **17** (1999), 271–274.

[62] M. A. Unger, H.-P. Chou, T. Thorsen, A. Scherer and S. R. Quake, Monolithic microfabricated valves and pumps by multilayer soft lithography, *Science*, **288** (2000), 113–116.

[63] Dawn E. Kataoka and Sandra M. Troian, Patterning liquid flow on the microscopic scale, *Nature*, **402** (1999), 794–797.

[64] Douglas B. Chrisey, The power of direct writing, *Science*, **289** (2000), 797–881.

[65] Seunghun Hong and Chad A. Mirkin, A nanoplotter with both parallel and serial writing capabilities, *Science*, **288** (2000), 1808–1811.

[66] Gongwen Peng, Feng Qiu, Valeriy V. Ginzburg, David Jasnow and Anna C. Balazs, Forming supramolecular networks from nanoscale rods in binary, phase-separating mixtures, *Science*, **288** (2000), 1802–1804.

[67] David H. Gracias, Joe Tien, Tricia L. Breen, Carey Hsu and George M. Whitesides, Forming electrical networks in three dimensions by self-assembly, *Science*, **289** (2000), 1170–1172.

[68] Mei Li, Heimo Schnablegger and Stephen Mann, Coupled synthesis and self-assembly of nanoparticles to give structures with controlled organization, *Nature*, **402** (1999), 393–395.

[69] Tadashi Okamoto, Tomohiro Suzuki and Nobuko Yamamoto, Microarray fabrication with covalent attachment of DNA using bubble jet technology, *Nature Biotechnology*, **18** (2000), 438–441.

[70] Edwin W. H. Jager, Olle Inganäds and Ingemar Lundström, Microrobots for micro-meter-size objects in aqueous media: potential tools for single-cell manipulation, *Science*, **288** (2000), 2335–2338.

[71] Michael J. B. Krieger, Jean-Bernard Billeter and Laurent Keller, Ant-like task allocation and recruitment in cooperative robots, *Nature*, **406** (2000), 992–995.

[72] Hod Lipson and Jordan B. Pollack, Automatic design and manufacture of robotic lifeforms, *Nature*, **406** (2000), 974–978.
Rodnay Brooks, From robot dreams to reality, *Nature*, **406** (2000), 945–947.

[73] Michael Knoblauch, Julian M. Hibberd, John C. Gray and Aart J. E. van Bel, A galinstan expansion femtosysringe for micro-injection of eukaryotic organelles and prokaryotes, *Nature Biotechnology*, **17** (1999), 906–909.

[74] Brian M. Cullum and Tuan Vo-Dinh, The development of optical nanosensors for biological measurements, *TIBTECH*, **18** (2000), 388–393.

[75] Nily Dan, Synthesis of hierarchical materials, *TIBTECH*, **18** (2000), 370–374.

[76] John Cumings and A. Zettl, Low-friction nanoscale linear bearing realized from multiwall carbon nanotubes, *Science*, **289** (2000), 602–604.

[77] Thomas Rueckes, Kyoungha Kim, Ernesto Joselevich, Greg Y. Tseng, Chin-Li Cheung and Charles M. Lieber, Carbon nanotube-based nonvolatile random access memory for molecular computing, *Science*, **289** (2000), 94–97.

[78] Rolf Schuster, Viola Kirchner, Philippe Allongue and Gerhard Ertl, Electro-chemical micromachining, *Science*, **289** (2000), 98–101.

[79] Bernard Yurke, Andrew J. Turberfield, Allen P. Mills, Jr, Friedrich C. Simmel and Jennifer L. Neumann, A DNA-fuelled molecular machine made of DNA, *Nature*, **406** (2000), 605–608.

[80] R. C. Hayward, D. A. Saville and I. A. Aksay, Electrophoretic assembly of colloidal crystals with optically tunable micro-patterns, *Nature*, **404** (2000), 56–59.

[81] Hongyou Fan, Yunfeng Lu, Aaron Stump, Scott T. Reed, Tom Baer, Randy Schunk, Victor Perez-Luna, Gabriel P. Lopez and C. Jeffrey Brinker, Rapid prototyping of patterned functional nanostructures, *Nature*, **405** (2000), 56–60.

[82] Sandra R. Whaley, D. S. English, Evelyn L. Hu, Paul F. Barbara and Angela M. Belcher, Selection of peptides with semiconductor binding specificity for directed nanocrystal assembly, *Nature*, **405** (2000), 665–668.
Chad A. Mirkin and T. Andrew Taton, Semiconductors meet biology, *Nature*, **405** (2000), 626–627.

[83] G. P. Lopinski, D. D. M. Wayner and R. A. Wolkow, Self-directed growth of molecular nanostructures on silicon, *Nature*, **406** (2000), 48–51.

[84] Nae-Lih Wu, Sze-Yen Wang and I. A. Rusakova, Inhibition of crystallite growth in the sol–gel synthesis of nanocrystalline metal oxides, *Science*, **285** (1999), 1375–1377.

[85] David J. Beebe, Jeffrey S. Moore, Joseph M. Bauer, Qing Yu, Robin H. Liu, Chelladurai Devadoss and Byung-Ho Jo, Functional hydrogel structures for autonomous flow control inside micro-fluidic channels, *Nature*, **404** (2000), 588–590.

[86] J. Fritz, M. K. Baller, H. P. Lang, H. Rothuizen, P. Vettiger, E. Meyer, H.-J. Güntherodt, Ch. Gerber and J. K. Gimzewski, Translating biomolecular recognition into nanomechanics, *Science*, **288** (2000), 316–318.

[87] Veronica Bermudez, Nathalie Capron, Torsten Gase, Francesco G. Gatti, Francois Kaijzar, David A. Leigh, Francesco Zerbetto and Songwei Zhang, Influencing intramolecular motion with an alternating electric field, *Nature*, **406** (2000), 608–611.

[88] James C. Weaver, Timothy E. Vaughan and R. Dean Astumian, Biological sensing of small field differences by magnetically sensitive chemical reactions, *Nature*, **405** (2000), 707–709.

[89] N. C. Seeman, DNA engineering and its application to nanotechnology, *TIBTECH*, **17** (1999), 437–443.

[90] D. Faulhammer, A. R. Cukras, R. J. Lipton and L. F. Landweber, Molecular computation: RNA solutions to chess problems, *Proceedings of the National Academy of Sciences USA*, **97** (2000), 1385–1389.

[91] J. Chen and D. H. Wood, Computation with biomolecules, *Proceedings of the National Academy of Sciences USA*, **97** (2000), 1328–1330.

[92] K. Sakamoto, H. Gouzu, K. Komiya, D. Kiga, S. Yokoyama, T. Yokomori and M. Hagiya, Molecular computation by DNA hairpin formation, *Science*, **288** (2000), 1223–1226.

[93] Qinghua Liu, Liman Wang, Anthony G. Frutos, Anne E. Condon, Robert M. Corn and Lloyd M. Smith, DNA computing on surfaces, *Nature*, **403** (2000), 175–179.

[94] Mitsunori Ogihara and Animesh Ray, DNA computing on a chip, *Nature*, **403** (2000), 143–144.

[95] Robert P. Lanza, Jose B. Cibelli and Michael D. West, Human therapeutic cloning, *Nature Medicine*, **5** (1999), 975–977.
      A. W. S. Chan, T. Dominko, C. M. Luetjens, E. Neuber, C. Martinovich, L. Hewitson, C. R. Simerly, G. P. Schatten, Clonal Propagation of Primate Offspring by Embryo Splitting, *Science*, **287** (2000), 317–321.

[96] L. R. Abeydeera *et al.*, Development and Viability of Pig Oocytes Matured in A Protein-Free Medium Containing Epidermal Growth Factor, *Theriogenology*, **54** (5) (2000), 787–797.
      Randall S. Prather, Pigs is pigs, *Science*, **289** (2000), 1886–1887.

[97] Mindy Tsai, Jochen Wedemeyer, Soula Ganiatsas, See-Ying Tam, Leonard I. Zon and Stephen J. Galli, *In vivo* immunological function of mast cells derived from embryonic stem cells: an approach for the rapid analysis of even embryonic lethal mutations in adult mice *in vivo*, *Proceedings of the National Academy of Sciences USA*, **97** (2000), 9186–9190.

[98] Oliver Brüstle, Kimberly N. Jones, Randall D. Learish, Khaled Karram, Khalid Choudhary, Otmar D. Wiestler, Ian D. Duncan, Ronald D. G. McKay, Embryonic Stem Cell-Derived Glial Precursors: A Source of Myelinating Transplants, *Science*, **285** (1999), 754–756.

[99] Ingrid Wickelgren, Rat Spinal Cord Function Partially Restored, *Science*, **286** (1999), 1826–1827.

[100] Henry M. Eppich, Russell Foxall, Kate Gaynor, David Dombkowski, Nobuyuki Miura, Tao Cheng, Sandra Silva-Arrieta, Richard H. Evans, Joseph A. Mangano, Frederic I. Preffer and David T. Scadden, Pulsed electric fields for selection of hematopoietic cells and depletion of tumor cell contaminants, *Nature Biotechnology*, **18** (2000), 882–887.

[101] D. Perry, Patients' voices: the powerful sound in the stem cell debate, *Science*, **287** (2000), 1423.
      J. M. W. Slack, Stem cells in epithelial tissues, *Science*, **287** (2000), 1431–1433.
      F. H. Gage, Mammalian neural stem cells, *Science*, **287** (2000), 1433–1438.
      D. van der Kooy and S. Weiss, Why stem cells, *Science*, **287** (2000), 1439–1441.
      P. J. Hines, B. A. Purnell and J. Marx, Stem cells branch out, *Science*, **287** (2000), 1417.

[102] Benjamin E. Reubinoff, Martin F. Pera, Chui-Yee Fong, Alan Trounson and Ariff Bongso, Embryonic stem cell lines from human blastocysts: somatic differentiation *in vitro*, *Nature Biotechnology*, **18** (2000), 399–404. *Nature Genetics*, **24** (2000), 372ff.

[103] Motonari Kondo, David C. Scherer, Toshihiro Miyamoto, Angela C. King, Koichi Akashi, Kazuo Sugamura and Irving L. Weissman, Cell-fate conversion of lymphoid-committed progenitors by instructive actions of cytokines, *Nature*, **407** (2000), 383–386.

[104] Maite Lewin, Nadia Carlesso, Ching-Hsuan Tung, Xiao-Wu Tang, David Cory, David T. Scadden and Ralph Weissleder, Tat peptide-derivatized magnetic nanoparticles allow *in vivo* tracking and recovery of progenitor cells, *Nature Biotechnology*, **18** (2000), 410–414.

[105] T. Cheng, N. Rodrigues, H. Shen, Y.-G. Yang, D. Dombkowski, M. Sykes and D. T. Scadden, Hematopoietic stem cell quiescence maintained by p21(cip1/waf1), *Science*, **287** (2000), 1804–1808.

[106] E. Marshall, New genetic tricks to rejuvenate ailing livers, *Science*, **287** (2000), 1185–1186.

[107] K. L. Rudolph, S. Chang, M. Millard, N. Schreiber-Agus and R. A. DePinho, Inhibition of experimental liver cirrhosis in mice by telomerase gene delivery, *Science*, **287** (2000), 1253–1258.

[108] Naoya Kobayashi, Toshiyoshi Fujiwara, Karen A. Westerman, Yusuke Inoue, Masakiyo Sakaguchi, Hirofumi Noguchi, Masahiro Miyazaki, Jin Cai, Noriaki Tanaka, Ira J. Fox, Philippe Leboulch, Prevention of Acute Liver Failure in Rats with Reversibly Immortalized Human Hepatocytes, *Science*, **287** (2000), 1258–1262.

[109] K. J. McCreath, J. Howcroft, K. H. S. Campbell, A. Colman, A. E. Schnieke and A. J. Kind, Production of gene-targeted sheep by nuclear transfer from cultured somatic cells, *Nature*, **405** (2000), 1066–1069.

[110] Akira Onishi, Masaki Iwamoto, Tomiji Akita, Satoshi Mikawa, Kumiko Takeda, Takashi Awata, Hirohumi Hanada and Anthony C. F. Perry, Pig cloning by microinjection of fetal fibroblast nuclei, *Science*, **289** (2000), 1188–1190.

[111] C. Kubota, H. Yamakuchi, J. Todoroki, K. Mizoshita, N. Tabara, M. Barber and X. Yang, Six cloned calves produced from adult fibroblast cells after long-term culture, *Proceedings of the National Academy of Sciences USA*, **97** (2000), 990–995.

[112] Yoko Kato, Tetsuya Tani, Yusuke Sotomaru, Kazuo Kurokawa, Jun-ya Kato, Hiroshi Doguchi, Hiroshi Yasue and Yukio Tsunoda, Eight calves cloned from somatic cells of a single adult, *Science*, **282** (1998), 2095–2098.

[113] Irina A. Polejaeva, Shu-Hung Chen, Todd D. Vaught, Raymond L. Page, Mullins, Suyapa Ball, Yifan Dai, Jeremy Boone, Shawn Walker, David L. Ayares, Alan Colman and Keith H. S. Campbell, Cloned pigs produced by nuclear transfer from adult somatic cells, *Nature*, **407** (2000), 86–90.

[114] Andy Coghlan, Cloning without embryos, *New Scientist*, **165** (No. 2223) (2000), 4.

[115] Michael Thomas, Lianqing Yang and Peter J. Hornsby, Formation of functional tissue from transplanted adrenocortical cells expressing telomerase reverse transcriptase, *Nature Biotechnology*, **18** (2000), 39–42.
J. W. Shay and W. E. Wright, The use of telomerized cells for tissue engineering, *Nature Biotechnology*, **18** (2000), 22–23.
Floyd E. Bloom, Breakthroughs 1999, *Science*, **286** (1999), 2267.
Gretchen Vogel, Capturing the Promise of Youth, *Science*, **286** (1999), 22438–2239.

[116] Robert P. Lanza, Jose B. Cibelli and Michael D. West, Prospects for the use of nuclear transfer in human transplantation, *Nature Biotechnology*, **17** (1999), 1171–1174.

[117] I. L. Weissman, Translating stem and progenitor cell biology to the clinic: barriers and opportunities, *Science*, **287** (2000), 1442–1446.

[118] D. Ferber, Growing human corneas in the lab, *Science*, **286** (1999), 2051–2052.

[119] May Griffith, Rosemarie Osborne, Rejean Munger, Xiaojuan Xiong, Charles J. Doillon, Noelani L. C. Laycock, Malik Hakim, Ying Song, Mitchell A. Watsky, Functional Human Corneal Equivalents Constructed from Cell Lines, *Science*, **286** (1999), 2169–2172.

[120] L. E. Niklason, J. Gao, W. M. Abbott, K. K. Hirschi, S. Houser, R. Marini and R. Langer, Functional arteries grown *in vitro*, *Science*, **284** (1999), 489–493.

[121] Jeffrey S. Schechner, Anjali K. Nath, Lian Zheng, Martin S. Kluger, Christopher C. W. Hughes, M. Rocio Sierra-Honigmann, Marc I. Lorber, George Tellides, Michael Kashgarian, Alfred L. M. Bothwell and Jordan S. Pober, *In vivo* formation of complex micro-vessels lined by human endothelial cells in an immunodeficient mouse, *Proceedings of the National Academy of Sciences USA*, **97** (2000), 9191–9196.

[122] Ioannis V. Yannas, Synthesis of organs: *in vitro* or *in vivo*?, *Proceedings of the National Academy of Sciences USA*, **97** (2000), 9354–9356.

[123] V. Tropepe, B. L. K. Coles, B. J. Chiasson, D. J. Horsford, A. J. Elia, R. R. McInnes and D. van der Kooy, Retinal stem cells in the adult mammalian eye, *Science*, **287** (2000), 2032–2036.

[124] Sang-Hun Lee, Nadya Lumelsky, Lorenz Studer, Jonathan M. Auerbach and Ron D. McKay, Efficient generation of midbrain and hindbrain neurons from mouse embryonic stem cells, *Nature Biotechnology*, **18** (2000), 675–679.

[125] Diana L. Clarke, Clas B. Johansson, Johannes Wilbertz, Biborka Veress, Erik Nilsson, Helena Karlström, Urban Lendahl and Jonas Frisen, Generalized potential of adult neural stem cells, *Science*, **288** (2000), 1660–1663.

[126] Sujata Kale, Sybil Biermann, Claire Edwards, Catherine Tarnowski, Michael Morris and Michael William Long, Three-dimensional cellular development is essential for *ex vivo* formation of human bone, *Nature Biotechnology*, **18** (2000), 954–958.

[127] Herve Petite, Veronique Viateau, Wassila Bensaid, Alain Meunier, Cindy de Pollak, Marianne Bourguignon, Karim Oudina, Laurent Sedel and Genevieve Guillemin, Tissue-engineered bone regeneration, *Nature Biotechnology*, **18** (2000), 959–963.

[128] Frank Oberpenning, Jun Meng, James J. Yoo and Anthony Atala, *De novo* reconstitution of a functional mammalian urinary bladder by tissue engineering, *Nature Biotechnology*, **17** (1999), 149–155.

[129] Susan Bonner-Weir, Monica Taneja, Gordon C. Weir, Krystyna Tatarkiewicz, Ki-Ho Song, Arun Sharma and John J. O'Neill, *In vitro* cultivation of human islets from expanded ductal tissue, *Proceedings of the National Academy of Sciences USA*, **97** (2000), 7999–8004.

[130] Gene C. Webb, Murtaza S. Akbar, Choongjian Zhao and Donald F. Steiner, Expression profiling of pancreatic β-cells: glucose regulation of secretory and metabolic pathways, *Proceedings of the National Academy of Sciences USA*, **97** (2000), 5773–5778.

[131] Toshio Imaizumi, Karen L. Lankford, Willis V. Burton, William L. Fodor and Jeffery D. Kocsis, Xeno-transplantation of transgenic pig olfactory ensheathing cells promotes axonal regeneration in rat spinal cord, *Nature Biotechnology*, **18** (2000), 949–953.

[132] Toru Kondo and Martin Raff, Oligodendrocyte precursor cells reprogrammed to become multipotential CNS stem cells, *Science*, **289** (2000), 1754–1757.

[133] Luc J. W. van der Laan, Christopher Lockey, Bradley C. Griffeth, Francine S. Frasier, Carolyn A. Wilson, David E. Onions, Bernard J. Hering, Zhifeng Long, Edward Otto, Bruce E. Torbett and Daniel R. Salomon, Infection by porcine endogenous retrovirus after islet xenotransplantation in SCID mice, *Nature*, **207** (2000), 90–94.

[134] R. P. Lanza, J. B. Cibelli and M. D. West, Prospects for the use of nuclear transfer in human transplantation, *Nature Biotechnology*, **17** (1999), 1171–1174.

[135] Megan J. Munsie, Anna E. Michalska, Carmel M. O'Brien, Alan O. Trounson, Martin F. Perta and Peter S. Mountford, Isolation of pluripotent embryonic stem cells from reprogrammed adult mouse somatic cell nuclei, *Current Biology*, **10** (2000), 989–992.

[136] Michael Thomas, Lianqing Yang and Peter J. Hornsby, Formation of functional tissue from transplanted adrenocortical cells expressing telomerase reverse transcriptase, *Nature Biotechnology*, **18** (2000), 39–42.

[137] Jerry W. Shay and Woodring E. Wright, The use of telomerized cells for tissue engineering, *Nature Biotechnology*, **18** (2000), 22–23.

[138] W. Mark Saltzman, Delivering tissue regeneration, *Nature Biotechnology*, **17** (1999), 534–535.

[139] Lonnie D. Shea, Elizabeth Smiley, Jeffrey Bonadio and David J. Mooney, DNA delivery from polymer matrices for tissue engineering, *Nature Biotechnology*, **17** (1999), 551–554.

[140] Mark C. Poznansky, Richard H. Evans, Russell B. Foxall, Ivona T. Olszak, Anita H. Piascik, Kelly E. Hartman, Christian Brander, Thomas H. Meyer, Mark J. Pykett, Karissa T. Chabner, Spyros A. Kalams, Michael Rosenzweig and David T. Scadden, Efficient generation of human T cells from a tissue-engineered thymic organoid, *Nature Biotechnology*, **18** (2000), 729–734.

[141] T. Lebestky, T. Chang, V. Hartenstein and U. Banerjee, Specification of *Drosophila* hematopoietic lineage by conserved transcription factors, *Science*, **288** (2000), 146–149.

[142] A. Colman and A. Kind, Therapeutic cloning: concepts and practicalities, *TIBTECH*, **18** (2000), 192–196.

[143] I. Cerina, Biochips: from science to business, *transkript*, **3** (6) (2000), 30–32.

[144] J. N. Cha, G. D. Stucky, D. E. Morse and T. J. Deming, Biomimetic synthesis of ordered silica structures mediated by block copolypeptides, *Nature*, **403** (2000), 289–292.

[145] A. Katz and M. E. Davis, Molecular imprinting of bulk, micro-porous silica, *Nature*, **403** (2000), 286–289.

[146] S. Schultz, D. R. Smith, J. J. Mock and D. A. Schultz, Single-target molecule detection with nonbleaching multicolor optical immunolabels, *Proceedings of the National Academy of Sciences USA*, **97** (2000), 996–1001.

[147] Tuan Vo-Dinh, Jean-Pierre Alarie, Brian M. Cullum and Guy D. Griffin, Antibody-based nanoprobe for measurement of a fluorescent analyte in a single cell, *Nature Biotechnology*, **218** (2000), 764–767.

[148] T. Andrew Taton, Chad A. Mirkin and Robert L. Letsinger, Scanometric DNA array detection with nanoparticle probes, *Science*, **289** (2000), 1757–1760.

[149] J. Hodgson, Crystal gazing the new biotechnologies, *Nature Biotechnology*, **18** (2000), 29–31.

[150] T. Peakman and Y. Bonduelle, Steering a course through the technology maze, *Drug Discovery Today*, **5** (2000), 337–343.

[151] J. Miller, Navigating the technology maze, *Drug Discovery Today*, **5** (2000), 265–266.

[152] C. Gubser and P. Hiscocks, How much of a threat is virtual pharma? *Scrip Magazine*, (November 1999), 41–42.

[153] T. G. Klopack, Strategic outsourcing: balancing the risks and the benefits, *Drug Discovery Today*, **5** (2000), 157–160.

[154] C. Sander, Genomic medicine and the future of health care, *Science*, **287** (2000), 1977–1978.

[155] M. Xu, G. Qiu, Z. Jiang, E. von Hofe and R. E. Humphreys, Genetic modulation of tumor antigen presentation, *TIBTECH*, **18** (2000), 167–172.

[156] Ramon Alemany, Cristina Balague and David T. Curiel, Replicative adenoviruses for cancer therapy, *Nature Biotechnology*, **18** (2000), 723–727.

[157] Jason C. Schense, Jecelyne Bloch, Patrick Aebischer and Jeffrey A. Hubbell, Enzymatic incorporation of bioactive peptides into fibrin matrices enhances neurite extension, *Nature Biotechnology*, **18** (2000), 415–419.

[158] A. Mountain, Gene therapy: the first decade, *TIBTECH*, **18** (2000), 119–128.

[159] G. Thurston, C. Suri, K. Smith, J. McClain, T. Sato, G. D. Yancopoulos and D. M. McDonald, Leakage-resistant blood vessels in mice transgenically overexpressing angiopoietin-1, *Science*, **286** (1999), 2511–2514.

[160] R. Draghia-Akli, M. L. Fiorotto, L. A. Hill, P. B. Malone, D. R. Deaver and R. J. Schwartz, Myogenic expression of an injectable protease-resistant growth hormone-releasing hormone augments long-term growth in pigs, *Nature Biotechnology*, **17** (1999), 1179–1183.

[161] K. M. L. Gaensler, G. Tu, S. Bruch, D. Liggitt, G. S. Lipshutz, A. Metkus, M. Harrison, T. D. Heath and R. J. Debs, Fetal gene transfer by transuterine injection of cationic liposome–DNA complexes, *Nature Biotechnology*, **17** (1999), 1188–1192.

[162] Ningya Shi and William M. Pardridge, Noninvasive gene targeting to the brain, *Proceedings of the National Academy of Sciences USA*, **97** (2000), 7567–7572.

[163] Toshihiro Takenaka, Gary J. Murray, Gangjian Qin, Jane M. Quirk, Toshio Ohshima, Pankaj Qasba, Kelly Clark, Ashok B. Kulkarni, Roscoe O. Brady and Jeffry A. Medin, Long-term enzyme correction and lipid reduction in multiple organs of primary and secondary transplanted Fabry mice receiving transduced bone marrow, *Proceedings of the National Academy of Sciences USA*, **97** (2000), 7515–7520.

[164] Paul Blezinger, Jijun Wang, Margaret Gondo, Abraham Quezada, Dorothy Mehrens, Martha French, Arun Singhal, Sean Sullivan, Alain Rolland, Rob Ralston and Wang Min, Systemic inhibition of tumor growth and tumor metastases by intramuscular administration of the endostatin gene, *Nature Biotechnology*, **17** (2000), 343–348.

[165] W. R. A. Brown, P. J. Mee and M. H. Shen, Artificial chromosomes: ideal vectors?, *TIBTECH*, **18** (2000), 218–223.

[166] Chad, Stefao Rivella, John Callegari, Glenn Heller, Karen M. L. Gaensler, Lucio Luzzatto and Michel Sadelain, Therapeutic haemoglobin synthesis in β-thalassemic mice expressing lentivirus-encoded human β-globin, *Nature*, **406** (2000), 82–86.

[167] Christian P. Kalberer, Robert Pawliuk, Suzan Imren, Thomas Bachelot, Ken J. Takekoshi, Mary Fabry, Connie J. Eaves, Irving M. London, R. Keith Humphries and Philippe Leboulch, Preselection of retrovirally transduced bone marrow avoids subsequent stem cell gene silencing and age-dependent extinction of expression of human β-globin in engrafted mice, *Proceedings of the National Academy of Sciences USA*, **97** (2000), 5411–5415.

[168] David W. Emery, Evangelia Yannaki, Julie Tubb and George Stamatoyannopoulos, A chromatin insulator protects retrovirus vectors from chromosomal position effects, *Proceedings of the National Academy of Sciences USA*, **97** (2000), 9150–9155.

[169] Tsuyoshi Tanabe, Tomoko Kuwabara, Masaki Warashina, Kenzaburo Tani, Kazunari Taira and Shigetaka Asano, Oncogene inactivation in a mouse model, *Nature*, **406** (2000), 473.

[170] Aubrey D. N. J. de Grey, Mitochondrial gene therapy: an arena for the biomedical use of inteins, *TIBTECH*, **18** (2000), 394–399.

[171] Huatao Guo, Michael Karberg, Meredith Long, J. P. Jones III, Bruce Sullenger and Alan M. Lambowitz, Group II introns designed to insert into therapeutically relevant DNA target sites in human cells, *Science*, **289** (2000), 452–457.

[172] Vitali Alexeev, Olga Igoucheva, Alla Domashenko, George Cotsarelis and Kyonggeun Yoon, Localized *in vivo* genotypic and phenotypic correction of the albino mutation in skin by RNA–DNA oligonucleotide, *Nature Biotechnology*, **18** (2000), 43.

[173] O. A. Kolesnikova, N. S. Entelis, H. Mireau, T. D. Fox, R. P. Martin and I. A. Tarassov, Suppression of mutations in mitochondrial DNA by tRNAs imported from the cytoplasm, *Science*, **289** (2000), 1931–1933.

[174] Jean-Baptiste Latouche and Michael Sadelain, Induction of human cytotoxic T lymphocytes by artificial antigen-presenting cells, *Nature Biotechnology*, **18** (2000), 405–409.

[175] Michael A. Chattergoon, J. Joseph Kim, Joo-Sung Yang, Tara M. Robinson, Daniel J. Lee, Tzvete Dentchev, Darren M. Wilson, Velpandi Ayyavoo and David B. Weiner, Targeted antigen delivery to antigen-presenting cells including dendritic cells by engineered Fas-mediated apoptosis, *Nature Biotechnology*, **18** (2000), 974–979.

[176] Alevtina Domashenko, Sonya Gupta and George Cotsarelis, Efficient delivery of transgenes to human hair follicle progenitor cells using topical lipoplex, *Nature Biotechnology*, **18** (2000), 420–423.

[177] Silvia M. Kreda, Raymond J. Pickles, Eduardo R. Lazarowski and Richard C. Boucher, G-protein-coupled receptors as targets for gene transfer vectors using natural small-molecule ligands, *Nature Biotechnology*, **18** (2000), 635–640.

[178] Richard J. Bartlett, Sabine Stockinger, Melvin M. Denis, William T. Bartlett, Luca Inverardi, T. T. Le, Nguyen thi Man, Glenn E. Morris, Daniel J. Bogan, Janet Metcalf-Bogan and Joe N. Kornegay, *In vivo* targeted repair of a point mutation in the canine dystrophin gene by a chimeric RNA/DNA oligonucleotide, *Nature Biotechnology*, **18** (2000), 615–622.

[179] Andrea W. Bledsoe, Cheryl A. Jackson, Sylvia McPherson and Casey D. Morrow, Cytokine production in motor neurons by poliovirus replicon vector gene delivery, *Nature Biotechnology*, **18** (2000), 964–969.

[180] Yoshikazu Yonemitsu, Christopher Kitson, Stefano Ferrari, Raymond Farley, Uta Griesenbach, Diane Judd, Rachel Steel, Philippe Scheid, Jie Zhu, Peter K. Jeffery, Atsushi Kato, Mohammad K. Hasan, Yoshiyuki Nagai, Ichiro Masaki, Masayuki Fukumura, Mamoru Hasegawa, Duncan M. Geddes and Eric W. F. W. Alton, Efficient gene transfer to airway epithelium using recombinant Sendai virus, *Nature Biotechnology*, **18** (2000), 970–973.

[181] Zahida Parveen, Anna Krupetsky, Martin Engel-Städter, Klaus Cichutek, Roger J. Pomerantz and Ralph Dornburg, Spleen necrosis virus-derived C-type retroviral vectors for gene transfer to quiescent cells, *Nature Biotechnology*, **18** (2000), 623–629.

[182] Ziying Yan, Yulong Zhang, Dongsheng Duan and John F. Engelhardt, *Trans*-splicing vectors expand the utility of adeno-associated virus for gene therapy, *Proceedings of the National Academy of Sciences USA*, **97** (2000), 6716–6721.

[183] Duncan Graham-Rowe, Mini motivators – Bionic nerves are battling muscle wasting after a stroke, *New Scientist*, **164** (No. 2216) (1999), 5.

[184] Ian Sample, Crystal fix – Biodegradable rods could help mend broken bones, *New Scientist*, **164** (No. 2216) (1999), 17.

[185] T. T. Ton-that, D. Doron, B. S. Pollard, J. Bacher and H. B. Pollard, *In vivo* bypass of hemophilia A coagulation defect by Factor XIIa implant, *Nature Biotechnology*, **18** (2000), 289–295.

[186] Robert F. Service, Scanners Get a Fix on Lab Animals, *Science*, **286** (1999), 2261–2263.

[187] Cecilia H. Jiang, Joe Z. Tsien, Peter G. Schultz, Yinghe Hu, The effects of aging on gene expression in the hypothalamus and cortex of mice, *Proceedings of the National Academy of Sciences USA* **98** (2001), 1930–1934.

[188] Cheol-Koo Lee, Roger G. Klopp, Richard Weindruch and Thomas A. Prolla, Gene expression profile of aging and its retardation by caloric restriction, *Science*, **285** (1999), 1390–1393.

[189] A. S. Kamath-Loeb, E. Johansson, P. M. J. Burgers and L. A. Loeb, Functional interaction between the Werner syndrome protein and DNA polymerase δ, *Proceedings of the National Academy of Sciences USA*, **97** (2000), 4603–4608.

[190] Joseph Landry, Ann Sutton, Stefan T. Tafrov, Ryan C. Heller, John Stebbins, Lorraine Pillus and Rolf Sternglanz, The silencing protein SIR2 and its homologs are NAD-dependent protein deacetylases, *Proceedings of the National Academy of Sciences USA*, **97** (2000), 5807–5811.

[191] Deborah J. Burks, Jalme Font de Mora, Markus Schubert, Dominic J. Withers, Martin G. Myers, Heather H. Towery, Shari L. Altamuro, Carrie L. Flint and Morris F. White, IRS-2 pathways integrate female reproduction and energy homeostasis, *Nature*, **407** (2000), 377–382.

[192] Su-Ju Lin, Pierre-Antoine Defossez and Leonard Guarente, Requirement of NAD and SIR2 for lifespan extension in *Saccharomyces cerevisiae*, *Science*, **289** (2000), 2126–2128.

[193] Jens C. Brüning, Dinesh Gautam, Deborah J. Burks, Jennifer Gillette, Markus Schubert, Paul C. Orban, Rüdiger Klein, Wilhelm Krone, Dirk Müller-Wieland and C. Ronald Kahn, Role of brain insulin receptor in control of body weight and reproduction, *Science*, **289** (2000), 2122–2125.

[194] Derek J. Symula, Kelly A. Frazer, Yukihiko Ueda, Patrice Denefle, Mary E. Stevens, Zhi-En Wang, Richard Locksley, Edward M. Rubin, Functional screening of an asthma QTL in YAC transgenic mice, *Nature Genetics*, **23** (1999), 241–244.

[195] Anne Frary, T. Clint Nesbitt, Amy Frary, Silvana Grandillo, Esther van der Knaap, Bin Cong, Jiping Liu, Jaroslaw Meller, Ron Elber, Kevin B. Alpert and Steven D. Tanksley, *fw2.2*: a quantitative trait locus key to the evolution of tomato fruit size, *Science*, **289** (2000), 85–88.

[196] John C. Clapham, Jonathan R. S. Arch, Helen Chapman, Andrea Haynes, Carolyn Lister, Gary B. T. Moore, Valerie Plercy, Sabrina A. Carter, Ines Lehner, Stephen A. Smith, Lee J. Beeley, Robert J. Godden, Nicole Herrity, Mark Skehel, K. Kumar Changani, Paul D. Hockings, David G. Reid, Sarah M. Squires, Jonathan Hatcher, Brenda Trail, Judy Latcham, Sohalla Rastan, Alexander J. Harper, Susana Cadenas, Julie A. Buckingham, Martin D. Brand and Alejandro Abuin, Mice overexpressing human uncoupling protein-3 in skeletal muscle are hyperphagic and lean, *Nature*, **406** (2000), 415–418.

[197] Jo Whelan, Reversing age-related and diabetic cardiovascular disease, *Drug Discovery Today*, **5** (2000), 272–273.

[198] Donald Metcalf, Christopher J. Greenhalgh, Elizabeth Viney, Tracy A. Wilson, Robyn Starr, Nicos A. Nicola, Douglas J. Hilton and Warren S. Alexander, Gigantism in mice lacking suppressor of cytokine signaling-2, *Nature*, **405** (2000), 1069–1073.

[199] Thomas M. Loftus, Donna E. Jaworsky, Gojeb L. Frehywot, Craig A. Townsend, Gabriele V. Ronnett, M. Daniel Lane and Francis P. Kuhajda, Reduced food intake and body weight in mice treated with fatty acid synthase inhibitors, *Science*, **288** (2000), 2379–2381.

[200] K. P. White, S. A. Rifkin, P. Hurban and D. S. Hogness, Microarray analysis of *Drosophila* development during metamorphosis, *Science*, **286** (1999), 2179–2184.

[201] D. H. Ly, D. J. Lockhart, R. A. Lerner and P. G. Schultz, Mitotic misregulation and human aging, *Science*, **287** (2000), 2486–2492.

[202] Martijn E. T. Dolle, Wendy K. Snyder, Jan A. Gossen, Paul H. M. Lohman and Jan Vijg, Distinct spectra of somatic mutations accumulated with age in mouse heart and small intestine, *Proceedings of the National Academy of Sciences USA*, **97** (2000), 8403–8408.

[203] Bey-Dih Chang, Keiko Watanabe, Eugenia V. Broude, Jing Fang, Jason C. Poole, Tatiana V. Kalinichenko and Igor B. Roninson, Effects of p21(Waf1/Cip1/Sdi1) on cellular gene expression: implications for carcinogenesis, senescence, and age-related diseases, *Proceedings of the National Academy of Sciences USA*, **97** (2000), 4291–4296.

[204] Shin-Ichiro Imai, Christopher M. Armstrong, Matt Kaeberlein and Leonard Guarente, Transcriptional silencing and longevity protein Sir2 is an NAD-dependent histone deacetylase, *Nature*, **403** (2000), 795–800.

[205] John J. Wyrick, Frank C. P. Holstege, Ezra G. Jennings, Helen C. Causton, David Shore, Michael Grunstein, Eric S. Lander and Richard A. Young, Chromosomal landscape of nucleosome-dependent gene expression and silencing in yeast, *Nature*, **402** (1999), 418–421.

[206] Itamar Simon, Toyoaki Tenzen, Benjamin E. Reubinoff, Dahlia Hillman, John R. McCarrey and Howard Cedar, Asynchronous replication of imprinted genes is established in the gametes and maintained during development, *Nature* 401 (1999), 929–932.

[207] Mark Pearson, Roberta Carbone, Carla Sebastiani, Mario Cioce, Marta Fagioli, Shin'ichi Saito, Yuichiro Higashimoto, Ettore Appella, Saverio Minucci, Pier Paolo Pandolfi and Pier Guiseppe Pelicci, PML regulates p53 acetylation and premature senescence induced by oncogenic Ras, *Nature*, **406** (2000), 207–210.

[208] Jing Wang, Gregory J. Hannon and David H. Beach, Risky immortalization by telomerase, *Nature*, **405** (2000), 755–756.

[209] David Kipling and Harry Rubin, Telomere-dependent senescence [Correspondence], *Nature Biotechnology*, **17** (1999), 313–314.

[210] David W. Walker, Gawain McColl, Nicole L. Jenkins, Jennifer Harris and Gordon J. Lithgow, Evolution of lifespan in *C. elegans*, *Nature*, **405** (2000), 296–297.

[211] Shawn Ahmed and Jonathan Hodgkin, MRT-2 checkpoint protein is required for germline immortality and telomere replication in *C. elegans*, *Nature*, **403** (2000), 159–164.

[212] Robert P. Lanza, Jose B. Cibelli, Catherine Blackwell, Vincent J. Cristofalo, Mary Kay Francis, Gabriela M. Baerlocher, Jennifer Mak, Michael Schertzer, Elizabeth A. Chavez, Nancy Sawyer, Peter Lansdorp and Michael West, Extension of cell life-span and telomere length in animals cloned from senescent somatic cells, *Science*, **288** (2000), 665–669.

[213] Teruhiko Wakayama, Yoichi Shinkai, Kellie L. K. Tamashiro, Hiroyuki Niida, D. Caroline Blanchard, Robert J. Blanchard, Atsuo Ogura, Kentaro Tanemura, Makoto Tachibana, Anthony C. F. Perry, Diana F. Colgan, Peter Mombaerts and Ryuzo Yanagimachi, Cloning of mice to six generations, *Nature*, **407** (2000), 318–319.

[214] Anne E. Pitts and David R. Corey, Inhibition of human telomerase by 2′-*O*-methyl-RNA, *Proceedings of the National Academy of Sciences USA*, **95** (1998), 11549–11554.

[215] Steven E. Artandi, Sandy Chang, Shwu-Luan Lee, Scott Alson, Geoffrey J. Gottlieb, Lynda Chin and Ronald A. DePinho, Telomere dysfunction promotes non-reciprocal translocations and epithelial cancers in mice, *Nature*, **406** (2000), 641–645.

[216] Simon Melov, Joanne Ravenscroft, Sarwatt Malik, Matt S. Gill, David W. Walker, Peter E. Clayton, Douglas C. Wallace, Bernard Malfroy, Susan R. Doctrow and Gordon J. Lithgow, Extension of lifespan with superoxide dismutase/catalase mimetics, *Science*, **289** (2000), 1567–1569.

Structures of EUK-8 and EUK-134: K. Baker, C. Bucay Marcus, K. Huffman, H. Kruk, B. Malfroy, S. R. Doctrow, Synthetic combined superoxide dismutase/catalase mimics are protective as a delayed treatment in a rat stroke model: a key role for reactive oxygen species in ischemic brain injury, *Journal of Pharmacology and Experimental Therapeutics*, **284** (1998), 215–221.
Effects in mouse ALS model: www.eukarion.com

[217] Ash A. Alizadeh, Michael B. Eisen, R. Eric Davis, Chi Ma, Izidore S. Lossos, Andreas Rosenwald, Jennifer C. Boldrick, Hajeer Sabet, Truc Tan, Xib Yu, John I. Powell, Liming Yang, Gerald E. Marti, Troy Moore, James Hudson, Jr, Lishend Lu, David B. Lewis, Robert Tibshirani, Gavin Sherlock, Wing C. Chan, Timothy C. Greiner, Dennis D. Weisenburger, James O. Armitage, Roger Warnke, Ronald Levy, Wyndham Wilson, Michael R. Grever, John C. Byrd, David Botstein, Patrick O. Brown and Louis M. Staudt, Distinct types of diffuse large B-cell lymphoma identified by gene expression profiling, *Nature*, **403** (2000), 503–511.

[218] M. Bittner, P. Meltzer, Y. Chen, Y. Jiang, E. Seftor, M. Hendrix, M. Radmacher, R. Simo, Z. Yakhini, A. Ben-Dor, N. Sampas, E. Dougherty, E. Wang, F. Marincola, C. Gooden, J. Lueders, A. Glatfelter, P. Pollck, J. Carpten, E. Gillanders, D. Leja, K. Dietrich, C. Beaudry, M. Berens, D. Alberts, V. Sondak, N. Hayward and J. Trent, Molecular classification of cutaneous malignant melanoma by gene expression profiling, *Nature*, **406** (2000), 536–540.

[219] Brad St Croix, Carlo Rago, Victor Velculescu, Giovanni Traverso, Katherine E. Romans, Elizabeth Montgomery, Anita Lal, Gregory J. Riggins, Christoph Lengauer, Bert Vogelstein and Kenneth W. Winzler, Genes expressed in human tumor endothelium, *Science*, **289** (2000), 1197–1202.

[220] Charles M. Perou, Therese Sorlie, Michael B. Eisen, Matt van de Rijn, Stefanie S. Jeffrey, Christian A. Rees, Jonathan R. Pollack, Douglas T. Ross, Hilde Johnsen, Lars A. Akslen, Öystein Fluge, Alexander Pergamenschikov, Cheryl Williams, Shirley X. Zhu, Per E. Lönning, Anne-Lise Börresen-Dale, Patrick O. Brown and David Botstein, Molecular portraits of human breast tumors, *Nature*, **406** (2000), 747–752.

[221] J. Drews, Drug discovery today – and tomorrow, *Drug Discovery Today*, **5** (2000), 2–4.

[222] S. L. Schreiber, Target-oriented and diversity-oriented organic synthesis in drug discovery, *Science*, **287** (2000), 1964–1969.

[223] Martin J. Valler and Darren Green, Diversity screening versus focussed screening in drug discovery, *Drug Discovery Today*, **5** (2000), 286–293.

[224] J. B. Gibbs, Mechanism-based target identification and drug discovery in cancer research, *Science*, **287** (2000), 1969–1973.

[225] Thomas Schindler, William Bornmann, Patricia Pellicena, W. Todd Miller, Bayard Clarkson and John Kuriyan, Structural mechanism for STI-571 inhibition of Abelson tyrosine kinase, *Science*, **289** (2000), 1938–1942.

[226] Anthony C. Bishop, Jeffrey A. Ubersax, Dejah T. Petsch, Dina P. Matheos, Nathanael S. Gray, Justin Blethrow, Elji Shimizu, Joe Z. Tsien, Peter G. Schultz, Mark D. Rose, John L. Wood, David O. Morgan and Kevan M. Shokat, A chemical switch for inhibitor-sensitive alleles of any protein kinase, *Nature*, **407** (2000), 395–401.

[227] William Shakespeare, Michael Yang, Regine Bohacek, Franklin Cerasoli, Karin Stebbins, Raji Sundaramoorthi, Mihai Azimioara, Chi Vu, Selvi Pradeepan, Chester Metcalf, III, Chad Haraldson, Taylor Merry, David Dalgarno, Surinder Narula, Marcos Hatada, Xiaode Lu, Marie Rose van Schravendijk, Susan Adams, Shelia Violette, Jeremy Smith, Wie Guan, Catherine Bartlett, Jay Herson, John Iuliucci, Manfred Weigele and Tomi Sawyer, Structure-based design of an osteoclast-selective, nonpeptide Src homology 2 inhibitor with *in vivo* antiresorptive activity, *Proceedings of the National Academy of Sciences USA*, **97** (2000), 9373–9378.

[228] Marcus D. Ballinger, Venkatakrishna Shyamala, Louise D. Forrest, Maja Deuter-Reinhard, Laura V. Doyle, Jian-xin Wang, Lootsee Panganiban-Lustan, Jennifer R. Stratton, Gerald Apell, Jill A. Winter, Michael V. Doyle, Steven Rosenberg and W. Michael Kavanaugh, Semirational design of a potent, artificial agonist of fibroblast growth factor receptors, *Nature Biotechnology*, **17** (1999), 1199.

[229] Atsuko Yoneda, Masahiro Asada, Yuko Oda, Masashi Suzuki and Toru Imamura, Engineering of an FGF–proteoglycan fusion protein with heparin-independent, mitogenic activity, *Nature Biotechnology*, **18** (2000), 641–644.

[230] Sang-Hyun Park and Ronald T. Raines, Genetic selection for dissociative inhibitors of designated protein–protein interactions, *Nature Biotechnology*, **18** (2000), 847–851.

[231] Zhihong Guo, Demin Zhou and Peter G. Schultz, Designing small-molecule switches for protein–protein interactions, *Science*, **288** (2000), 2042–2045.

[232] Anna M. Kapp, Aseem Z. Ansari, Mark Ptashne and Peter B. Dervan, Activation of gene expression by small molecule transcription factors, *Proceedings of the National Academy of Sciences USA*, **97** (2000), 3930–3935.

[233] Sakari Hietanen, Sonia Lain, Eberhard Krausz, Christine Blattner and David P. Lane, Activation of p53 in cervical carcinoma cells by small molecules, *Proceedings of the National Academy of Sciences USA*, **97** (2000), 8501–8506.

[234] M. Faria, C. D. Wood, L. Perrouault, J. S. Nelson, A. Winter, M. R. H. White, C. Helene and C. Giovannangeli, Targeted inhibition of transcription elongation in cells mediated by triplex-forming oligonucleotides, *Proceedings of the National Academy of Sciences USA*, **97** (2000), 3862–3867.

[235] Daniel A. Erlanson, Andrew C. Braisted, Darren R. Raphael, Mike Randal, Robert M. Stroud, Eric M. Gordon and James A. Wells, Site-directed ligand discovery, *Proceedings of the National Academy of Sciences USA*, **97** (2000), 9367–9372.

[236] Rene F. Ketting, Ronald H. A. Plasterk, A genetic link between co-suppression and RNA interference in *C. elegans*, *Nature*, **404** (2000), 296–298.

[237] Jason R. Kennerdell and Richard W. Carthew, Heritable gene silencing in *Drosophila* using double-stranded RNA, *Nature Biotechnology*, **18** (2000), 896–898.

[238] James C. Clemens, Carolyn A. Worby, Nancy Simonson-Leff, Marco Muda, Tomohiko Maehama, Brian A. Hemmings and Jack E. Dickson, Use of double-stranded RNA interference in *Drosophila* cell lines to dissect signal transduction pathways, *Proceedings of the National Academy of Sciences USA*, **97** (2000), 6499–6503.

[239] Chiou-Fen Chuang and Elliott M. Meyerowitz, Specific and heritable genetic interference by double-stranded RNA in *Arabidopsis thaliana*, *Proceedings of the National Academy of Sciences USA*, **97** (2000), 4985–4990.

[240] K. Asish Xavier, Paul S. Eder and Tony Giordano, RNA as a drug target: methods for biophysical characterization and screening, *TIBTECH*, **18** (2000), 349–356.

[241] Erik A. Schultes and David P. Bartel, One sequence, two ribozymes: implications for the emergence of new ribozyme folds, *Science*, **289** (2000), 448–452.

[242] Koichi Ito, Makiko Uno and Yoshikazu Nakamura, A tripeptide,anticodon' deciphers stop codons in messenger RNA, *Nature*, **403** (2000), 680–684.

[243] William H. Miller, Richard M. Keenan, Robert N. Willette, Michael W. Lark, Identification and *in vivo* efficacy of small-molecule antagonists of integrin $\alpha\beta$ (the vitronectin receptor), *Drug Discovery Today*, **5** (2000), 397–408.

[244] Li Sun and Gerald McMahon, Inhibition of tumor angiogenesis by synthetic receptor tyrosine kinase inhibitors, *Drug Discovery Today*, **5** (2000), 344–353.

[245] Kathryn M. Koeller and Chi-Huey Wong, Emerging themes in medicinal glycoscience, *Nature Biotechnology*, **18** (2000), 835–841.

[246] Nigel R. A. Beeley, Can peptides be mimicked?, *Drug Discovery Today*, **5** (2000), 354–363.

[247] Fiorenza Falcioni, Kouichi Ito, Damir Vidovic, Charles Belunis, Robert Campbell, Steven J. Berthel, David R. Bolin, Paul B. Gillespie, Nicholas Huby, Gary L. Olson, Ramakanth Sarabu, Jeanmarie Guenot, Vincent Madison, Jürgen Hammer, Francesco Sinigaglia, Michael Steinmetz and Zoltan A. Nagy, Peptidomimetic compounds that inhibit antigen presentation by autoimmune disease-associated class II major histocompatibility molecules, *Nature Biotechnology*, **17** (1999), 562–567.

[248] Kevin FitzGerald, *In vitro* display technologies – new tools for drug discovery, *Drug Discovery Today*, **5** (2000), 253–258.

[249] Eskil Söderlind, Leif Strandberg, Pernilla Jirholt, Norihiro Kobayashi, Vessela Alexeiva, Anna-Maria Aberg, Anna Nilson, Bo Jansson, Mats Ohlin, Christer Wingren, Lena Danielsson, Roland Carlsson and Carl A. K. Borrebaeck, Recombining germline-derived CDR sequences for creating diverse single-framework antibody libraries, *Nature Biotechnology*, **18** (2000), 852–856.

[250] Tajib Mirzabekov, Harry Kontos, Michael Farzan, Wayne Marasco and Joseph Sodroski, Paramagnetic proteo-liposomes containing a pure, native and oriented seven-membrane segment protein, CCR5, *Nature Biotechnology*, **18** (2000), 649–654.

[251] Charles Parnot, Sabine Bardin, Stephanie Miserey-Lenkei, Denis Guedin, Pierre Corvol and Eric Clauser, Systematic identification of mutations that constitutively activate the angiotensin II type 1A receptor by screening a randomly mutated cDNA library with an original pharmacological bioassay, *Proceedings of the National Academy of Sciences USA*, **97** (2000), 7615–7620.

[252] Anne B. Satterthwaite, Fiona Willis, Prim Kanchanastit, David Fruman, Lewis C. Cantley, Cheryl D. Helgason, R. Keith Humphries, Clifford A. Lowell, Melvin Simon, Michael Leitges, Alexander Tarakhovsky, Thomas F. Tedder, Ralf Lesche, Hong Wu and Owen N. Witte, A sensitized genetic system for the analysis of murine B lympho-cyte signal transduction pathways dependent on Bruton's tyrosine kinase, *Proceedings of the National Academy of Sciences USA*, **97** (2000), 6687–6692.

[253] Eric V. Shusta, Philip D. Holler, Michele C. Kieke, David M. Kranz and K. Dane Wittrup, Directed evolution of a stable scaffold for T-cell receptor engineering, *Nature Biotechnology*, **18** (2000), 754–759.

[254] Gordon C. K. Roberts, Applications of NMR in drug discovery, *Drug Discovery Today*, **5** (2000), 230–240.

[255] Peter J. Eddershaw, Alan P. Beresford and Martin K. Bayliss, ADME/PK as part of a rational approach to drug discovery, *Drug Discovery Today*, **5** (2000), 409–414.

[256] Richard A. Lovett, Toxicologists brace for genomics revolution, *Science*, **289** (2000), 536–537.

[257] Nicholas A. Meanwell and Mark Krystal, Respiratory syncytical virus: recent progress towards the discovery of effective prophylactic and therapeutic agents, *Drug Discovery Today*, **5** (2000), 241–252.

[258] Noriaki Shimizu, Kotaro Sugimoto, Jianwei Tang, Takeyuki Nishi, Iwao Sato, Masaki Hiramoto, Shin Aizawa, Mamoru Hatakeyama, Reiko Ohba, Hideaki Hatori, Tatsufumi Yoshikawa, Fumihiko Suzuki, Akira Oomori, Hirotoshi Tanaka, Haruma Kawaguchi, Hajime Watanabe and Hiroshi Handa, High-performance affinity beads for identifying drug receptors, *Nature Biotechnology*, **18** (2000), 877–881.

[259] Mary J. Cismowski, Aya Takesono, Chienling Ma, Jeffrey S. Lizano, Xiaobing Xie, Hans Fuernkranz, Stephen M. Lanier and Emir Duzic, Genetic screens in yeast to identify mammalian nonreceptor modulators of G-protein signaling, *Nature Biotechnology*, **17** (1999), 878–883.

[260] H. Breithaupt, The new antibiotics: can novel antibacterial treatments combat the rising tide of drug-resistant infections?, *Nature Biotechnology*, **17** (1999), 1165–1169.

[261] D. A. Cowan, Microbial genomes – the untapped resource, *TIBTECH*, **18** (2000), 14–16.

[262] Doloressa Gleba, Nikolai V. Borisjuk, Ludmyla G. Borysjuk, Ralf Kneer, Alexander Poulev, Marina Skarzhinskaya, Slavik Dushenkov, Sithes Logendra, Yuri Y. Gleba and Ilya Raskin, Use of plant roots for phytoremediation and molecular farming, *Proceedings of the National Academy of Sciences USA*, **96** (1999), 5973–5977.

[263] Nikolai V. Borisjuk, Ludmyla G. Borisjuk, Sithes Logendra, Frank Peterson, Yuri Gleba and Ilya Raskin, Production of recombinant proteins in plant root exudates, *Nature Biotechnology*, **17** (1999), 466–469.

[264] M. R. Rondon, R. M. Goodman and J. Jadelsman, The Earth's bounty: assessing and accessing soil microbial diversity, *TIBTECH*, **17** (1999), 403–409.

[265] W. R. Strohl, The role of natural products in a modern drug discovery program, *Drug Discovery Today*, **5** (2000), 39–41.

[266] Charles Gerday, Mohamed Aittaleb, Mostafa Bentahir, Jean-Pierre Chessa, Paule Claverie, Tony Collins, Salvino D'Amico, Joelle Dumont, Genevieve Garsoux, Daphne Georlette, Anne Hoyoux, Thierry Lonhienne, Marie-Alice Meuwis, Georges Feller, Cold-adapted enzymes: from fundamentals to biotechnology, *TIBTECH*, **18** (2000), 103–107.

[267] Alan Harvey, Strategies for discovering drugs from previously unexplored natural products, *Drug Discovery Today*, **5** (2000), 294–300.

[268] G. S. Shen, R. T. Layer and R. T. McCabe, Conopeptides: from deadly venoms to novel therapeutics, *Drug Discovery Today*, **5** (2000), 98–106.

[269] W. van Eden, Stress proteins as targets for anti-inflammatory therapies, *Drug Discovery Today*, **5** (2000), 115–120.

[270] P. Fredericks and J. Dankert, *Journal of Experimental Zoology*, **287** (2000), 340ff.

[271] Fernando S. Santiago, Harry S. Lowe, Mary M. Kavurma, Colin N. Chesterman, Andrew Baker, David G. Atkins, Levon M. Khachichian *et al.*, New DNA enzyme targeting Egr-1 mRNA inhibits vascular smooth muscle proliferation and regrowth after injury, *Nature Medicine*, **5** (1999), 1264–1269. *Nature Medicine,* **5** (1999), 1438. (Correction)

[272] E. Finkel, DNA Cuts its teeth – as an enzyme, *Science*, **286** (1999), 2441–2442.

[273] A. Foster, H. A. Coffey, M. J. Morin and F. Rastinejad, Pharmacological rescue of mutant p53 conformation and function, *Science*, **286** (1999), 2507–2510.

[274] M. Ostermeier, J. H. Shim and S. J. Benkovic, A combinatorial approach to hybrid enzymes independent of DNA homology, *Nature Biotechnology*, **17** (1999), 1205–1209.

[275] Chia-Chun J. Chang, Teddy T. Chen, Brett W. Cox, Glenn N. Dawes, Willem P. C. Stemmer, Juha Punnonen and Phillip A. Patten, Evolution of a cytokine using DNA family shuffling, *Nature Biotechnology*, **17** (2000), 793–797.

[276] J. M. Fostel and P. A. Lartey, Emerging novel antifungal agents, *Drug Discovery Today*, **5** (2000), 25–32.

[277] G. Cutrona, E. M. Carpaneto, M. Ulivi, S. Roncella, O. Landt, M. Ferrarini and L. C. Boffa, Effects in live cells of a c-*myc* anti-gene PNA linked to a nuclear localization signal, *Nature Biotechnology*, **18** (2000), 300–303.

[278] M. Schapira, B. M. Raaka, H. H. Samuels and R. Abagyan, Rational discovery of novel nuclear hormone receptor antagonists, *Proceedings of the National Academy of Sciences USA*, **97** (2000), 1008–1013.

[279] Benjamin Bader, Karsten Kuhn, David J. Owen, Herbert Waldmann, Alfred Wittinghofer and Jürgen Kuhlmann, Bioorganic synthesis of lipid-modified proteins for the study of signal transduction, *Nature*, **403** (2000), 223–226.

[280] Pavel I. Kitov, Joanna M. Sadowska, George Mulvey, Glen D. Armstrong, Hong Ling, Navraj S. Pannu, Randy J. Read and David R. Bundle, Shiga-like toxins are neutralized by tailored multivalent carbohydrate ligands, *Nature*, **403** (2000), 669–672.

[281] C. Kendall Stover, Paul Warrener, Donald R. VanDevanter, David R. Sherman, Taraq M. Arain, Michael H. Langhorne, Scott W. Anderson, J. Andrew Towell, Ying Yuan, David N. McMurray, Barry N. Kreiswirth, Clifton E. Barry and William R. Baker, A small-molecule nitroimidazopyran drug candidate for the treatment of tuberculosis, *Nature*, **405** (2000), 962–966.

[282] Anna P. Tretiakova, C. Scott Little, Kenneth J. Blank and Bradford A. Jameson, Rational design of cytotoxic T-cell inhibitors, *Nature Biotechnology*, **18** (2000), 984–988.

[283] Hong Zhang, Jesse Cook, Jeffrey Nickel, Rosie Yu, Kimberley Stecker, Kathleen Myers and Nicholas M. Dean, Reduction of liver Fas expression by an antisense oligonucleotide protects mice from fulminant hepatitis, *Nature Biotechnology*, **18** (2000), 862–867.

[284] S. E. Driver, G. S. Robinson, J. Flanagan, W. Shen, L. E. H. Smith, D. W. Thomas and P. C. Roberts, Oligonucleotide-based inhibition of embryonic gene expression, *Nature Biotechnology*, **17** (1999), 1184–1187.

[285] Y. Wang and S. Huang, Adenovirus technology for gene manipulation and functional studies, *Drug Discovery Today*, **5** (2000), 10–16.

[286] Y. Zhu, M. C. Jong, K. SA. Frazer, E. Gong, R. M. Krauss, J.-F. Cheng, D. Boffelli and E. M. Rubin, Genomic interval engineering of mice identifies a novel modulator of triglyceride production, *Proceedings of the National Academy of Sciences USA*, **97** (2000), 1137–1142.

[287] A. Grishok, H. Tabara and C. C. Mello, Genetic requirements for inheritance of RNAi in *C. elegans*, *Science*, **287** (2000), 2494–2497.

[288] Ami Mankodi, Eric Logigian, Linda Callahan, Carolyn McClain, Robert White, Don Henderson, Matt Krym and Charles A. Thornton, Myotonic dystrophy in transgenic mice expressing an expanded CUG repeat, *Science*, **289** (2000), 1769–1772.

[289] Mary Ellen Domeier, Daniel P. Morse, Scott W. Knight, Michael Potereiko, Brenda L. Bass and Susan E. Mango, A link between RNA interference and nonsense-mediated decay in *Caenorhabditis elegans*, *Science*, **289** (2000), 1928–1930.

[290] Michael S. Neuberger and James Scott, RNA editing AIDS antibody diversification?, *Science*, **289** (2000), 1705–1706.
M. Muramatsu *et al.*, *Cell*, **102** (2000), 553ff.
P. Revy *et al.*, *Cell*, **102** (2000), 565ff.

[291] T. Raveh, H. Berissi, M. Eisenstein, T. Spivak and A. Kimchi, A functional genetic screen identifies regions at the C-terminal tail and death-domain of death-associated protein kinase that are critical for its proapoptotic activity, *Proceedings of the National Academy of Sciences USA*, **97** (2000), 1572–1577.

[292] Ji Li, Ildiko Sarosi, Xiao-Qiang Yan, Sean Morony, Casey Capparelli, Hong-Lin Tan, Susan McCabe, Robin Elliott, Sheila Scully, Gwyneth Van, Stephen Kaufman, Shao-Chieh Juan, Yu Sun, John Tarpley, Laura Martin, Kathleen Christensen, James McCabe, Paul Kostenuik, Hailing Hsu, Frederick Fletcher, Colin R. Dunstan, David L. Lacey, William J. Boyle, RANK is the intrinsic hematopoietic cell surface receptor that controls osteoclastogenesis and regulation of bone mass and calcium metabolism, *Proceedings of the National Academy of Sciences USA*, **97** (2000), 1566–1571.

[293] A. Y. Louie, M. M. Hübner, E. T. Ahrens, U. Rothbächer, R. Moats, R. E. Jsacobs, S. E. Fraser and T. J. Meadde, *In vivo* visualization of gene expression using magnetic resonance imaging, *Nature Biotechnology*, **18** (2000), 321–325.

[294] D. J. Ecker and R. H. Griffey, RNA as a small-molecule drug target: doubling the value of genomics, *Drug Discovery Today*, **4** (1999), 420–429.

[295] Stephen K. Burley, Steven C. Almo, Jeffrey B. Bonanno, Malcolm Capel, Mark R. Chance, Terry Gaasterland, Dawei Lin, Andrej Sali, F. William Studier, Subramanyam Swaminathan, Structural genomics: beyond the Human Genome Project, *Nature Genetics*, **23** (1999), 151–157.

[296] Richard Neutze, Remco Wouts, David van der Spoel, Edgar Weckert and Janos Hajdu, Potential for biomolecular imaging with femtosecond X-ray pulses, *Nature*, **406** (2000), 752–757.

[297] H. Loferer, A. Jacobi, A. Posch, C. Gauss, S. Meier-Eweret and B. Seizinger, Integrated bacterial genomics for the discovery of novel antimicrobials, *Drug Discovery Today*, **5** (2000), 107–114.

[298] P. D. Karp, M. Krummenacker, S. Paley and J. Wagg, Integrated pathway-genome databases and their role in drug discovery, *TIBTECH*, **17** (1999), 275–281.

[299] S. Zozulya, M. Lioubin, R. J. Hill, C. Abram and M. L. Gishizky, Mapping signal transduction pathways by phage display, *Nature Biotechnology*, **17** (1999), 1193–1198.

[300] J. D. Klemm, S. L. Schreiber and G. R. Crabtree, Dimerization as a regulatory mechanism in signal transduction, *Annual Review of Immunology*, **16** (1998), 569–59.

[301] S. J. Spengler, Bioinformatics in the information age, *Science*, **287** (2000), 1221–1222.

[302] D. E. Clark and S. D. Pickett, Computational methods for the prediction of drug-likeness, *Drug Discovery Today*, **5** (2000), 49–58.

[303] Matteo Pellegrini, Edward M. Marcotte, Michael J. Thompson, David Eisenberg, Todd O. Yeates, Assigning protein functions by comparative genome analysis: Protein phylogenetic profiles, *Proceedings of the National Academy of Sciences USA* 96 (1999), 4285–4288.
Sarah A. Teichmann and G. Mitchison, Computing protein function, *Nature Biotechnology*, **18** (2000), 27.

[304] Anton J. Enright, Ioannis Iliopoulos, Nikos C. Kyrpides and Christos A. Ouzounis, Protein interaction maps for complete genomes based on gene fusion events, *Nature*, **402** (2000), 86–90.

[305] Edward M. Marcotte, Matteo Pellegrini, Michael J. Thompson, Todd O. Yeates and David Eisenberg, A combined algorithm for genome-wide prediction of protein function, *Nature*, **402** (2000), 83–86.

[306] Harmen J. Bussemaker, Hao Li and Eric D. Siggia, Building a dictionary for genomes: identification of presumptive regulatory sites by statistical analysis, *Proceedings of the National Academy of Sciences USA*, **97** (2000), 10096–10100.

[307] Orly Alter, Patrick O. Brown and David Botstein, Singular value decomposition for genome-wide expression data processing and modeling, *Proceedings of the National Academy of Sciences USA*, **97** (2000), 10101–10106.

[308] D. B. Searls, Using bioinformatics in gene and drug discovery, *Drug Discovery Today*, **5** (2000), 135–143.

[309] Christopher J. Roberts, Bryce Nelson, Matthew J. Marton, Roland Stoughton, Michael R. Meyer, Holly A. Bennett, Yudong D. He, Hongyue Dai, Wynn L. Walker, Timothy R. Hughes, Mike Tyers, Charles Boone, Stephen H. Friend, Signaling and Circuitry of Multiple MAPK pathways Revealed by a Matrix of Global Gene Expression Profiles, *Science*, **287** (2000), 873–880.

[310] D. B. Kell and R. D. King, On the optimization of classes for the assignment of un-identified reading frames in functional genomics programmes: the need for machine learning, *TIBTECH* **18** (2000), 93–98.

[311] J. Skolnick and J. S. Fetrow, From genes to protein structure and function: novel applications of computational approaches in the genomic era, *TIBTECH*, **18** (2000), 34–39.

[312] G. G. Loots, R. M. Locksley, C. M. Blankespoor, Z. E. Wang, W. Miller, E. M. Rubin and K. A. Frazer, Identification of a coordinate regulator of interleukins 4, 13, and 5 by cross-species comparisons, *Science*, **288** (2000), 136–140.

[313] Attila Becskei and Luis Serrano, Engineering stability in gene networks by auto-regulation, *Nature*, **405** (2000), 590–593.

[314] Richard C. Strohman, Organization becomes cause in the matter, *Nature Bio-technology*, **18** (2000), 575–576.

[315] Harold J. Morowitz, Jennifer D. Kostelnik, Jeremy Yang and George D. Cody, The origin of intermediary metabolism, *Proceedings of the National Academy of Sciences USA*, **97** (2000), 7704–7708.
Peter Schuster, Taming combinatorial explosion, *Proceedings of the National Academy of Sciences USA*, **97** (2000), 7678–7680.

[316] Gaell Mainguy, Maria Luz Montesinos, Brigitte Lesaffre, Branco Zevnik, Mika Karasawa, Rashmi Kothary, Wolfgang Wurst, Alain Prochiantz and Michael Volovitch, An induction gene trap for identifying a homeoprotein-regulated locus, *Nature Biotechnology*, **18** (2000), 746–749.

[317] D. Pineda, J. Gonzalez, P. Callaets, K. Ikeo, W. J. Gehring and E. Salo, Searching for the prototypic eye genetic network: *sine oculis* is essential for eye regeneration in planarians, *Proceedings of the National Academy of Sciences USA*, **97** (2000), 4525–4529.

[318] Michael Y. Galperin and Eugene V. Koonin, Who's your neighbor? New computational approaches for functional genomics, *Nature Biotechnology*, **18** (2000), 609–613.

[319] Martin Fussenegger, James E. Bailey and Jeffrey Varner, A mathematical model of caspase function in apoptosis, *Nature Biotechnology*, **18** (2000), 768–774.

[320] J. S. Edwards and B. O. Palsson, The *Escherichia coli* MG1655 *in silico* metabolic genotype: Its definition, characteristics, and capabilities, *Proceedings of the National Academy of Sciences USA*, **97** (2000), 5528–5533.

[321] Richard E. Lenski, Charles Ofria, Travis C. Collier and Christoph Adami, Genome complexity, robustness and genetic interactions in digital organisms, *Nature*, **400** (1999).

[322] Neal S. Holter, Madhusmita Mitra, Amos Maritan, Marek Cieplak, Jayanth R. Banavar and Nina V. Federoff, Fundamental patterns of underlying gene expression profiles: simplicity from complexity, *Proceedings of the National Academy of Sciences USA*, **97** (2000), 8409–8414.

[323] Christoph Adami, Charles Ofria and Travis C. Collier, Evolution of biological complexity, *Proceedings of the National Academy of Sciences USA*, **97** (2000), 4463–4468.

[324] J. A. Thomson and J. S. Odorico, Human embryonic stem cell and embryonic germ cell lines, *TIBTECH*, **18** (2000), 53–57.

[325] Daphne W. Bell, Jennifer M. Varley, Tara E. Szydlo, Deborah H. Kang, Doke C. R. Wahrer, Kristen E. Shannon, Marcie Lubratovich, Sigitas J. Verselis, Kurt J. Isselbacher, Joseph F. Fraumeni, Jillian M. Birch, Frederick P. Li, Judy E. Garber, Daniel A. Haber, Heterozygous Germ Line hCHK2 Mutations in Li–Fraumeni Syndrome, *Science*, **286** (1999), 2528–2431.

[326] Karen Cichowski, T. Shane Shih, Earlene Schmitt, Sabrina Santiago, Karlyne Reilly, Margaret E. McLaughlin, Roderick T. Bronson, Tyler Jacks, Mouse models of tumor development in neurofibromatosis type 1, *Science*, **287** (1999), 2172–2176;
Kristine S. Vogel, Laura J. Klesse, Susana Velasco-Miguel, Kimberley Meyers, Elizabeth J. Rushing, Luis F. Parada, Mouse Tumor Model for Neurofibromatosis Type 1, *Science*, **286** (1999), 2176–2179.

[327] Adriana Donovan, Alison Brownelle, Yi Zhou, Jennifer Shepard, Stephen J. Pratt, John Moynihan, Barry H. Paw, Anna Drejer, Bruce Barut, Agustin Zapata, Terence C. Law, Carlo Brugnara, Samuel E. Lux, Geraldine S. Pinkus, Jack L. Pikus, Paul D. Kingsley, James Palis, Mark D. Fleming, Nancy C. Andrews and Leonard I. Zon, Positional cloning of zebrafish *ferroportin1* identifies a conserved vertebrate iron exporter, *Nature*, **403** (2000), 776–781.

[328] Carl J. Neumann and Christiane Nuesslein-Vollhard, Patterning of the zebrafish retina by a wave of Sonic Hedgehog activity, *Science*, **289** (2000), 2137–2139.

[329] Kevin J. Lee, Paula Dietrich and Thomas M. Jessell, Genetic ablation reveals that the roof plate is essential for dorsal interneuron specification, *Nature*, **403** (2000), 734–740.

[330] Taeko Ichise, Masanobu Kano, Kouichi Hashimoto, Dai Yanagihara, Kazuki Nakao, Ryuichi Shigemoto, Motoya Katsuki and Atsu Aiba, *mGluR1* in cerebellar Purkinje cells essential for long-term depression, synapse elimination, and motor coordination, *Science*, **288** (2000), 1832–1835.

[331] Hiroshi Ishimoto, Akira Matsumoto and Teiichi Tanimura, Molecular identification of a taste receptor gene for trehalose in *Drosophila*, *Science*, **289** (2000), 116–119.

[332] Thomas Günther, Zhou-Feng Chen, Jaesang Kim, Matthias Priemel, Johannes M. Rueger, Michael Amling, Jane M. Moseley, T. John Martin, David J. Anderson and Gerard Karsenty, Genetic ablation of parathyroid glands reveals another source of para-thyroid hormone, *Nature*, **406** (2000), 199–203.

[333] Mel B. Feany and Welcome W. Bender, A *Drosophila* model of Parkinson's disease, *Nature*, **404** (2000), 394–398.

[334] David Schneider and Mohammed Shahabuddin, Malaria parasite development in a *Drosophila* model, *Science*, **288** (2000), 2376–2379.

[335] P. Cluzel, M. Surette and S. Leibler, An ultrasensitive bacterial motor revealed by monitoring signaling proteins in single cells, *Science*, **287** (2000), 1652–1655.

[336] G. D. Meredith, C. E. Sims, J. S. Soughayer and N. L. Allbritton, Measurement of kinase activation in single mammalian cells, *Nature Biotechnology*, **18** (2000), 309–312.

[337] R. J. Davenport, G. J. L. Wuite, R. Landick and C. Bustamante, Single-molecule study of transcriptional pausing and arrest by *E. coli* RNA polymerase, *Science*, **287** (2000), 2497–2500.

[338] Gljs J. L. Wulte, Steven B. Smith, Mark Young, David Keller and Carlos Bustamante, Single-molecule studies of the effect of template tension on T7 DNA polymerase activity, *Nature*, **404** (2000), 103–106.

[339] G. G. Lennon, High-throughput gene expression analysis for drug discovery, *Drug Discovery Today*, **5** (2000), 59–66.

[340] A. P. Watt, D. Morrison and D. C. Evans, Approaches to higher-throughput pharmacokinetics (HTPK) in drug discovery, *Drug Discovery Today*, **5** (2000), 17–24.

[341] H. P. Nestler and R. Liu, Combinatorial libraries: studies in molecular recognition, *Combinatorial Chemistry and High Throughput Screening*, **1** (1998), 113–126.

[342] C. Karan and B. L. Miller, Dynamic diversity in drug discovery: putting small-molecule evolution to work, *Drug Discovery Today*, **5** (2000), 67–75.

[343] D. G. Powers and D. L. Coffen, Convergent automated parallel synthesis, *Drug Discovery Today*, **4** (1999), 377–383.

[344] Richard A. Houghten, Darcy B. Wilson and Clemencia Pinilla, Drug discovery and vaccine development using mixture-based synthetic combinatorial libraries, *Drug Discovery Today*, **5** (2000), 276–285.

[345] C. J. Welch, M. N. Protopopova and G. A. Baht, Microscale synthesis and screening of combinatorial libraries of new chromatographic stationary phases (Special Publication), *Royal Society of Chemistry*, **235** (1999), 129–138.

[346] D. S. Tan, M. A. Foley, M. D. Shair and S. L. Schreiber, Stereoselective synthesis of over two million compounds having structural features both reminiscent of natural products and compatible with miniaturized cell-based assays, *Journal of the American Chemical Society*, **120** (1998), 8565–8566. S. L. Schreiber, Chemical genetics resulting from a passion for synthetic organic chemistry, *Bioorganic and Medicinal Chemistry*, **6** (1998), 1127–1152.

[347] Jennifer L. Harris, Bradley J. Backes, Francesco Leonetti, Sami Mahrus, Jonathan A. Ellman, Charles S. Craik, Rapid and general profiling of protease specificity by using combinatorial fluorogenic substrate libraries, *Proceedings of the National Academy of Sciences USA*, **97** (2000), 7754–7759.

[348] C. W. Harwig, D. J. Gravert, K. D. Janda, Soluble polymers: new options in both traditional and combinatorial synthesis, *CHEMTRACTS Organic Chemistry*, **12** (1999), 1–26. Nicholas W. Hird, Automated synthesis: new tools for the organic chemist, *Drug Discovery Today,* **4** (1999), 265–274. Dominique Gorse, Anthony Rees, Michel Kaczorek, Roger Lahana, Molecular diversity and its analysis, *Drug Discovery Today,* **4** (1999), 257–264.

[349] G. R. Rosania, Y.-T. Chang, O. Perez, D. Sutherlin, H. Dong, D. J. Lockhart and P. G. Schultz, Myoseverin, a microtubule-binding molecule with novel cellular effects, *Nature Biotechnology*, **18** (2000), 304–308. D. G. Hall, Combinatorial chemistry gives cell biology some muscle, *Nature Bio-technology*, **18** (2000), 261–262.

[350] G. R. Lenz, H. M. Nash and S. Jindal, Chemical ligands, genomics and drug discovery, *Drug Discovery Today*, **5** (2000), 145–156.

[351] J. Rademann and G. Jung, Integrating combinatorial synthesis and bioassays, *Science*, **287** (2000), 1947–1948.

[352] Paul Beroza and Mark J. Suto, Designing chiral libraries for drug discovery, *Drug Discovery Today*, **5** (2000), 364–372.

[353] Andrew R. Leach and Michael M. Hann, The *in silico* world of virtual libraries, *Drug Discovery Today*, **5** (2000), 326–336.

[354] Raymond J. Cho, Michael Mindrinos, Daniel R. Richards, Ronald J. Sapolsky, Mary Anderson, Eliana Drenkard, Julia Dewdney, T. Lynne Reuber, Melanie Stammers, Nancy Federspiel, Athanasios Theologis, Wie-Hsien Yang, Earl Hubbell, Melinda Au, Edward Y. Chung, Deval Lashkari, Bertrand Lemieux, Caroline Dean, Robert J. Lipshutz, Frederick M. Ausubel, Ronald W. Davis, Peter J. Oefner, Genome-wide mapping with biallelic markers in *Arabidopsis thaliana*, *Nature Genetics*, **23** (1999), 203–207.

[355] Hai Yan, Kenneth W. Kinzler and Bert Vogelstein, Genetic testing – present and future, *Science*, **289** (2000), 1890–1892.

[356] A. M. Garvin, K. C. Parker and L. Haff, MALDI-TOF based mutation detection using tagged *in vitro* synthesized peptides, *Nature Biotechnology*, **18** (2000), 95–97.

[357] F. J. Steemers, J. A. Ferguson and D. R. Walt, Screening unlabeled DNA targets with radomly ordered fiber-optic gene arrays, *Nature Biotechnology*, **18** (2000), 91–94.

[358] David A. Campbell, AnaMaria Valdes and Nigel Spurr, Making drug discovery a SN(i)P, *Drug Discovery Today*, **5** (2000), 388–396.

[359] Dale R. Pfost, Michael T. Boyce-Jacino and Denis M. Grant, A SNPshot: pharmacogenetics and the future of drug therapy, *TIBTECH*, **18** (2000), 334–338.

[360] F. Schweighoffer, *et al.*, Qualitative gene profiling: a novel tool in genomics and pharmacogenomics that deciphers messenger RNA isoforms diversity, *Pharmacogenomics*, **1** (2000), 187–197.

[361] Fors, L. *et al.*, Large-scale SNP scoring from unamplified genomic DNA, *Pharmacogenomics*, **1** (2000), 219–229.

[362] Victor Lyamichev, Andrea L. Mast, Jeff G. Hall, James R. Prudent, Michael W. Kaiser, Tsetska Takova, Robert W. Kwiatkowski, Tamara J. Sander, Monika de Arruda, David A. Arco, Bruce P. Neri and Mary Ann D. Brow, Polymorphism identification and quantitative detection of genomic DNA by invasive cleavage of oligonucleotide probes, *Nature Biotechnology*, **17** (1999), 292–296.

[363] Jeff G. Hall, Peggy S. Eis, Scott M. Law, Luis P. Reynaldo, James R. Prudent, David J. Marshall, Hatim T. Allawi, Andrea L. Mast, James E. Dahlberg, Robert W. Kwiatkowski, Monika de Arruda, Bruce P. Neri and Victor I. Lyamichev, Sensitive detection of DNA polymorphisms by the serial invasive signal amplification reaction, *Proceedings of the National Academy of Sciences USA*, **97** (2000), 8272–8277.

[364] David Voss, Gene express, *New Scientist*, **164** (No. 2208) (1999), 40–43.

[365] C. Aston, B. Mishra and D. C. Schwartz, Optical mapping and its potential for large-scale screening projects, *TIBTECH*, **17** (1999), 297–302.

[366] A. Meller, L. Nivon, E. Brandin, J. Golovchenko and D. Branton, Rapid nanopore discrimination between single polynucleotide molecules, *Proceedings of the National Academy of Sciences USA*, **97** (2000), 1079–1084.

[367] D. W. Deamer and M. Akeson, Nanopores and nucleic acids: prospects for ultrarapid sequencing, *TIBTECH*, **18** (2000), 147–151.

[368] J. Han and H. G. Craighead, Separation of long DNA molecules in a micro-fabricated entropic trap array, *Science*, **288** (2000), 1026–1029.

[369] Adam T. Woolley, Chantal Guillemette, Chin Li Cheung, David E. Housman and Charles M. Lieber, Direct haplotyping of kilobase-size DNA using carbon nanotube probes, *Nature Biotechnology*, **18** (2000), 760–763.

[370] Sydney Brenner, Maria Johnson, John Bridgham, George Golda, David H. Lloyd, Davida Johnson, Shujun Luo, Sarah McCurdy, Michael Foy, Mark Ewan, Rithy Roth, Dave George, Sam Eletr, Glann Albrecht, Eric Vermaas, Steven R. Williams, Keith Moon, Timothy Burcham, Michael Pallas, Robert B. DuBridge, Kames Kirchner, Karen Fearon, Jen-i Mao and Kevin Corcoran, Gene expression analysis by massively parallel signature sequencing (MPSS) on microbead arrays, *Nature Biotechnology*, **18** (2000), 630–634.

[371] Shaorong Liu, Hongji Ren, Qiufeng Gao, David J. Roach, Robert T. Loder, Jr, Thomas M. Armstrong, Qinglu Mao, Iuliu Blaga, David L. Barker and Stevan B. Jovanovich, Automated parallel DNA sequencing on multiple channel microchips, *Proceedings of the National Academy of Sciences USA*, **97** (2000), 5369–5374.

[372] David W. Oldach, Charles F. Delwiche, Kjetill S. Jakobsen, Torstein Tengs, Ernest G. Brown, Jason W. Kempton, Eric F. Schaefer, Holly A. Bowers, Howard B. Glasgow, Jr, JoAnn M. Burkholder, Karen A. Steidinger and Parke A. Rublee, Heteroduplex mobility assay-guided sequence discovery: elucidation of the small subunit (18S) rDNA sequences of *Pfiesteria piscicida* and related dinoflagellates from complex algal culture and environmental sample DNA pools, *Proceedings of the National Academy of Sciences USA*, **97** (2000), 4303–4308.

Douglas W. Selinger, Kevin J. Cheung, Rui Mei, Erik M. Johannson, Craig S. Richmond, Frederick R. Blattner, David J. Lockhart, George M. Church, RNA expression analysis using a 30 base pair resolution *Escherichia coli* genome array, *Nature Biotechnology*, **18** (2000), 1262–1268.

[373] I. Wickelgren, Mining the genome for drugs, *Science*, **285** (1999), 998–1001.

[374] J. D. Keasling, Gene-expression tools for the metabolic engineering of bacteria, *TIBTECH*, **17** (1999), 452–460.

[375] S. Schuster, D. A. Fell and T. Dandekar, A general definition of metabolic pathways useful for the systematic organization and analysis of complex metabolic networks, *Nature Biotechnology*, **18** (2000), 326–332.

[376] Gregor Daum, Bernd Bartenbach and Mike Battrum, Praxisrelevante Simulations-werkzeuge für mehrstufige Batch-Prozesse, *Chemie Ingenieur Technik*, **71** (1999), 1253–1261.

[377] M. Ortega Lorenzo, C. J. Baddeley, C. Muryn and R. Raval, Extended surface chirality from supramolecular assemblies of adsorbed chiral molecules, *Nature*, **404** (2000), 376–379.

[378] Varda Mann, Mark Harker, Iris Pecker and Joseph Hirschberg, Metabolic engineering of astaxanthin production in tobacco flowers, *Nature Biotechnology*, **18** (2000), 888–892.

[379] Manuela Albrecht, Shinichi Takaichi, Sabine Steiger, Zheng-Yu Wang and Gerhard Sandmann, Novel hydroxy-carotenoids with improved antioxidative properties produced by gene combination in *Escherichia coli*, *Nature Biotechnology*, **18** (2000), 843–846.

Gerhard Sandmann, Manuela Albrecht, Georg Schnurr, Oliver Knörzer, Peter Böger, The biotechnological potential and design of novel carotenoids by gene combination in *Escherichia coli*, *TIBTECH*, **17** (1999), 233–237.

[380] Claudia Schmidt-Dannert, Daisuke Umeno and Frances H. Arnold, Molecular breeding of carotenoid biosynthetic pathways, *Nature Biotechnology*, **18** (2000), 750–753.

[381] K. A. Ward, Transgene-mediated modifications to animal biochemistry, *TIBTECH*, **18** (2000), 99–102.

[382] G. Bulfield, Farm animal biotechnology, *TIBTECH*, **18** (2000), 10–13.

[383] J. M. Staub, B. Garcia, J. Graves, P. T. J. Hajdukiewicz, P. Hunter, N. Nehra, V. Paradkar, M. Schlittler, J. A. Carroll, D. Spatola, D. Ward, G. Ye and D. A. Russell, High-yield production of a human therapeutic protein in tobacco chloroplasts, *Nature Biotechnology*, **18** (2000), 333–338.

[384] H. Horvath, J. Huang, O. Wong, E. Kohl, T. Okita, C. G. Kannangara and D. von Wettstein, The production of recombinant proteins in transgenic barley grains, *Proceedings of the National Academy of Sciences USA*, **97** (2000), 1914–1919.

[385] Xudong Ye, Salim Al-Babili, Andreas Klöti, Jing Zhang, Paoloa Lucca, Peter Beyer, Ingo Potrykus, Engineering the Provitamin A (β-Carotene) Biosynthetic Pathway into (Carotenoid-Free) Rice Endosperm, *Science*, **287** (2000), 303–305.

[386] Oliver May, Peter T. Nguyen and Frances H. Arnold, Inverting enantioselectivity by directed evolution of hydantoinase for improved production of L-methionine, *Nature Biotechnology*, **18** (2000), 317–320.

[387] M. M. Altamirano, J. M. Blackburn, C. Aguayo and A. R. Fersht, Directed evolution of new catalytic activity using the α/β-barrel scaffold, *Nature*, **403** (2000), 617–622.

[388] K. Kuchner and F. H. Arnold, Directed evolution of enzyme catalysts [Review], *Trends in Biotechnology*, **15** (1997), 523–530.

[389] Joel R. Cherry, Michael H. Lamsa, Palle Schneider, Jesper Vind, Allan Svendsen, Aubrey Jones and Anders H. Petersen, Directed evolution of a fungal peroxidase, *Nature Biotechnology*, **17** (1999), 379–384.

[390] Jon E. Ness, Mark Welch, Lori Giver, Manuel Bueno, Joel R. Cherry, Torben V. Borchert, Willem P. C. Stemmer and Jeremy Minshull, DNA shuffling of subgenomic sequences of subtilisin, *Nature Biotechnology*, **17** (1999), 893–896.

[391] Lance P. Encell, Daniel M. Landis and Lawrence A. Loeb, Improving enzymes for cancer therapy, *Nature Biotechnology*, **17** (2000), 143–147.

[392] Robin J. Gouka, Casper Gerk, Paul J. J. Hooykaas, Paul Bundock, Wouter Musters, C. Theo Verrips and Marcel J. A. de Groot, Transformation of *Aspergillus awamori* by *Agrobacterium tumefaciens*-mediated homologous recombination, *Nature Bio-technology*, **17** (1999), 598–601.

[393] M. Sosio, F. Giusino, C. Cappellano, E. Bossi, A. M. Puglia and S. Donadio, Artificial chromosomes for antibiotic-producing actinomycetes, *Nature Bio-technology*, **18** (2000), 343–345.

[394] J. Rosamund and A. Allsop, Harnessing the power of the genome in the search for new antibiotics, *Science*, **287** (2000), 1973–1976.

[395] Javier Velasco, Jose Luis Adrio, Miguel Angel Moreno, Bruno Diez, Gloria Soler and Jose Luis Barredo, Environmentally safe production of 7-aminodeacetoxy-cephalo-sporanic acid (7-ADCA) using recombinant strains of *Acremonium chrysogenum*, *Nature Biotechnology*, **18** (2000), 857–861.

[396] Nicholas Judson and John J. Mekalanos, TnAraOut, a transposon-based approach to identify and characterize essential bacterial genes, *Nature Biotechnology*, **18** (2000), 740–745.

[397] T. Ashton Cropp, Dennis J. Wilson and K. A. Reynolds, Identification of a cyclo-hexylcarbonyl CoA biosynthetic gene cluster and application in the production of doramectin, *Nature Biotechnology*, **18** (2000), 980–983.

[398] John W. Trauger, Rahul M. Kohll, Henning D. Mootz, Mohamed A. Marahiel and Christopher T. Walsh, Peptide cyclization catalysed by the thioester domain of tyrocidine synthetase, *Nature*, **407** (2000), 215–218.

[399] Peter J. Belshaw, Christopher T. Walsh and Torsten Stachelhaus, Aminoacyl-CoAs as probes of condensation domain selectivity in nonribosomal peptide synthesis, *Science*, **284** (1999), 486–489.

[400] Rajesh S. Gokhale, Stuart Y. Tsuji, David E. Cane and Chaitan Khosla, Dissecting and exploiting intermodular communication in polyketide synthases, *Science*, **284** (1999), 482–485.

[401] Jonathan Kennedy, Karine Auclair, Steven G. Kendrew, Cheonseok Park, John C. Vederas and C. Richard Hutchinson, Modulation of polyketide synthase activity by accessory proteins during lovastatin biosynthesis, *Science*, **284** (1999), 1368–1372.

[402] Nobutaka Funa, Yasuo Ohnishi, Isao Fujii, Masaaki Shibuya, Yutaka Ebizuka and Suehara Horinouchi, A new pathway for polyketide synthesis in microorganisms, *Nature*, **400** (2000), 897–899.

[403] Nicolai I. Burzlaff, Peter J. Rutledge, Ian J. Clifton, Charles M. H. Hensgens, Michael Pickford, Robert M. Adlington, Peter L. Roach and Jack E. Baldwin, The reaction cycle of isopenicillin N synthase observed by X-ray diffraction, *Nature*, **401** (1999), 721–724.

[404] L. Tang, S. Shah, L. Chung, J. Carney, L. Katz, C. Khosla and B. Julien, Cloning and heterologous expression of the epothilone gene cluster, *Science*, **287** (2000), 640–642.

[405] R. T. Lorenz and G. R. Cysewski, Commercial potential for *Haematococcus* microalgae as a natural source of astaxanthin, *TIBTECH*, **18** (2000), 160–167.

[406] M. Juza, M. Mazzotti and M. Morbidelli, Simulated moving-bed chromatography and its application to chirotechnology, *TIBTECH*, **18** (2000), 108–118.

[407] Myriam M. Altamirano, Consuelo Garcia, Lourival D. Possani and Alan R. Fersht, Oxidative refolding chromatography: folding of the scorpion toxin Cn5, *Nature Bio-technology*, **17** (2000), 187–191.

[408] David W. Wood, Wie Wu, Georges Belfort, Victoria Derbyshire and Marlene Belfort, A genetic system yields self-cleaving inteins for bioseparations, *Nature Bio-technology*, **17** (1999), 889–892.

[409] Oliver Hofstetter, Heike Hofstetter, Meir Wilchek, Volker Schurig and Bernard S. Green, Chiral discrimination using an immunosensor, *Nature Biotechnology*, **17** (1999), 371–374.

[410] Lidietta Giorno and Enrico Drioli, Biocatalytic membrane reactors: applications and perspectives, *TIBTECH*, **18** (2000), 339–349.

[411] Lydia P. Ooijkaas, Frans J. Weber, Reinetta M. Buitelaar, Johannes Tramper and Arjen Rinzema, Defined media and inert supports: their potential as solid-state fermentation production systems, *TIBTECH*, **18** (2000), 356–360.

[412] O. Doblhoff-Dier and R. Bliern, Quality control and assurance from the development to the production of biopharmaceuticals, *TIBTECH*, **17** (1999), 266–270.

[413] F. Hesse and R. Wagner, Developments and improvements in the manufacturing of human therapeutics with mammalian cell cultures, *TIBTECH*, **18** (2000), 173–180.

[414] M. Susana Levy, Ronan D. O'Kennedy, Parviz Ayazi-Shamlou and Peter Dunhill, Biochemical engineering approaches to the challenges of producing pure plasmid DNA, *TIBTECH*, **18** (2000), 296–305.

[415] Guilherme N. M. Ferreira, Gabriel A. Monteiro, Duarte M. F. Prazeres and Joaquim M. S. Cabral, Down-stream processing of plasmid DNA for gene therapy and DNA vaccine applications, *TIBTECH*, **18** (2000), 380–388.

[416] B. Harris, Exploiting antibody-based technologies to manage environmental pollution, *TIBTECH*, **17** (1999), 290–296.
John D. Coates, Robert T. Anderson, Emerging techniques for anaerobic bio-remediation of contaminated environments, *TIBTECH*, **18** (2000), 408–412.

[417] Tillman U. Gerngross, Can biotechnology move us toward a sustainable society?, *Nature Biotechnology*, **17** (1999), 541–544.
Bernhard Zechendorf, Sustainable development: how can biotechnology contribute?, *TIBTECH*, **17** (1999), 219–225.

[418] The Safety in Biotechnology Working Party of the European Federation of Biotechnology, Safe Biotechnology 10: DNA content of biotechnological process waste, *TIBTECH*, **18** (2000), 141–146.

[419] Cristiano Nicolella, Mark C. M. van Loosdrecht and Sef J. Heijnen, Particle-based biofilm reactor technology, *TIBTECH*, **18** (2000), 312–320.

[420] D. Perry, Patients' voices: the powerful sound in the stem cell debate, *Science*, **287** (2000), 1423.
F. E. Young, A time for restraint, *Science*, **287** (2000), 1424.
N. Lenoir, Europe confronts the embryonic stem cell research challenge, *Science*, **287** (2000), 1425–1427.

[421] M. K. Cho, D. Magnus, A. L. Caplan, D. McGee and the Ethics of Genomics Group, Ethical considerations in synthesizing a minimal genome, *Science*, **286** (1999), 2087–2090.

[422] J. Kinderlerer, Is a European convention on the ethical use of modern biotechnology needed?, *TIBTECH*, **18** (2000), 87–90.

[423] C. A. Hutchison, III, S. N. Peterson, S. R. Gill, R. T. Cline, O. White, C. M. Fraser, H. O. Smith and J. C. Venter, Global transposon mutagenesis and a minimal mycoplasma genome, *Science*, **286** (1999), 2165–2169.
J. C. Polkinghorne, Ethical issues in biotechnology, *TIBTECH*, **18** (2000), 8–10.

[424] J. Harris, Intimations of Immortality, *Science*, **288** (2000), 59.

[425] S.-I. Imai, C. M. Armstrong, M. Kaeberlein and L. Guarente, Transcriptional silencing and longevity protein Sir2 is an NAD-dependent histone deacetylase, *Nature*, **403** (2000), 795–800.

[426] R. Breaker, G. Joyce and S. Santoro (1994), The Scripps Research Institute.

[427] S. W. Michnik and F. H. Arnold, 'Itching' for new strategies in protein engineering, *Nature Biotechnology*, **17** (1999), 1159–1180.

[428] Janet M. Thornton, Annabel E. Todd, Duncan Milburn, Neera Borkakoti and Christine A. Orengo, From structure to function: approaches and limitations, *Nature Structural Biology, Structural Genomics Supplement*, **7** (2000), 991–994.

[429] Steven E. Brenner, Target selection for structural genomics, *Nature Structural Biology, Structural Genomics Supplement*, **7** (2000), 967–969.

[430] Aled M. Edwards, Cheryl H. Arrowsmith, Dinesh Christendat, Akil Dharamsi, James D. Friesen, Jack F. Greenblatt and Masoud Vedadi, Protein production: feeding the crystallographers and NMR spectroscopists, *Nature Structural Biology, Structural Genomics Supplement*, **7** (2000), 970–972.

[431] Enrique Abola, Peter Kuhn, Thomas Earnest and Raymond C. Stevens, Automation of X-ray crystallography, *Nature Structural Biology, Structural Genomics Supplement*, **7** (2000), 973–977.

[432] Victor S. Lamzin and Anastassis Perrakis, Current state of automated crystallographic data analysis, *Nature Structural Biology, Structural Genomics Supplement*, **7** (2000), 978–981.

[433] Gaetano T. Montelione, Deyou Zheng, Yuanpeng J. Huang, Kristin C. Gunsalus and Thomas Szyperski, Protein NMR spectroscopy in structural genomics, *Nature Structural Biology, Structural Genomics Supplement*, **7** (2000), 982–985.

[434] Roberto Sanchez, Ursula Pieper, Francisco Melo, Narayanan Eswar, Marc A. Marti-Renom, M. S. Madhusudhan, Nebosja Mirkovic and Andrej Sali, Protein structure modeling for structural genomics, *Nature Structural Biology, Structural Genomics Supplement*, **7** (2000), 986–990.

[435] Jennifer A. Doudna, Structural genomics of RNA, *Nature Structural Biology, Structural Genomics Supplement*, **7** (2000), 954–956.

[436] Thomas C. Terwilliger, Structural genomics in North America, *Nature Structural Biology, Structural Genomics Supplement*, **7** (2000), 935–939.

[437] Udo Heinemann, Structural genomics in Europe: slow start, strong finish?, *Nature Structural Biology, Structural Genomics Supplement*, **7** (2000), 940–942.

[438] Shigeyuki Yokoyama, Hiroshi Hirota, Takanori Kigawa, Takashi Yabuki, Mikako Shirouzu, Takaho Terada, Yutaka Ito, Yo Matsuo, Yutaka Kuroda, Yoshifumi Nishimura, Yoshimasa Kyogoku, Kunio Miki, Ryoji Masui and Seiki Kuramitsu, Structural genomics projects in Japan, *Nature Structural Biology, Structural Genomics Supplement*, **7** (2000), 943–945.

[439] Sarah Dry, Sean McCarthy and Tim Harris, Structural genomics in the biotechnology sector, *Nature Structural Biology, Structural Genomics Supplement*, **7** (2000), 946–949.

[440] Alison Abbott, Structures by numbers, *Nature*, **408** (2000), 130–132.

[441] Vicki L. Nienaber, Paul L. Richardson, Vered Klighofer, Jennifer J. Bouska, Vincent L. Giranda and Jonathan Greer, Discovering novel ligands for macromolecules using X-ray crystallographic screening, *Nature Biotechnology*, **18** (2000), 1105–1108.

[442] Arnold L. Demain, Microbial biotechnology, *TIBTECH*, **18** (2000), 26–31.

[443] Simon Ostergaard, Lisbeth Olsson, Mark Johnston and Jens Nielsen, Increasing galactose consumption by *Saccharomyces cerevisiae* through metabolic engineering of the *GAL* gene regulatory network, *Nature Biotechnology*, **18** (2000), 1283–1286.

[444] David McCaskill and Rodney Croteau, Strategies for bioengineering the development and metabolism of glandular tissues in plants, *Nature Biotechnology*, **17** (1999), 31–36.

[445] A. Kusnadi, Z. L. Nikolov and J. A. Howard, Production of recombinant proteins in transgenic plants: practical considerations, *Biotechnology and Bioengineering*, **56** (1997), 473–484.

[446] Glynis Giddings, Gordon Allison, Douglas Brooks and Adrian Carter, Transgenic plants as factories for biopharmaceuticals, *Nature Biotechnology*, **18** (2000), 1151–1155.

[447] Liz J. Richter, Yasmin Thanavala, Charles J. Arntzen and Hugh S. Mason, Production of hepatitis B surface antigen in transgenic plants for oral immunization, *Nature Bio-technology*, **18** (2000), 1167–1171.

[448] Oliver Fiehn, Joachim Kopka, Peter Dörmann, Thomas Altmann, Richard N. Trethe-wey and Lothar Willmitzer, Metabolite profiling for plant functional genomics, *Nature Biotechnology*, **18** (2000), 1157–1161.

[449] David M. Horn, Roman A. Zubarev and Fred W. McLafferty, Automated *de novo* sequencing of proteins by tandem high-resolution mass spectrometry, *Proceedings of the National Academy of Sciences USA*, **97** (2000), 10313–10317.

[450] William J. Coates, David J. Hunter and William S. MacLachlan, Successful implementation of automation in medicinal chemistry, *Drug Discovery Today*, **5** (2000), 521–527.

[451] Henning Vollert, Birgit Jordan and Irvin Winkler, Wandel in der Wirkstoffsuche – Ultra-High-Through-put-Screening-Systeme in der Pharmaindustrie, *LABORWELT*, **1** (2000), 5–10.

[452] Martin Winter, Robotik und Automatisierungskonzepte in der Kombinatorischen Chemie – Synthese- und Pipettierroboter, *LABORWELT*, **1** (2000), 25–29.

[453] Odilo Müller, Lab-on-a-Chip-Technologie – Revolution im (bio-)analytischen Labor, *LABORWELT*, **1** (2000), 36–38.

[454] Holger Eickhoff, Eckhard Nordhoff, Wilfried Nietfeld, Dolores Cahill, Martin Horn, Konrad Büssow and Hans Lehrach, Vom Gen zur Proteinstruktur – Hochparallele Ansätze zur Expressions-, Protein- und Strukturanalyse, *LABORWELT*, **1** (2000), 19–23.

[455] Egon Amann, Genom-Diagnose, SNPs und Pharmakogenetik – Einzug in das diagnostische Routinelabor?, *LABORWELT*, **1** (2000), 11–18.

[456] Tamas Balla, Identification of intracellular signaling domains, *New Technologies for Life sciences: A Trends Guide*, December (2000), 51–55.

[457] Andrew W. Mulvaney, C. Ian Spencer, Steven Culliford, John J. Borg, Stephen G. Davies and Roland Z. Koslowski, Cardiac chloride channels: physiology, pharma-cology and approaches for identifying novel modulators of activity, *Drug Discovery Today*, **5** (2000), 492–505.

[458] Jeffrey J. Clare, Simon N. Tate, Malcolm Nobbs and Mike A. Romanos, Voltage-gated sodium channels as therapeutic targets, *Drug Discovery Today*, **5** (2000), 506–520.

[459] Ricky K. Soong, George D. Bachand, Hercules P. Neves, Anatoli G. Olkhovets, Harold G. Craighead and Carlo D. Montemagno, Powering an inorganic nanodevice with a bio-molecular motor, *Science*, **290** (2000), 1555–1558.

[460] David J. Beebe, Jeffrey S. Moore, Qing Yu, Robin H. Liu, Mary L. Kraft, Byung-Ho Jo and Chelladurai Devadoss, Microfluidic tectonics: a comprehensive construction platform for microfluidic systems, *Proceedings of the National Academy of Sciences USA*, **97** (2000), 13488–13493.

[461] Jürgen Drews, *Quo vadis*, biotech? (Part 1), *Drug Discovery Today*, **5** (2000), 547–553.

[462] Jürgen Drews, *Quo vadis*, biotech? (Part 2), *Drug Discovery Today*, **6** (2001), 21–26.

[463] David R. Gilbert, Michael Schroeder and Jacques van Helden, Interactive visualization and exploration of relationships between biological objects, *TIBTECH*, **18** (2000), 487–494.

[464] Joshua B. Tenenbaum, Vin de Silva and John C. Langford, A global framework for nonlinear dimensionality reduction, *Science*, **290** (2000), 2319–2323.

[465] Sam T. Roweis and Lawrence K. Saul, Nonlinear dimensionality reduction by local linear embedding, *Science*, **290** (2000), 2323–2326.

[466] Paolo P. Saviotti, Marie-Angele deLooze, Sylvie Michelland and David Catherine, The changing marketplace of bioinformatics, *Nature Biotechnology*, **18** (2000), 1247–1249.

[467] Teresa K. Attwood, The Babel of bioinformatics, *Science*, **290** (2000), 471–473.

[468] Christopher Ahlberg, Visual exploration of HTS data bases: bridging the gap between chemistry and biology, *Drug Discovery Today*, **4** (1999), 370–376.

[469] Nicholas J. Hrib and Norton P. Peet, Chemoinformatics: are we exploiting this new science?, *Drug Discovery Today*, **5** (2000), 483–485.

[470] John R. S. Newman, Ethan Wolf and Peter S. Kim, A computationally directed screen identifying interacting coiled coils from *Saccharomyces cerevisiae*, *Proceedings of the National Academy of Sciences USA*, **97** (2000), 13203–13208.

[471] Karen E. Nelson, Ian T. Paulsen, John F. Heidelberg and Claire M. Fraser, Status of genome projects for nonpathogenic bacteria and archaea, *Nature Biotechnology*, **18** (2000), 1049–1054.

[472] Bernhard Palsson, The challenges of *in silico* biology, *Nature Biotechnology*, **18** (2000), 1147–1150.

[473] Christophe H. Schilling, Stefan Schuster, Bernhard O. Palsson, Reinhart Heinrich, Metabolic Pathway Analysis: Basic Concepts and Scientific Applications in the Post-genomic Era, *Biotechnology Progress*, **15** (1999), 296–303.

[474] Aris Persidis, The business of pharmacogenomics, *Nature Biotechnology*, **16** (1998), 209–210 (reprinted in *Nature Biotechnology*, **18** (Suppl.) (2000), IT40–IT42.

[475] Joseph A. Rininger, Vincent A. DiPippo and Bonnie E. Gould-Rothberg, Differential gene expression technologies for identifying surrogate markers of drug efficacy and toxicity, *Drug Discovery Today*, **5** (2000), 560–568.

[476] Allan D. Roses, Pharmacogenetics and the practice of medicine, *Nature*, **405** (2000), 857–865.

[477] Paul L. Gourley, Mark F. Gourley, Biocavity lasers for biomedicine, *TIBTECH*, **18** (2000), 443–448.

[478] M. Werner, S.-P. Heyn and Th. Köhler, Microsystems in health care markets: recent achievements and future trends, *mst*news, **4** (2000), 4–7.

[479] Patrick Griss, Helene Andersson and Göran Stemme, BioMEMS activities at S3, *mst*news, **4** (2000), 24–25.

[480] Gert Blankenstein, Microfluidic devices for biomedical applications, *mst*news, **4** (2000), 14–15.

[481] A. Offenhäuser and W. Knoll, Living cells as functional elements of a hybrid electronic/bio-microsystem, *mst*news, **4** (2000), 26–28.

[482] Johan Wessberg, Christopher R. Stambaugh, Jerald D. Kralik, Pamela D. Beck, Marl Laubach, John K. Chapin, Jung Kim, S. James Biggs, Mandayam A. Srinivasan, Miguel A. L. Nicolelis, Real-time prediction of hand trajectory by ensembles of cortical neurons in primates, *Nature*, **408** (2000), 361–365.
Sandro Mussa-Ivaldi, Real brains for real robots, *Nature* **408** (2000), 305–306.

[483] Kathleen M. O'Craven, Paul E. Downing and Nancy Kanwisher, fMRI evidence for objects as the units of attentional selection, *Nature*, **401** (1999), 584–587.

[484] Andro Zangaladze, Charles M. Epstein, Scott T. Grafton and K. Sathian, Involvement of visual cortex in tactile discrimination of orientation, *Nature*, **401** (1999), 587–591.

[485] Mark Hallett, Transcranial magnetic stimulation and the human brain, *Nature*, **406** (2000), 147–150.

[486] Harley I. Kornblum, Dalia M. Araujo, Alexander J. Annala, Keith J. Tatsukawa, Michael E. Phelps and Simon R. Cherry, *In vivo* imaging of neuronal activation and plasticity in the rat brain by high resolution positron emission tomography (microPET), *Nature Biotechnology*, **18** (2000), 655–660.

[487] Alexey Terskikh, Arkady Fradkov, Galina Ermakova, Andrey Zaraisky, Patrick Tan, Andrey V. Kajava, Xiaoning Zhao, Sergey Lukyanov, Mikhail Matz, Stuart Kim, Irving Weissman and Paul Siebert, 'Fluorescent Timer': protein that changes color with time, *Science*, **290** (2000), 1585–1588.

[488] Christopher Walsh, Enabling the chemistry of life, *Nature*, **409** (2001), 226–231.

[489] Kathryn M. Koeller and Chi-Huey Wong, Enzymes for chemical synthesis, *Nature*, **409** (2001), 232–240.

[490] Alexander M. Klibanov, Improving enzymes by using them in organic solvents, *Nature*, **409** (2001), 241–246.

[491] Chaitan Khosla and Pehr B. Harbury, Modular enzymes, *Nature*, **409** (2001), 247–252.

[492] Frances H. Arnold, Combinatorial and computational challenges for biocatalyst design, *Nature*, **409** (2001), 253–257.

[493] David E. Benson, Michael S. Wisz and Homme W. Hellinga, Rational design of nascent metalloenzymes, *Proceedings of the National Academy of Sciences USA*, **97** (2000), 6292–6297.

[494] Eric W. Stawinski, Albion E. Baucom, Scott C. Lohr and Lydia M. Gregoret, Predicting protein function from structure: unique structural features of proteases, *Proceedings of the National Academy of Sciences USA*, **97** (2000), 3954–3958.

[495] Edmond Y. Lau, Kalju Kahn, Paul A. Bash and Thomas C. Bruice, The importance of reactant positioning in enzyme catalysis: a hybrid quantum mechanics/molecular mechanics study of a haloalkane dehalogenase, *Proceedings of the National Academy of Sciences USA*, **97** (2000), 9937–9942.

[496] Lutz Riechmann and Greg Winter, Novel folded protein domains generated by combinatorial shuffling of polypeptide segments, *Proceedings of the National Academy of Sciences USA*, **97** (2000), 10068–10073.

[497] Amelie Karlstrom, Guofu Zhong, Christoph Rader, Nicholas A. Larsen, Andreas Heine, Robertsa Fuller, Benjamin List, Fujie Tanaka, Ian A. Wilson, Carlos F. Barbas, III and Richard A. Lerner, Using antibody catalysis to study the outcome of multiple evolutionary trials of a chemical task, *Proceedings of the National Academy of Sciences USA*, **97** (2000), 3873–3883.

[498] A. Schmid, J. S. Dordick, B. Hauer, A. Kiener, M. Wubbolts and B. Witholt, Industrial bio-catalysis today and tomorrow, *Nature*, **409** (2001), 258–268.

[499] Tomoko Kuwabara, Masaki Warashina and Kazunari Raira, Allosterically controllable maxizymes cleave mRNA with high efficiency and specificity, *TIBTECH*, **18** (2000), 462–468.

[500] Hyotcherl Ihee, Vladimir A. Lobastov, Udo M. Gomez, Boyd M. Goodson, Ramesh Srinivasan, Chong-Yu Ruan and Ahmed H. Zewail, Direct imaging of transient molecular structures with ultrafast diffraction, *Science*, **291** (2001), 458–462.

[501] Sergey A. Piletsky, Susan Alcock and Anthony P. F. Turner, Molecular imprinting: at the edge of the third millennium, *TIBTECH*, **19** (2001), 9–12.

[502] Simon Ostergaard, Lisbeth Olsson, Jens Nielsen, Metabolic Engineering of *Saccharomyces cerevisiae*, *Microbiology and Molecular Biology Reviews*, **64** (2000), 34–50.

[503] Blaine A. Pfeifer, Suzanne J. Admiraal, Hugo Gramajo, David E. Cane, Chaitan Khosla, Biosynthesis of Complex Polyketides in a Metabolically Engineered Strain of *E. coli, Science*, **291** (2001), 1790–1792.

[504] L. P. Wackett, Environmental biotechnology, *TIBTECH*, **18** (2000), 19–21.

[505] Brent R. Stockwell, Frontiers in chemical genetics, *TIBTECH*, **18** (2000), 449–454.

[506] Bernhard Straub, Elisabeth Meyer, Peter Fromherz, *Nature Biotechnology*, **19** (2001), 121–124.

[507] Jean-Christophe Rain, Luc Selig, Hilde De Reuse, Veronique Battaglia, Celine Reverdy, Stephane Simon, Gerlinde Lenzen, Fabien Petel, Jerome Wojcik, Vincent Schächter, Y. Chemama, Agnes Labigne, Pierre Legrain, The protein-protein inter-action map of *Helicobacter pylori, Nature*, **490** (2001), 211–215.

[508] Nicole F. Mathon, Denise S. Malcolm, Marie C. Harrisingh, Lili Cheng, Alison C. Lloyd, Lack of Replicative Senescence in Normal Rodent Glia, *Science,* **291** (2001), 872–875.

[509] Dean G. Tang, Yasuhito M. Tokumoto, James A. Apperly, Alison C. Lloyd, Martin C. Raff, Lack of Replicative Senescence in Cultured Rat Oligodendryte Precursor Cells, *Science,* **291** (2001), 868–871.

[510] Jerry W. Shay, Woodring E. Wright, When Do Telomers Matter?, *Science,* **291** (2001), 839–840.

[511] Catherine A. Wolkow, Koutarou D. Kimura, Ming-Sum Lee, Gary Ruvkun, Regulation of *C. elegans* Life-Span by Insulinlike Signaling in the Nervous system, *Science,* **290** (2000), 147–150.

[512] Javier Apfeld, Cynthia Kenyon, Regulation of lifespan by sensory perception in *Caenorhabditis elegans*, *Nature*, **402** (1999), 804–809.

[513] Enrica Migliaccio, Marco Giorgio, Simonetta Mele, Giuliana Pelicci, Paolo Reboldi, Pier Paolo Pandolfi, Luisa Lanfrancone, Pier Giuseppe Pelicci, The p66^{shc} adaptor protein controls oxidative stress response and life span in mammals, *Nature,* **402** (1999), 309–313.

[514] Sige Zou, Sarah Meadows, Linda Sharp, Lily Y. Jan, Yuh Nung Jan, Genome-wide study of aging and oxidative stress response in *Drosophila melanogaster*, *Proceedings of the National Academy of Sciences USA* **97** (2000), 13726–13731.

[515] Blanka Rogina, Robert A. Reenan, Steven P. Nilsen, Stephen L. Helfand, Extended Life-Span Conferred by Cotransporter Gene Mutations in *Drosophila, Science,* **290** (2000), 2137–2140.

[516] Yoshiko Takahashi, Makoto Kuro-o, Fuyuki Ishikawa, Aging mechanisms, *Proceedings of the National Academy of Sciences USA,* **97** (2000), 12407–12408.

[517] Maarten H. K. Linskens, Junli Feng, William H. Andrews, Brett E. Enlow, Shahin M. Saati, Leath A. Tonkin, Walter D. Funk, Bryant Villeponteau, Cataloging altered gene expression in young and senescent cells using enhanced differential display, *Nucleic Acids Research,* **23** (1995), 3244–3251.

[518] Eric Dufour, Joceline Boulay, Vincent Rincheval, Annie Sainsard-Chanet, A causal link between respiration and senescence in *Podospora anserina*, *Proceedings of the National Academy of Sciences USA,* **97** (2000), 4138–4143.

[519] Thomas B. L. Kirkwood, Steven N. Austad, Why do we age?, *Nature,* **408** (2000), 233–238.

[520] Toren Finkel, Nikki J. Holbrook, Oxidants, oxidative stress and the biology of ageing, *Nature,* **408** (2000), 239–247.

[521] Ronald A. DePinho, The age of cancer, *Nature,* **408** (2000), 248–254.

[522] Leonard Guarente, Cynthia Kenyon, Genetic pathways that regulate ageing in model organisms, *Nature,* **408** (2000), 255–262.

[523] George M. Martin, Junko Oshima, Lessons from human progeroid syndromes, *Nature,* **408** (2000), 263–266.

[524] Leonard Hayflick, The future of ageing, *Nature,* **408** (2000), 267–269.

[525] Elizabeth Blackburn, Telomere states and cell fates, *Nature,* **408** (2000), 53–56.

[526] James R. Mitchell, Emily Wood, Kathleen Collins, A telomerase component is defective in the human disease dyskeratosis congenita, *Nature,* **402** (1999), 551–555.

[527] Tanya Jonassen, Pamela L. Larsen, Catherine F. Clarke, A dietary source of coenzyme Q is essential for growth of long-lived *Caenorhabditis elegans clk-1* mutants, *Proceedings of the National Academy of Sciences USA,* **98** (2001), 421–426.

[528] Markus Drescher, Michael Hentschel, Reinhard Kienberger, Gabriel Tempea, Christian Spielmann, Georg A. Reider, Paul B. Corkum, Ferenc Krausz, X-ray Pulses Approaching the Attosecond Frontier, *Science,* **291** (2001), 1923–1927.

[529] Albert M. Brouwer, Celine Frochot, Francesco G. Gatti, David A. Leigh, Loic Mottier, Francesco Paolucci, Sergio Roffia, George W. H. Wurpel, Photoinduction of Fast, Reversible Translational Motion in a Hydrogen-Bonded Molecular Shuttle, *Science,* **291** (2001), 2124–2128.

[530] Jean-Pierre Suvage, A Light-Driven Linear Motor at the Molecular Level, *Science,* **291** (2001), 2105–2106.

[531] Guanghua Wu, Haifeng Ji, Karolyn Hansen, Thomas Thundat, Ram Datar, Richard Cote, Michael F. Hagan, Arup K. Chakraborty, Arunava Majumdar, Origin of nano-mechanical cantilever motion generated from biomolecular interactions, *Proceedings of the National Academy of Sciences USA,* **98** (2001), 1560–1564.

[532] H. B. Chan, V. A. Aksyuk, R. N. Kleiman, D. J. Bishop, Federico Capasso, Quantum Mechanical Actuation of Microelectromechanical Systems by the Casimir Force, *Science,* **291** (2001), 1941–1944.

[533] Bin Zhao, Jeffrey S. Moore, David J. Beebe, Surface-Directed Liquid Flow Inside Microchannels, *Science,* **291** (2001), 1023–1026.

[534] A. Hatzor, P. S. Weiss, Molecular Rulers for Scaling Down Nanostructures, *Science,* **291** (2001), 1019–1020.
Philip G. Collins, Michael S. Arnold, Phaedon Avouris, Engineering Carbon Nanotubes and Nanotube Circuits Using Electrical Breakdown, *Science,* **292** (2001), 706–709.
Min Ouyang, Jin-Lin Huang, Chin Li Cheung, Charles M. Lieber, Energy Gaps in 'Metallic' Single-Walled Carbon Nanotubes, *Science,* **292** (2001), 702–705.
Mildred S. Dresselhaus, Burn and Interrogate, *Science,* **292** (2001), 650–651.

[535] Li-Qun Gu, Stephen Cheley, Hagan Bayley, Capture of a Single Molecule in a Nanocavity, *Science,* **291** (2001), 636–640.

[536] Hideo Ohno, Toward Functional Spintronics, *Science,* **291** (2001), 840–841.

[537] Yuji Matsumoto, Makoto Murakami, Tomoji Shono, Tetsuya Hasegawa, Tomoteru Fukumura, Masashi Kawasaki, Parhat Ahmet, Toyohiro Chikyow, Shin-ya Koshihara, Hideomi Koinuma, Room-Temperature Ferromagnetism in Transparent Transition Metal-Doped Titanium Dioxide, *Science,* **291** (2001), 854–856.

[538] Yu Huang, Xiangfeng Duan, Qingqiao Wie, Charles M. Lieber, Directed Assembly of One-Dimensional Nanostructures into Functional Networks, *Science,* **291** (2001), 630–633.

[539] Susan Daniel, Manoj K. Chaudhury, John C. Chen, Fast Drop Movements Resulting from a Phase Change on a Gradient Surface, *Science,* **291** (2001), 633–636.

[540] Darsh T. Wasan, Alex D. Nikolov, Howard Brenner, Droplets Speeding on Surfaces, *Science,* **291** (2001), 605–606.
M. Heuberger, M. Zäch, N. D. Spencer, Density Fluctuations Under Confinement: When Is a Fluid Not a Fluid?, *Science,* **292** (2001), 905–908.

[541] Yi Cui, Charles M. Lieber, Functional Nanoscale Electronic Devices Assembled Using Silicon Nanowire Building Blocks, *Science,* **291** (2001), 851–853.

[542] M. W. J. Prins, W. J. J. Welters, J. W. Weekamp, Fluid Control in Multichannel Structures by Electrocapillary Pressure, *Science,* **291** (2001), 277–280.

[543] Jan Genzer, Kirill Efimenko, Creating Long-Lived Superhydrophobic Polymer Surfaces Through Mechanically Assembled Monolayers, *Science,* **290** (2000), 2130–2133.

[544] A. Prasanna de Silva, David B. Fox, Thomas S. Moody, Sheenagh M. Weir, The development of molecular fluorescent switches, *TIBTECH,* **19** (2001), 29–34.

[545] S. Komiyama, O. Astafiev, V. Antonov, T. Kutsuwa, H. Hirai, A single-photon detector in the far-infrared range, *Nature,* **403** (2000), 405–407.

[546] Leo Kouwenhoven, One photon seen by one electron, *Nature,* **403** (2000), 374–375.

[547] K. Schwab, E. A. Henriksen, J. M. Worlock, M. L. Roukes, Measurement of the quantum of thermal conductance, *Nature,* **404** (2000), 974–977.

[548] Leo P. Kouwenhoven, Lisbeth C. Venema, Heat flow through nanobridges, *Nature,* **404** (2000), 943–944.

[549] C. Joachim, J. K. Gimzewski, A. Aviram, Electronics using hybrid-molecular and mono-molecular devices, *Nature,* **408** (2000), 541–548.

[550] David I. Gittins, Donald Bethell, David J. Schiffrin, Richard J. Nichols, A nanometre-scale electronic switch consisting of a metal cluster and redox-addressable groups, *Nature,* **408** (2000), 67–69.

[551] Dan Feldheim, Flipping a molecular switch, *Nature,* **408** (2000), 45–46.

[552] Nagatoshi Koumura, Robert W. J. Zijlstra, Richard A. van Delden, Nobuyuki Harada, Ben L. Feringa, Light-driven monodirectional molecular motor, *Nature,* **401** (1999), 152–155.

[553] T. Ross Kelly, Harshani De Silva, Richard A. Silva, Unidirectional rotary motion in a molecular system, *Nature,* **401** (1999), 150–152.

[554] Dhaval A. Doshi, Nicola K. Huesing, Mengcheng Lu, Hongyou Fan, Yunfeng Lu, Kelly Simmons-Potter, B. G. Potter Jr., Alan J. Hurd, C. Jeffrey Brinker, Optically Defined Multifunctional Patterning of Photosensitive Thin-Film Silica Mesophases, *Science,* **290** (2000), 107–111.

[555] H. Sirringhaus, T. Kawase, R. H. Friend, T. Shimoda, M. Inbasekaran, W. Wu, E. P. Woo, High-Resolution Inkjet Printing of All-Polymer Transistor Circuits, *Science,* **290** (2000), 2123–2126.

[556] T. Thurn-Albrecht, J. Schotter, G. A. Kästle, N. Emley, T. Shibauchi, L. Krusin-Elbaum, K. Guarini, C. T. Black, M. T. Tuominen, T. P. Russell, Ultrahigh-Density Nanowire Arrays Grown in Self-Assembled Diblock Copolymer Templates, *Science,* **290** (2000), 2126–2129.

[557] Peng Jiang, Jane F. Bertone, Vicki L. Colvin, A Lost-Wax Approach to Monodisperse Colloids and Their Crystals, *Science,* **291** (2001), 453–457.

[558] Thomas E. Mallouk, Stretching the Mold, *Science,* **291** (2001), 443–444.

[559] Wenonah Vercoutere, Stephen Winters-Hilt, Hugh Olsen, David Deamer, David Haussler, Mark Akeson, Rapid discrimination among individual DNA hairpin molecules at single-nucleotide resolution using an ion channel, *Nature Biotechnology,* **19** (2001), 248–252.

[560] Fernando Patolsky, Amir Lichtenstein, Itamar Willner, Detection of single-based DNA mutations by enzyme-amplified electronic transduction, *Nature Biotechnology,* **19** (2001), 253–257.

[561] Bonnie E. Gould Rothberg, Mapping a role for SNPs in drug development, *Nature Biotechnology,* **19** (2001), 209–211.

[562] Heiko O. Jacobs, George M. Whitesides, Submicrometer Patterning of Charge in Thin-Film Electrets, *Science,* **291** (2001), 1763–1766.

[563] Michael P. Washburn, Dirk Wolters, John R. Yates III, Large-scale analysis of the yeast proteome by multidimensional protein identification technology, *Nature Bio-technology,* **19** (2001), 242–247.
    Trey Ideker, Vesteinn Thorsson, Jeffrey A. Ranish, Rowan Christmas, Jeremy Buhler, Jimmy K. Eng, Roger Bumgarner, David R. Goodlett, Ruedi Aebersold, Leroy Hood, Integrated Genomic and Proteomic Analyses of a Systematically Perturbed Metabolic Network, *Science,* **292** (2001), 929–934.

[564] Tom Misteli, Protein Dynamics: Implications for Nuclear Architecture and Gene Expression, *Science,* **291** (2001), 843–847.

[565] Jiwei Yang, Usha Nagavarapu, Kenneth Relloma, Michael D. Sjaastad, William C. Moss, Antonino Passaniti, G. Scott Herron, Telomerized human microvasculature is functional *in vivo*, *Nature Biotechnology,* **19** (2001), 219–224.

[566] Sophie Petit-Zeman, Regenerative medicine, *Nature Biotechnology,* **19** (2001), 201–206.

[567] David Bailey, Edward Zanders, Philip Dean, The end of the beginning for genomic medicine, *Nature Biotechnology,* **19** (2001), 207–209.

[568] Fumihiko Sato, Takahashi Hashimoto, Akira Hachiya, Ken-ichi Tamura, Kum-Boo Choi, Takashi Morishige, Hideki Fujimoto, Yasuyuki Yamada, Metabolic engineering of plant alkaloid biosynthesis, *Proceedings of the National Academy of Sciences USA,* **98** (2001), 367–372.

[569] Zhiyong Luo, Qisheng Zhang, Yoji Oderaotoshi, Dennis P. Curran, Fluorous Mixture Synthesis: A fluorous-Tagging Strategy for the Synthesis and Separation of Mixtures of Organic Compounds, *Science,* **291** (2001), 1766–1769.

[570] Obadiah J. Plante, Emma R. Palmacci, Peter H. Seeburger, Automated Solid-Phase Synthesis of Oligosaccharides, *Science,* **291** (2001), 1523–1527.

[571] Piotr E. Marzalek, Hongbin Li, Julio M. Fernandez, Fingerprinting polysaccharides with single-molecule atomic force microscopy, *Nature Biotechnology,* **19** (2001), 258–262.

[572] Mary Sara McPeek, From mouse to human: Fine mapping of quantitative trait loci in a model organism, *Proceedings of the National Academy of Sciences USA,* **97** (2000), 12389–12390.

[573] Richard Mott, Christopher J. Talbot, Maria G. Turri, Allan C. Collins, Jonathan Flint, A method for fine mapping quantitative trait loci in outbred animal stocks, *Proceedings of the National Academy of Sciences USA,* **97** (2000), 12649–12654.

[574] R. W. Doerge, Bruce A. Craig, Model selection for quantitative trait locus analysis in polypolids, *Proceedings of the National Academy of Sciences USA,* **97** (2000), 7951–7956.

[575] Eyal Fridman, Tzili Pleban, Dani Zamir, A recombination hotspot delimits a wild-species quantitative trait locus for tomato sugar content to 484 bp within an invertase gene, *Proceedings of the National Academy of Sciences USA,* **97** (2000), 4718–4723.

[576] Dirk-Jan de Koning, Annemieke P. Rattink, Barbara Harlizius, Johan A. M. van Arendonk, E. W. Brascamp, Martien A. M. Groenen, Genome-wide scan for body composition in pigs reveals important role of imprinting, *Proceedings of the National Academy of Sciences USA,* **97** (2000), 7947–7950.

[577] C. Lonjou, K. Barnes, H. Chen, W. O. C. M. Cookson, K. A. Deichmann, I. P. Hall, J. W. Holloway, T. Laitinen, L. J. Palmer, M. Wjst, N. E. Morton, A first trial of retrospective collaboration for positional cloning in complex inheritance: Assay of the cytokine region on chromosome 5 by the Consortium on Asthma Genetics (COAG), *Proceedings of the National Academy of Sciences USA,* **97** (2000), 10942–10947.

[578] Linda M. Brzustowicz, Kathleen A. Hodgkinson, Eva W. C. Chow, William G. Honer, Anne S. Bassett, Location of a Major Susceptibility Locus for Familial Schizophrenia on Chromosome 1q21-q22, *Science,* **288** (2000), 678–682.

[579] Federico Sesti, Geoffrey W. Abbott, Jian Wie, Katherine T. Murray, Sanjeev Saksena, Peter J. Schwartz, Silvia G. Priori, Dan M. Roden, Alfred L. George, Jr., Steve A. N. Goldstein, A common polymorphism associated with antibiotic-induced cardiac arrhythmia, *Proceedings of the National Academy of Sciences USA,* **97** (2000), 10613–10618.

[580] Pamela M. Greenwood, Trey Sunderland, Judy L. Friz, Raja Parasuraman, Genetics and visual attention: Selective deficits in healthy adult carriers of the epsilon4 allele of the apolipoprotein *E* gene, *Proceedings of the National Academy of Sciences USA,* **97** (2000), 11661–11666.

[581] Richard Bellamy, Nulda Beyers, Keith P. W. J. McAdam, Cyril Ruwende, Robert Gie, Priscilla Samaai, Danite Bester, Mandy Meyer, Tumani Corrah, Matthew Collin, D. Ross Camidge, David Wilkinson, Eileen Hoal-van Helden, Hilton C. Whittle, William Amos, Paul van Helden, Adrian V. S. Hill, Genetic susceptibility to tuberculosis in Africans: A genome-wide scan, *Proceedings of the National Academy of Sciences USA,* **97** (2000), 8005–8009.

[582] John E. J. Rasko, Jean-Luc Battini, Leonid Kruglyak, David R. Cox, A. Dusty Miller, Precise gene localization by phenotypic assay of radiation hybrid cells, *Proceedings of the National Academy of Sciences USA,* **97** (2000), 7388–7392.

[583] Akihito Ozawa, Mark R. Band, Joshua H. Larson, Jena Donovan, Cheryl A. Green, James E. Womack, Harris A. Lewin, Comparative organization of cattle chromosome 5 revealed by comparative mapping by annotation and sequence similarity and radiation hybrid mapping, *Proceedings of the National Academy of Sciences USA,* **97** (2000), 4150–4155.

[584] Adam C. Bell, Adam G. West, Gary Felsenfeld, Insulators and Boundaries: Versatile Regulatory Elements in the Eukaryotic Genome, *Science,* **291** (2001), 447–450.

[585] Ekaterina Muravyova, Anton Golovnin, Elena Gracheva, Aleksander Parshikov, Tatiana Belenkaya, Vincenzo Pirrotta, Pavel Georgiev, Loss of Insulator Activity by Paired Su(Hw) Chromatin Insulators, *Science,* **291** (2001), 495–498.

[586] Haini N. Cai, Ping Shen, Effects of cis Arrangement of Chromatin Insulators on Enhancer-Blocking Activity, *Science,* **291** (2001), 493–495.

[587] Gary J. Lye, John M. Woodley, Application of in situ product-removal techniques to biocatalytic processes, *TIBTECH,* **17** (1999), 395–402.

[588] Min Jiang, Jubin Ryu, Monika Kiraly, Kyle Duke, Valerie Reinke, Stuat K. Kim, Genome-wide analysis of developmental and sex-regulated gene expression profiles in *Caenorhabditis elegans*, *Proceedings of the National Academy of Sciences USA,* **98** (2001), 218–223.

[589] Yao-Cheng Li, Tzu-Hao Cheng, Marc R. Gartenberg, Establishment of Transcriptional Silencing in the Absence of DNA Replication, *Science,* **291** (2001), 650–653.

[590] Ann L. Kirchmaier, Jasper Rine, DNA Replication-Independent Silencing in *S. cerevisiae, Science,* **291** (2001), 646–650.

[591] Jeffrey S. Smith, Jef D. Boeke, Is S Phase Important for transcriptional silencing?, *Science,* **291** (2001), 608–609.

[592] Marina N. Nikiforova, James R. Stringer, Ruthann Blough, Mario Medvedovic, James A. Fagin, Yuri E. Nikiforov, Proximity of Chromosomal Loci That Participate in Radiation-Induced Rearrangements in Human Cells, *Science,* **290** (2000), 138–141.

[593] Cesar Llave, Kristin D. Kasschau, James C. Carrington, Virus-encoded suppressor of post-transcriptional gene silencing targets a maintenance step in the silencing pathway, *Proceedings of the National Academy of Sciences USA,* **97** (2000), 13401–13406.

[594] Wolfgang Mayer, Alain Niveleau, Jörn Walter, Reinald Fundele, Thomas Haaf, Demethylation of the zygotic paternal genome, *Nature* **403** (2000), 501–502.

[595] Frank Lyko, Bernard H. Ramsahoye, Rudolf Jaenisch, DNA methylation in *Droso-phila melanogaster, Nature,* **408** (2000), 538–540.

[596] Jeremy S. Edwards, Rafael U. Ibarra, Bernhard O. Palsson, *In silico* predictions of *Escherichia coli* metabolic capabilities are consistent with experimental data, *Nature Biotechnology,* **19** (2001), 125–130.

[597] Sharon K. Powell, Michele A. Kaloss, Anne Pinkstaff, Rebecca McKee, Irina Burimski, Michael Pensiero, Edward Otto, Willem P. C. Stemmer, Nay-Wie Soong, Breeding of retroviruses by DNA shuffling for improved stability and processing yields, *Nature Biotechnology,* **18** (2000), 1279–1282.

[598] Kevin Walker, Rodney Croteau, Taxol biosynthesis: Molecular cloning of a benzoyl-CaA:taxane 2alpha-O-benzoyltransferase cDNA from *Taxus* and functional expression in *Escherichia coli, Proceedings of the National Academy of Sciences USA,* **97** (2000), 13591–13596.

[599] W. Christian Wigley, Rheas D. Stidham, Nathan M. Smith, John F. Hunt, Philip J. Thomas, Protein solubility and folding monitored *in vivo* by structural complementation of a genetic marker protein, *Nature Biotechnology,* **19** (2001), 131–136.

[600] Eileen E. M. Furlong, David Profitt, Matthew P. Scott, Automated sorting of live transgenic embryos, *Nature Biotechnology,* **19** (2001), 153–156.

[601] Kevin Eggan, Hidenori Akutsu, Konrad Hochedlinger, William Rideout III, Ryuzo Yanagimachi, Rudolf Jaenisch, X-Chromosome Inactivation in Cloned Mouse Embryos, *Science*, **290** (2000), 1578–1581.

[602] S. Gronthos, M. Mankani, J. Brahim, P. Gehron Robey, S. Shi, Postnatal human dental pulp stem cells (DPSCs) in vitro and in vivo, *Proceedings of the National Academy of Sciences USA*, **97** (2000), 13625–13630.

[603] Robert M. Lavker, Tung-Tien Sun, Epidermal stem cells: Properties, markers, and location, *Proceedings of the National Academy of Sciences USA*, **97** (2000), 13473–13475.

[604] Martin J. Cohn, Giving limbs a hand, *Nature*, **406** (2000), 953–954.

[605] Eva Mezey, Karen J. Chandross, Gyöngyi Harta, Richard A. Maki, Scott R. McKercher, Turning Blood into Brain: Cells Bearing Neuronal Antigens Generated in Vivo from Bone Marrow, *Science*, **290** (2000), 1779–1782.

[606] Jeff Betthauser, Erik Forsberg, Monica Augenstein, Lynette Childs, Kenneth Eilertsen, Joellyin Enos, Todd Forsythe, Paul Golueke, Gail Jurgella, Richard Koppang, Tiffany Lesmeister, Kelly Mallon, Greg Mell, Pavla Misica, Marvin Pace, Martha Pfister-Genskow, Nikolai Strelchenko, Gary Voelker, Steven Watt, Simon Thompson, Michael Bishop, Production of cloned pigs from in vitro systems, *Nature Biotechnology*, **18** (2000), 1055–1059.

[607] Alan Dove, Milking the genome for profit, *Nature Biotechnology*, **18** (2000), 1045–1048.

[608] Mark Westhusin, Jorge Piedrahita, Three little pigs worth the huff and puff?, *Nature Biotechnology*, **18** (2000), 1144–1145.

[609] Mark Noble, Can neural stem cells be used to track down and destroy migratory brain tumor cells while also providing a means of repairing tumor-associated damage, *Proceedings of the National Academy of Sciences USA*, **97** (2000), 12393–12395.

[610] Karen S. Aboody, Alice Brown, Nikolai G. Rainov, Kate A. Bower, Shaoxiong Liu, Wendy Yang, Juan E. Small, Ulrich Herrlinger, Vaclav Ourednik, Peter McL. Black, Xandra O. Brakefield, Evan Y. Snyder, Neural stem cells display extensive tropism for pathology in adult brains: Evidence from intracranial gliomas, *Proceedings of the National Academy of Sciences USA*, **97** (2000), 12846–12851.

[611] Maya Schuldiner, Ofra Yanuka, Joseph Itskovitz-Eldor, Douglas A. Melton, Nissim Benvenisty, Effects of eight growth factors on the differentiation of cells derived from human embryonic stem cells, *Proceedings of the National Academy of Sciences USA*, **97** (2000), 11307–11312.

[612] Koichi Akashi, David Traver, Toshihiro Miyamoto, Irving L. Weissman, A clonogenic common myeloid progenitor that gives rise to all myeloid lineages, *Nature*, **404** (2000), 193–197.

[613] Malcolm R. Alison, Richard Poulsom, Rosemary Jefferey, Amar P. Dhillon, Alberto Quaglia, Joe Jacob, Marco Novelli, Grant Prentice, Jill Williamson, Nicholas A. Wright, Hepatocytes from non-hepatic adult stem cells, *Nature*, **406** (2000), 257.

[614] Ron McKay, Stem cells – hype and hope, *Nature* **406** (2000), 361–364.

[615] Diana L. Clarke, Clas B. Johansson, Johannes Wilbertz, Biborka Veress, Erik Nilsson, Helena Karlström, Urban Lendahl, Jonas Frisen, Generalized Potential of Adult Neural Stem Cells, *Science*, **288** (2000), 1660–1663.

[616] Robert L. Phillips, Robin E. Ernst, Brian Brunk, Natalia Ivanova, Mark A. Mahan, Julia K. Deanehan, Kateri A. Moore, G. Christian Overton, Ihor R. Lemischka, The Genetic Program of Hematopoietic Stem Cells, *Science*, **288** (2000), 1635–1640.

[617] Nobuaki Kikyo, Paul A. Wade, Dmitry Guschin, Hui Ge, Alan P. Wolffe, Active Remodeling of Somatic Nuclei in Egg Cytoplasm by the Nucleosomal ATPase ISWI, *Science*, **289** (2000), 2360–2362.

[618] Qiang Tong, Gökhan Dalgin, Haiyan Xu, Chao-Nan Ting, Jeffrey M. Leiden, Gökhan S. Hotamisligil, Function of GATA Transcription Factors in Preadipocyte-Adipocyte Transition, *Science*, **290** (2000), 134–138.

[619] Ting Xie, Allan C. Spradling, A Niche Maintaining Germ Line Stem Cells in the Drosophila Ovary, *Science*, **290** (2000), 328–330.

[620] Zhongtao Zhang, Rongguang Zhang, Andrzej Joachimiak, Joseph Schlessinger, Xiang-Peng Kong, Crystal structure of human stem cell factor: Implication for stem cell factor receptor dimerization and activation, *Proceedings of the National Academy of Sciences USA,* **97** (2000), 7732–7737.

[621] Ognjenka Goga Vukmirovic, Shirley M. Tilghman, Exploring genome space, *Nature,* **405** (2000), 820–822.

[622] Randall T. Peterson, Brian A. Link, John E. Dowling, Stuart L. Schreiber, Small molecule developmental screens reveal the logic and timing of vertebrate development, *Proceedings of the National Academy of Sciences USA,* **97** (2000), 12965–12969.

[623] Gary P. Kobinger, Daniel J. Weiner, Qian-Chun Yu, James M. Wilson, *Filovirus*-pseudotyped lentiviral vector can efficiently and stably transduce airway epithelia *in vivo, Nature Biotechnology,* **19** (2001), 225–230.

[624] Roberto P. Stock, Alejandro Olvera, Ricardo Sanchez, Andreas Saralegui, Sonia Scarfi, Rosana Sanchez-Lopez, Marco A. Ramos, Lidia C. Boffa, Umberto Benatti, Alejandro Alagon, Inhibition of gene expression in *Entamoeba histolytica* with antisense peptide nucleic acid oligomers, *Nature Biotechnology,* **19** (2001), 231–234.

[625] Marianne D. De Backer, Bart Nelissen, Marc Logghe, Jasmine Viaene, Inge Loonen, Sandy Vandonink, Ronald de Hoogt, Sylviane Dewaele, Fermin A. Simons, Peter Verhasselt, Greet Vanhoof, Roland Contreras, Walter H. M. L. Luyten, An antisense-based functional genomics approach for identification of genes critical for growth of *Candida albicans, Nature Biotechnology,* **19** (2001), 235–241.

[626] Dominique Sanglard, Integrated antifungal drug discovery in *Candida albicans, Nature Biotechnology,* **19** (2001), 212–213.

[627] Paul A. Wender, Dennis J. Mitchell, Kanaka Pattabiraman, Erin T. Pelkey, Lawrence Steinman, Jonathan B. Rothbard, The design, synthesis, and evaluation of molecules that enable or enhance cellular uptake: Peptoid molecular transporters, *Proceedings of the National Academy of Sciences USA,* **97** (2000), 13003–13008.

[628] Ting-Fung Chan, John Carvalho, Linda Riles, X. F. Steven Zheng, A chemical genomics approach toward understanding the global functions of the target of rapamycin protein (TOR), *Proceedings of the National Academy of Sciences USA,* **97** (2000), 13227–13232.

[629] Michelle A. Blaskovich, Qing Lin, Frederic L. Delarue, Jiazhi Sun, Hyung Soon Park, Domenico Coppola, Andrew D. Hamilton, Said M. Sebti, Design of GBF-111, a platelet-derived growth factor binding molecule with antiangiogenic and anticancer activity against human tumors in mice, *Nature Biotechnology,* **18** (2000), 1065–1070.

[630] Mark J. Olsen, Daren Stephens, Devin Griffiths, Patrick Daugherty, George Georgiou, Brent L. Iverson, Function-based isolation of novel enzymes from a large library, *Nature Biotechnology,* **18** (2000), 1071–1074.

[631] John V. Frangioni, Leah M. LaRiccia, Lewis C. Cantley, Marc R. Montminy, Minimal activators that bind to the KIX domain of p300/CBP identified by phage display screening, *Nature Biotechnology,* **18** (2000), 1080–1085.

[632] Bing Wang, Juan Li, Xiao Xiao, Adeno-associated virus vector carrying human minidystrophin genes effectively ameliorates muscular dystrophy in mdx mouse model, *Proceedings of the National Academy of Sciences USA,* **97** (2000), 13714–13719.

[633] Jean-Marie Lehn, Alexey V. Eliseev, Dynamic Combinatorial Chemistry, *Science,* **291** (2001), 2331–2332.

[634] Brian F. Volkman, Doron Lipson, David E. Wemmer, Dorothee Kern, Two-State Allosteric Behaviour in a Single-Domain Signaling Protein, *Science,* **291** (2001), 2429–2433.

[635] Matthias Buck, Michael K. Rosen, Flipping a Switch, *Science,* **291** (2001), 2329–2330.

[636] Huib Caron, Barbera van Schaik, Merlijn van der Mee, Frank Baas, Gregory Riggins, Peter van Sluis, Marie-Christine Hermus, Ronald van Asperen, Kathy Boon, P. A. Voute, Siem Heisterkamp, Antoine van Kampen, Rogier Versteeg, The Human Transcriptome Map: Clustering of Highly Expressed Genes in Chromosomal Domains, *Science,* **291** (2001), 1289–1292.

Junaid Ziauddin, David M. Sabatini, Microarrays of cells expressing defined cDNAs, *Nature,* **411** (2001), 107–110.

[637] Richard T. Wood, Michael Mitchell, John Sgouros, Thomas Lindahl, Human DNA Repair Genes, *Science,* **291** (2001), 1284–1289.

[638] L. Aravind, Vishva M. Dixit, Eugene V. Koonin, Apoptotic Molecular Machinery: Vastly Increased Complexity in Vertebrates Revealed by Genome Comparisons, *Science,* **291** (2001), 1279–1284.

[639] Joseph Alper, Searching for Medicine's Sweet Spot, *Science,* **291** (2001), 2338–2343.

[640] Pamela Sears, Chi-Huey Wong, Toward Automated Synthesis of Oligosaccharides and Glycoproteins, *Science,* **291** (2001), 2344–2350.

[641] Anne Dell, Howard R. Morris, Glycoprotein Structure Determination by Mass Spectrometry, *Science,* **291** (2001), 2351–2356.

[642] Carolyn R. Bertozzi, Laura L. Kiessling, Chemical Glycobiology, *Science,* **291** (2001), 2357–2363.

[643] Ari Helenius, Markus Aebi, Intracellular Functions of N-Linked Glycans, *Science,* **291** (2001), 2364–2369.

[644] Pauline M. Rudd, Tim Elliott, Peter Cresswell, Ian A. Wilson, Raymond A. Dwek, Glycosylation and the Immune System, *Science,* **291** (2001), 2370–2376.

[645] Lance Wells, Keith Vosseller, Gerald W. Hart, Glycosylation of Nucleocytoplasmic Proteins: Signal Transduction and O-GlcNAc, *Science,* **291** (2001), 2376–2378.

[646] Neil J. Risch, Searching for genetic determinants in the new millenium, *Nature,* **405** (2000), 847–856.

[647] David Eisenberg, Edward M. Marcotte, Ioannis Xenarios, Todd O. Yeates, Protein function in the post-genomic era, *Nature,* **405** (2000), 823–826.

[648] Hyun Chul Lee, Su-Jin Kim, Kyung-Sup Kim, Hang-Cheol Shin, Ji-Won Yoon, Remission in models of type 1 diabetes by gene therapy using a single-chain insulin analogue, *Nature,* **408** (2000), 483–488.

[649] Marina Cavazzano-Calvo, Salima Hacein-Bey, Genevieve de Saint Basile, Fabian Gross, Eric Yvon, Patrick Nusbaum, Francoise Selz, Christophe Hue, Stephanie Certain, Jean-Laurent Casanova, Philippe Bousso, Francoise Le Deist, Alain Fischer, Gene Therapy of Human Severe Combined Immunodeficiency (SCID)-X1 Disease, *Science,* **288** (2000), 669–672.

[650] Bruce D. Klugherz, Peter L. Jones, Xiumin Cui, Weiliam Chen, Nicolas F. Meneveau, Suzanne DeFelice, Jeanne Connolly, Robert L. Wilensky, Robert J. Levy, Gene delivery from a DNA controlled release stent in porcine coronary arteries, *Nature Biotechnology,* **18** (2000), 1181–1184.

[651] Richard Wade-Martins, Robert E. White, Hiroshi Kimura, Peter R. Cook, Michael R. James, Stable correction of a genetic deficiency in human cells by an episome carrying a 115 kb genomic transgene, *Nature Biotechnology,* **18** (2000), 1311–1314.

[652] Ryohei Yasuda, Hiroyuki Noji, Masasuke Yoshida, Kazuhiko Kinosita Jr., Hiroyasu Itoh, Resolution of distinct rotational substeps by millisecond kinetic analysis of F1-ATPase, Nature, 410 (2001), 898–904.

[653] Vincenzo Balzani, Alberto Credi, Francisco M. Raymo, J. Fraser Stoddart, Künstliche molekulare Maschinen, *Angewandte Chemie*, **112** (2000), 3484–3530.

[654] David J. Clancy, David Gems, Lawrence G. Harshman, Sean Oldham, Hugo Stocker, Ernst Hafen, Sally J. Leevers, Linda Partridge, Extension of Life-Span by Loss of CHICO, a *Drosophila* Insulin Receptor Substrate Protein, *Science,* **292** (2001), 104–106.
Yi-Jyun Lin, Laurent Seroude, Seymour Benzer, Extended Life-Span and Stress Resistance in the *Drosophila* Mutant *methuselah, Science,* **282** (1998), 943–946.

[655] Tatar, A. Kopelman, D. Epstein, M.-P. Tu, C.-M. Yin, R. S. Garofalo, A Mutant *Drosophila* Insulin Receptor Homolog that Extends Life-Span and Impairs Neuro-endocrine Function, *Science,* **292** (2001), 107–110.

[656] Evelyn Strauss, Growing Old Together, *Science,* **292** (2001), 41–43.

[657] Adam Curtis, Chris Wilkinson, Nanotechniques and approaches in biotechnology, *TIBTECH,* **19** (2001), 97–101.

[658] Amy C. Groth, Eric C. Olivares, Bhaskar Thyagarajan, Michele P. Calos, A phage integrase directs efficient site-specific integration in human cells, *Proceedings of the National Academy of Sciences USA,* **97** (2001), 5995–6000.

[659] Alistair S. Irvine, Peter K. E. Trinder, David L. Laughton, Helen Ketteringham, Ruth H. McDermott, Sophie C. H. Reid, Adrian M. R. Haines, Abdu Amir, Rhonda Husain, Rajeev Doshi, Lawrence S. Young, Andrew Mountain, Efficient nonviral transfection of dendritic cells and their use for in vivo immunization, *Nature Biotechnology,* **18** (2000), 1273–1278.

[660] Aubrey D. N. J. de Grey, Mitochondrial gene therapy: an arena for the biomedical use of inteins, *TIBTECH,* **18** (2000), 394–399.

[661] Donald Orlic, Jan Kajstura, Stefano Chimenti, Igor Jakoniuk, Stacie M. Anderson, Baosheng Li, James Pickel, Ronald McKay, Bernardo Nadal-Ginard, David M. Bodine, Annarosa Leri, Piero Anversa, Bone marrow cells regenerate infarcted myo-cardium, *Nature,* **410** (2001), 701–705.
Teruhiko Wakayama, Viviane Tabar, Ivan Rodriguez, Anthony C. F. Perry, Lorenz Studer, Peter Mombaerts, Differentiation of Embryonic Stem Cell Lines Generated from Adult Somatic Cells by Nuclear Transfer, *Science* 292 (2001), 740–743.

[662] H. Stocker, E. Hafen, Genetic control of cell size, *Current Opinion in Genetics and Development,* **10** (2000), 529–535.
F. Wittwer, A. van Straaten, K. Keleman, B. J. Dickson, E. Hafen, Lilliputian: anAF4/FMR2-related protein that controls cell identitiy and cell growth, *Development,***128** (2001), 791–800.
R. Bohni, J. Riesgo-Escovar, S. Oldham, W. Brogiolo, H. Stocker, B. F. Andruss, K. Beckingham, E. Hafen, Autonomous control of cell and organ size by CHICO, a *Drosophila* homolog of verte-brate IRS1-4, *Cell,* **97** (1999), 865–875.
R. Burke, D. Nellen, M. Bellotto, E. Hafen, K. A. Senti, B. J. Dickson, K. Basler, Dispatched, a novel sterol-sensing domain protein dedicated to the release of cholesterol-modified hedgehog from signalling, *Cell,* **99** (1999), 803–815.

[663] Masaru Tomita, Whole-cell simulation: a grand challenge of the 21st century, *TIBTECH,* **19** (2001), 205–210.

[664] Erick Noensie, Harry Dietz, A strategy for disease gene identification through nonsense-mediated mRNA decay inhibition, *Nature Biotechnology,* **19** (2001), 434–439.

[665] John J. Harrington, Bruce Sherf, Stephen Rundlett, P. David Jackson, Rob Perry, Scott Cain, Christina Leventhal, Mark Thornton, Rakesh Ramachandran, Jessica Whittington, Laura Lerner, Dana Costanzo, Karen McElligott, Sherry Boozer, Robert Mays, Emery Smith, Neil Veloso, Alison Klika, Jennifer Hess, Kevin Cothren, Kalok Lo, Jason Offenbacher, Joel Danzig, Matt Ducar, Creation of genome-wide protein expression libraries using random activation of gene expression, *Nature Bio-technology,* **19** (2001), 440–445.

[666] Itamar Willner, Bilha Willner, Biomaterials integrated with electronic elements: en route to bioelectronics, *TIBTECH,* **19** (2001), 222–230.

[667] Kelvin H. Lee, Proteomics: a technology-driven and technology-limited discovery science, *TIBTECH,* **19** (2001), 217–222.

[668] Yoshiharu Ishii, Akihiko Ishijima, Toshio Yanagida, Single molecule nano-manipulation of biomolecules, *TIBTECH,* **19** (2001), 211–216.

[669] Volker Sieber, Carlos A. Martinez, Frances H. Arnold, Libraries of hybrid proteins from distantly related sequences, *Nature Biotechnology,* **19** (2001), 456–460.

[670] Steven A. Farber, Michael Pack, Shiu-Ying Ho, Iain D. Johnson, Daniel S. Wagner, Roland Dosch, Mary C. Mullins, H. Stewart Hendrickson, Elizabeth K. Hendrickson, Marnie E. Halpern, Genetic Analysis of Digestive Physiology Using Fluorescent Phospholipid Markers, *Science,* **292** (2001), 1385–1388.

[671] Maeve A. Caldwell, Xiaoling He, Neil Wilkie, Scott Pollack, George Marshall, Keith A. Wafford, Clive N. Svendsen, Growth factors regulate the survival and fate of cells derived from human neurospheres, *Nature Biotechnology,* **19** (2001), 475–479.

[672] Nadya Lumelsky, Oliver Blondel, Pascal Laeng, Ivan Velasco, Rea Ravin, Ron McKay, Differentiation of Embryonic Stem Cells to Insulin-Secreting Structures Similar to Pancreatic Islets, *Science*, **292** (2001), 1389–1394.

[673] Joost A. Kolkman, Willem P. C. Stemmer, Directed evolution of proteins by exon shuffling, *Nature Biotechnology*, **19** (2001), 423–428.

[674] Serguei P. Golovan, M. Anthony Hayes, John P. Phillips, Cecil W. Forsberg, Transgenic mice expressing bacterial phytase as a model for phosphorous pollution control, *Nature Biotechnology*, **19** (2001), 429–433.

[675] Sayda M. Elbashir, Jens Harborth, Winfried Lendeckel, Abdullah Yalcin, Klaus Weber, Thomas Tuschl, Duplexes of 21-nucleotide RNAs mediate RNA interference in cultured mammalian cells, *Nature*, **411** (2001), 494–498.

[676] Andrew G. Fraser, Ravi S. Kamath, Peder Zipperlen, Maruxa Martinez-Campos, Marc Sohrmann, Julie Ahringer, Functional genomic analysis of *C. elegans* chromo-some I by systematic RNA interference, *Nature*, **408** (2000), 325–330.

[677] Pierre Gönczy, Christophe Echeverri, Karen Oegema, Alan Coulson, Steven J. M. Jones, Richard R. Copley, John Duperon, Jeff Oegema, Michael Brehm, Etienne Cassin, Eva Hannak, Matthew Kirkham, Silke Pichler, Kathrin Flohrs, Anoesjka Goessen, Sebastian Leidel, Anne-Marie Alleaume, Cecille Martin, Nurhan Özlü, Peer Bork, Anthony A. Hyman, Functional genomic analysis of cell division in *C. elegans* using RNAi of genes on chromosome III, *Nature*, **408** (2000), 331–336.

[678] Dancia Wu-Scharf, Byeong-Ryool Jeong, Chaomei Zhang, Heriberto Cerutti, Trans-gene and Transposon Silencing in *Chlamydomonas reinhardtii* by a DEAH-Box RNA Helicase, *Science*, **290** (2000), 1159–1162.

[679] Shoshy Altuvia, E. Gerhart H. Wagner, Switching on and off with RNA, *Proceedings of the National Academy of Sciences USA*, **97** (2000), 9824–9826.

[680] Paolo Amedeo, Yoshiko Habu, Karin Afsar, Ortrun Mittelsten Scheid, Jerzy Paszkowski, Disruption of the plant gene *MOM* releases transcriptional silencing of methylated genes, *Nature*, **405** (2000), 203–206.

[681] Scott M. Hammond, Emily Bernstein, David Beach, Gregory J. Hannon, An RNA-directed nuclease mediates post-transcriptional gene silencing in *Drosophila* cells, *Nature*, **406** (2000), 293–296.

[682] Mathilde Fagard, Stephanie Boutet, Jean-Benoit Morel, Catherine Bellini, Herve Vaucheret, AGO1, QDE-2, and RDE-1 are related proteins required for post-trans-criptional gene silencing in plants, quelling in fungi, and RNA interference in animals, *Proceedings of the National Academy of Sciences USA*, **97** (2000), 11650–11654.

[683] Caterina Catalanotto, Gianluca Azzalin, Giuseppe Macino, Carlo Cogoni, Gene silencing in worms and fungi, *Nature*, **404** (2000), 245.

[684] Wolf Reik, Adele Murrell, Silence across the border, *Nature*, **405** (2000), 408–409.
Adam C. Bell, Gary Felsenfeld, Methylation of a CTFC-dependent boundary controls imprinted expression of the *Igf2* gene, *Nature*, **405** (2000), 482–485.
Amy T. Hark, Christopher J. Schoenherr, David J. Katz, Robert S. Ingram, John M. Levorse, Shirley M. Tilghman, CTFC mediates methylation-sensitive enhancer-blocking activity at the *H19/Igf2* locus, *Nature*, **405** (2000), 486–489.

[685] Jeffrey A. Bailey, Laura Carrel, Aravinda Chakravarti, Evan Eichler, Molecular evidence for a relationship between LINE-1 elements and X chromosome in-activation: The Lyon repeat hypothesis, *Proceedings of the National Academy of Sciences USA*, **97** (2000), 6634–6639.

[686] Joel N. Hirschhorn, Pamela Sklar, Kerstin Lindblad-Toh, Yin-Mei Lim, Melisa Ruiz-Gutierrez, Stacey Bolk, Bradley Langhorst, Steven Schaffner, Ellen Winchester, Eric S. Lander, SBE-TAGS: An array-based method for efficient single-nucleotide polymorphism genotyping, *Proceedings of the National Academy of Sciences USA*, **97** (2000), 12164–12169.

[687] Elizabeth M. Boon, Donato M. Ceres, Thomas G. Drummond, Michael G. Hill, Jacqueline K. Barton, Mutation detection by electrocatalysis at DNA-modified electrodes, *Nature Biotechnology*, **18** (2000), 1096–1100.

[688] Hongping Jiang, Dong-chul Kang, Deborah Alexandre, Paul B. Fisher, RaSH, a rapid subtraction hybridization approach for identifying and cloning differentially expressed genes, *Proceedings of the National Academy of Sciences USA*, **97** (2000), 12684–12689.

[689] A. A. Hill, C. P. Hunter, B. T. Tsung, G. Tucker-Kellogg, E. L. Brown, Genomic Analysis of Gene Expression in *C. elegans*, *Science*, **290** (2000), 809–812.

[690] Shu-Mei Dai, Hsiu-Hua Chen, Cheng Chang, Arthur D. Riggs, Steven D. Flanagan, Ligation-mediated PCR for quantitative in vivo footprinting, *Nature Biotechnology*, **18** (2000), 1108–1111.

[691] Eaves, I. A. *et al.*, The genetically isolated populations of Finland and Sardinia may not be a panacea for linkage disequilibrium mapping of common disease genes, *Nature Genetics*, **25** (2000), 320–323.

[692] Taillon-Miller, P. *et al.*, Juxtaposed regions of extensive and minimal linkage disequilibrium in human Xq25 and Xq28, *Nature Genetics*, **25** (2000), 324–328.

[693] Neto E. D. *et al.*, Shotgun sequencing of the human transcriptome with ORF expressed sequence tags, *Proceedings of the National Academy of Sciences USA*, **97** (2000), 3491–3496.

[694] Liu H. *et al.*, Development of multichannel devices with an array of electrospray tips for high-throughput mass spectrometry, *Anal. Chem.*, **72** (2000), 3303–3310.

[695] Atul J. Butte, Pablo Tamayo, Donna Slonim, Todd R. Golub, Isaac S. Kohane, Discovering functional relationships between RNA expression and chemotherapeutic susceptibility using relevance networks, *Proceedings of the National Academy of Sciences USA*, **97** (2000), 12182–12186.

[696] Emad M. El-Omar, Mary Carrington, Wong-Ho Chow, Kenneth E. L. McColl, Jay H. Bream, Howard A. Young, Jesus Herrera, Jolanta Lissowska, Chiu-Chin Yuan, Nathaniel Rothman, George Lanyon, Maureen Martin, Joseph F. Fraumeni Jr., Charles S. Rabkin, Interleukin-1 polymorphisms associated with increased risk of gastric cancer, *Nature*, **404** (2000), 398–402.

[697] Andreas Jenne, Jörg S. Hartig, Nicolas Piganeau, Andreas Tauer, Dmitry A. Samarsky, Michael R. Green, Julian Davies, Michael Famoluk, Rapid identification and characterization of hammerhead-ribozyme inhibitors using fluorescence-based technology, *Nature Biotechnology*, **19** (2001), 56–61.

[698] Leonie M. Raamsdonk, Bas Teusink, David Broadhurst, Nianshu Zhang, Andrew Hayes, Michael C. Walsh, Jan A. Berden, Kevin M. Brindle, Douglas B. Kell, Jem J. Rowland, Hans V. Westerhoff, Karel van Dam, Stephen G. Oliver, A functional genomics strategy that uses metabolome data to reveal the phenotype of silent mutations, *Nature Biotechnology*, **19** (2001), 45–50.

[699] Kazuhiko Nakatani, Shinsuke Sando, Isao Saito, Scanning of guanine-guanine mis-matches in DNA by synthetic ligands using surface plasmon resonance, *Nature Biotechnology*, **19** (2001), 51–55.

[700] Robert Jenison, Shao Yang, Ayla Haeberli, Barry Polisky, Interference-based detection of nucleic acid targets on optically coated silicon, *Nature Biotechnology*, **19** (2001), 62–65.

[701] Wie Zhou, Gennaro Galizia, Steven N. Goodman, Katharine E. Romans, Kenneth W. Kinzler, Bert Vogelstein, Michael A. Choti, Elizabeth A. Montgomery, Counting alleles reveals a connection between chromosome 18q loss and vascular invasion, *Nature Biotechnology*, **19** (2001), 78–81.

[702] David E. Kerr, Karen Plaut, A. John Bramley, Christine M. Williamson, Alistair J. Lax, Karen Moore, Kevin D. Wells, Robert J. Wall, Lysostaphin expression in mammary glands confers protection against staphylococcal infection in transgenic mice, *Nature Biotechnology*, **19** (2001), 66–70.

[703] Tatsuhiro Joki, Marcelle Machluf, Anthony Atala, Jianhong Zhu, Nicholas T. Seyfried, Ian F. Dunn, Toshiaki Abe, Rona S. Carroll, Peter McL. Black, Continuous release of endostatin from microencapsulated engineered cells for tumor therapy, *Nature Biotechnology*, **19** (2001), 35–39.

[704] Tracy-Ann Read, Dag R. Sorensen, Rupavathana Mahesparan, Per O. Enger, Rupert Timpl, Bjorn R. Olsen, Mari H. B. Hjelstuen, Olav Haraldseth, Rolf Bjerkvig, Local endostatin treatment of gliomas administered by microencapsulated producer cells, *Nature Biotechnology*, **19** (2001), 29–34.

[705] Gabriele Bergers, Douglas Hanahan, Cell factories for fighting cancer, *Nature Bio-technology*, **19** (2001), 20–21.

Antonia E. Stephen, Peter T. Masiakos, Dorry L. Segev, Joseph P. Vacanti, Patricia K.Donahue, David T. MacLaughlin, Tissue-engineered cells producing complex recombinant proteins inhibit ovarian cancer in vivo, *Proceedings of the National Academy of Sciences USA,* **98** (2001), 3214–3219.

[706] David T. Page, Sally Cudmore, Innovations in oral gene delivery: challenges and potentials, *Drug Discovery Today,* **6** (2001), 92–101.

[707] Malin Mejare, Leif Bülow, Metal-binding proteins and peptides in bioremediation and phyto-remediation of heavy metals, *TIBTECH*, **19** (2001), 67–73.

[708] Andreas Offenhäusser, Wolfgang Knoll, Cell-transistor hybrid systems and their potential applications, *TIBTECH*, **19** (2001), 62–66.

[709] John A. Rogers, Zhenan Bao, Kirk Baldwin, Ananth Dodabalapur, Brian Crone, V. R. Raju, Valerie Kuck, Howard Katz, Karl Amundson, Jay Ewing, Paul Drzaic, Paper-like electronic displays: Large-area rubber-stamped plastic sheets of electronics and microencapsulated electrophoretic inks, *Proceedings of the National Academy of Sciences USA,* **98** (2001), 4835–4840.

[710] Ralph G. Nuzzo, The future of electronics manufacturing is revealed in the fine print, *Proceedings of the National Academy of Sciences USA*, **98** (2001), 4827–4829.

Index

U.W.E.L.   NG RESOURCES